Velocity-Free Localization Methodology for Acoustic and Microseismic Sources

Longjun Dong · Xibing Li

Velocity-Free Localization Methodology for Acoustic and Microseismic Sources

Springer

Longjun Dong
School of Resources and Safety Engineering
Central South University
Changsha, China

Xibing Li
School of Resources and Safety Engineering
Central South University
Changsha, China

ISBN 978-981-19-8609-3 ISBN 978-981-19-8610-9 (eBook)
https://doi.org/10.1007/978-981-19-8610-9

This Springer imprint is published by the registered company Springer Nature Singapore Pte Ltd.
The registered company address is: 152 Beach Road, #21-01/04 Gateway East, Singapore 189721, Singapore

Preface

Microseismic/Acoustic Emission (MS/AE) source location is a technology using information of acoustic waves, such as arrival time and amplitude, to determine acoustic and microseismic sources. A large number of location methods have been proposed over the past hundred years. Most of the methods assume that the propagation path of acoustic wave is a ray, which means that the location accuracy is sensitive to the accuracy of wave velocity. In fact, the wave velocity changes with time and space during the propagation, which means that the pre-measured wave velocity hardly matches the actual wave velocity. Therefore, using pre-measured wave velocity may cause poor location accuracy.

Therefore, we proposed velocity-free localization methods for acoustic and microseismic sources. These methods do not require the pre-determination of wave velocity, which is a dynamically adjusted free real-time parameter. They solve the problem of large location errors caused by the difference between measured wave velocity and actual wave velocity in the source area, and greatly improve the positioning accuracy. They are suitable for complex structures where the wave velocity changes dynamically in time and space, such as mines, bridges, buildings, pavements, loaded mechanical structures, dams, geothermal mining, oil extraction, and other engineering fields.

This book presents the velocity-free source localization methods from basic theory to methodology and to application. It is intended to serve as an introduction to velocity-free localization for advanced students and professionals.

This book is arranged into twelve chapters. Chapter 1 introduces the progress in the development of localization methods and details the merits and demerits of different localization methods. The basic theory about Acoustic Emission (AE) source localization is demonstrated in Chap. 2. Chapter 3 investigates and discusses the factors affecting the accuracy of source localization. Chapters 4–6 present respectively analytical, iterative, and collaborative localization methods without the predetermination of velocity. Considering the complex structure in practical projects, a velocity-free method based on non-straight wave travel paths is introduced in Chap. 7. Chapters 8–12 demonstrate respectively the application of velocity-free localization methods in slope stability analysis, localization of the trapped people after an accident or disaster, localization of autonomous driverless vehicles in underground intelligent mines, the study on micro-crack mechanism and precursor of rock instability, and the prediction of rock instability risk.

Changsha, China

Longjun Dong
Xibing Li

Acknowledgements

The book is partially supported by the International Cooperation and Exchange of the National Natural Science Foundation of China (52161135301), National Science Foundation for Excellent Young Scholars of China (51822407), the Natural Science Foundation of China (51774327), the Special Fund for Basic Scientific Research Operations in Universities (2282020cxqd055), and the National Key Research and Development Program of China (2021YFC2900500).

I would like to express my sincere appreciation to my supervisor Prof. Xi-Bing Li for his support of Master's and Ph.D. studies. I would also like to thank my students Yong-Chao Chen, Si-Jia Deng, Qing-Chun Hu, Qiao-Mu Luo, Zhong-Wei Pei, Wei-Wei Shu, Dao-Yuan Sun, Zheng Tang, Qing Tao, Xiao-Jie Tong, Jian Wang, Xian-Han Yan, Long-Bin Yang, Yi-Han Zhang, and Wei Zou, who assisted me in carrying out the verification experiments and blasting tests of all methods.

Finally, I am grateful to my parents wife, and children for their encouragement and support.

Changsha, Hunan, China — Longjun Dong

Contents

About the Authors

Longjun Dong is a Professor and the Director of Department of Safety Science and Engineering, Central South University, China. His research interests include rock mechanics, engineering seismicity, and applied acoustics. He obtained Ph.D., M.Sc., and B.Sc. in 2013, 2009, and 2007, respectively. He was appointed as Professor, Associate Professor, and Lecturer at Central South University in 2017, 2015, and 2013, respectively. He was invited as a research assistant at Australian Center for Geomechanics (ACG) from 2012 to 2013.

He presided over 20 projects of the National Key Research and Development Program, International Cooperation and Exchange of the National Natural Science Foundation of China (NSFC), etc. He is the recipient of the NSFC for Excellent Young Scholars (2018), the Young Elite Scientist sponsored by the China Association for Science and Technology (CAST) (2016), the leading talent of Science and Technology Innovation of Hunan Province (2021), and the Distinguished Young Scholars Fund of Hunan Province (2018). He established the multi-source acoustic theory of rock mass, and developed a complete set of apparatuses, based on the velocity-free localization methodology, for geo-acoustic monitoring and hazard early warning. He has 166 publications in peer-reviewed journals, with 5009 citations and 42 h-index in Google Scholar, and has 51 Chinese patents. He is an IAAM Fellow, ISRM Member, and an IEEE Senior Member. He was invited to present over 10 Keynote Presentations and Invited Presentations at international conferences. He is a Topical Editor in Chief of Arabian Journal of Geosciences (Springer Nature); an Associate Editor of Journal of the Acoustical Society of America Express Letter, and Shock and Vibration; an Editorial Board Member of Safety Science (Elsevier), Soils and Foundations (Elsevier, the official journal of the Japanese Geotechnical Society), Scientific Reports (Springer Nature), and Chinese Journal of Rock Mechanics and Engineering; an Editorial Advisory Board Member of Archives of Mining Sciences (established by the Polish Academy of Sciences since 1956), and a guest editor of Induced Seismicity in Scientific Reports. He was selected into the list of "Highly Cited Chinese Researchers (Elsevier)" and the "Top 2% of Scientists in the World (Stanford University)".

Xibing Li is a professor of Central South University. He obtained Ph.D., M.Sc., and B.Sc. at Central South University of Technology in 1992, 1986, and 1983, respectively. He was invited as a visiting researcher at University of Missouri-Rolla, America, from July 1998 to August 1999, and served as a researcher at Nanyang Technological University, Singapore, from September 1999 to May 2001. His research interests include the theory and technology of resource mining under hidden danger, mining without waste and hazard, mining of seabed bedrock, and hard rock mining in depth. He has been engaging in teaching and research in the field of mining and geotechnical engineering for more than 40 years during which about 80 Ph.D. students graduated under his supervision. One of the Ph.D. graduates was nominated for the National Excellent Doctoral Thesis and another four were awarded the Excellent Doctoral Thesis of Hunan Province. He has presided over more than 50 programs, such as Key Program and Major Program of National Natural Science Foundation of China, National Natural Science Foundation for Distinguished Young Scholars, National Key Research and Development program, National Basic Research Program of China (973 Program), National High-tech Research and Development Program of China (863 Program). He is the recipient of the National Science Foundation for Distinguished Young Scholars (1996) and the Distinguished Professor of "Changjiang Scholar" by Ministry of Education of China (2000). He has authorized 114 national invention patents and published 8 monographs. He has more than 380 publications in peer-reviewed journals, with 13298 citations and h-index of 60. He was the vice chairman of the Chinese Society of Rock Mechanics & Engineering, and vice chairman of the National Group of the International Society of Rock Mechanics (ISRM). He is an Editor in chief of Journal of safety and sustainability, an Associate Editor of Journal of Central South University, and an Editorial Board Member of Chinese Journal of Rock Mechanics and Engineering. He has obtained 4 Second Prizes of National Science and Technology Progress Award, and 13 First Prizes (and above) of provincial and ministerial level, as well as industrial science and technology awards. He also obtained the nomination of the National Top Ten Outstanding Scientific and Technological Workers. He was selected into the list of "Highly Cited Researchers(Clarivate)", "Highly Cited Chinese Researchers (Elsevier)", and the "Career long impact" as well as "Single year impact" lists of the world's top 2% scientists (World's Top 2% Sciences released by Stanford University).

Chapter 1
Introduction

Seismic source localization is one of the most classic and fundamental problems in seismology, hence it is of great significance for researching the basic problems related to seismology and rock mechanics (e.g., the structure of seismic activity, the internal structure of the Earth, the geometry structure of the seismic source, the response of microseismic to the rock mass, rock support under seismic loads, and the evolution of energy in the rock mass) [1–6]. Furthermore, fast and accurate source localization is also crucial in rescue and relief work after earthquake disasters and mine earthquakes. The localization methods are also widely applied to the localization of microseismic sources in mines, defects in materials, and acoustic emission (AE) sources of indoor experiments [7–9]. The location result can be used to effectively analyze the evolution law of high stress area and explore the propagation of material cracks [10, 11, 138–140]. In addition, source location is the basis for other analysis, for example tomography [141–143]. Figure 1.1 demonstrates the definition of AE technology, microseismic technology, and seismic monitoring technology according to the fracture scale of the monitoring. The field of microseismic/acoustic emission (MS/AE) source localization and series of effective localization methods have been developed.

To detail the merits and demerits of different localization methods, this chapter is arranged in four sections. In Sect. 1.1, the origin of MS/AE source localization and early development are descripted briefly. For better explanation, the MS/AE source localization methods are divided into three categories which are analytical methods, iterative methods and emerging methods. The three kinds of methods are reviewed in Sects. 1.2–1.4. It is noted that the classification is not absolute because sometimes two or more methods have some correlations.

L. Dong and X. Li, *Velocity-Free Localization Methodology for Acoustic and Microseismic Sources*, https://doi.org/10.1007/978-981-19-8610-9_1

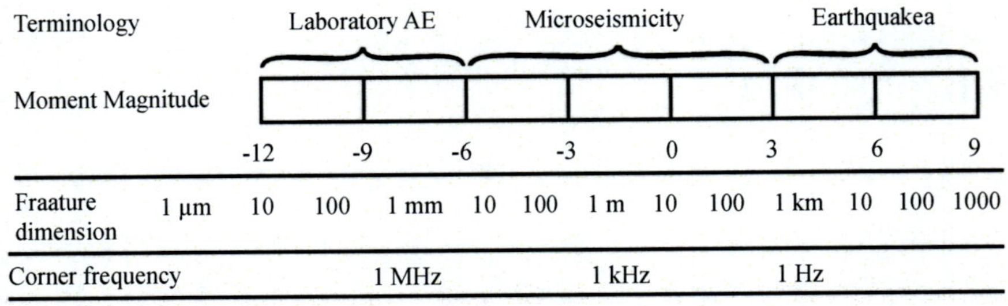

Fig. 1.1 Key aspects and parameters associated with rock fracture at different scales (from [12])

1.1 Origin and Early Development of MS/AE Source Localization

Due to the limitation of experimental equipment and computational tools, the method which was generally used to locate the sources in the early days was the geometry graphing method which impended the development of the MS/AE technique. The application and optimization of MS/AE source localization methods have been advanced greatly as a result of the rapid development of the computer technology (e.g., the Geiger localization method was proposed in 1910 [13, 14], but it was not widely used for source localization until the 1970s [15]). Based on Geiger method, more and more localization methods were developed, such as HYPOINVERSE [16], HYPOCENTER [17]. Then, joint inversion methods [18–20], and double-difference localization method (DD) [21] were developed. Since the least square method is introduced and the objective function is nonlinear, optimization theories, such as simplex method (SM) and Levenberg-Marquardt method (LM), were applied to solve the nonlinear equations and then obtain source position [20, 22].

In the past decades, various localization methods have been proposed and mainly divided into three categories which are detailed below.

1.2 Analytical Localization Methods

The coordinates and arrival time of each sensor is used in the analytical localization methods, to derive the analytical solutions with mathematical formulas. In 1928, Inglada [23] proposed an efficient and explicit method to solve the location equation. The advantage of the method lies in that fewer sensors (four) are needed to locate the AE. In addition, there is a unique analytic solution inside the sensor array. Essentially, the analytical method is also used to obtain the localization result in the Inglada method. Although the unique solution can be determined using the method, the wave velocity needs to be measured in advance. Christy [24] performed an iterative optimization of the Inglada method and achieved better location results than that of non-iteration. In 1968, Mogi [25] of the Japan Institute of Geological Survey carried out the transverse bending test of long strip samples with premeasured wave velocity, where two and four sensors were used for one-dimensional (1-D) and two-dimensional (2-D) localizations, respectively. It can be classified as a localization method of arrival-time difference with the small number of deployed sensors.

In 1970, Leighton and Blake [26] of the US Bureau of Mines developed an analytical localization method with simple steps, called USBM, which is widely applied in the localization of MS/AE source. In the method, the arrival times and coordinates of four sensors were recorded to achieve better localization accuracy. In 1974, Blake [27] then proved the uniqueness of the USBM solution. Godson and Bridgs [28] designed a 32-channel localization system, where USBM was applied to AE source localization with high accuracy. Strang [29] conducted an iterative study of the USBM and obtained a better localization effect compared to the non-iterative method.

Smith and Abel [30] developed a new analytical localization method using the sphere interpolation and least squares. Duraiswami et al. [31] proposed a set of analytical localization methods based on time difference localization. He argued that the localization solutions can be optimized using the L1 norm when the number of sensors is more than four. Brandstein et al. [32] presented a three-position method based on linear interpolation. Zhou et al. [33] proposed a closed-form method based on weight estimation, and improved location accuracy by using normal distribution to eliminate the outliers.

All the analytical localization methods mentioned above share the same disadvantage: the wave velocity needs to be determined in advance. In fact, in many practical engineering applications, the real-time wave velocities are unknown or difficult to measure due to the dynamic velocities and complex environments, which cause great temporal and spatial errors. Focusing on the vital issue, many scholars have proposed a series of optimized analytical localization methods to eliminate the effects of premeasured velocity on localization accuracy. Dong et al. [34] proposed the multi-sensor analytical localization method which is based on the logistic probability density function. The basic methodology of this multi-sensor analytical localization method is explained as follows: six sensors are selected randomly from m triggered sensors to combine $n = C_m^6$ groups of analytical solutions. The logistic probability density function is then applied to fit the whole solutions; the coordinates of the AE source are exactly the abscissas corresponding to the maximum value of the logistic probability density function. Figure 1.2 clarifies the localization process and highlights the proposed 3-D analytical localization method. After that, Zhou et al. [35] proposed a velocity-free localization using complete time difference of arrival (TDOA) measurements, which has higher noise immunity than the traditional methods. Rui et al. [36] applied DBSCAN principle to eliminate the effect of outliers on location accuracy and the velocity-free idea to avoid the error caused by pre-measured velocity.

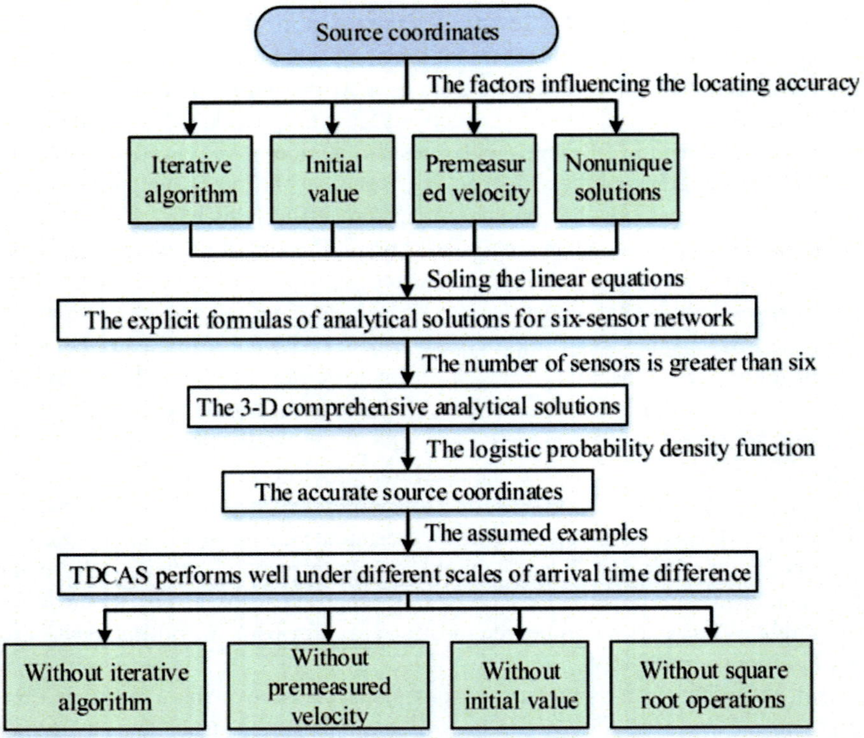

Fig. 1.2 Localization process and advantages of the proposed 3-D analytical localization method. To eliminate the effects of the iterative algorithm, initial value, premeasured velocity, and non-unique solutions, a 3-D comprehensive analytical localization method was proposed for random sensor networks. The localization processes for six sensors networks and random sensor networks are shown above. This method highlights four advantages: without iterative algorithm, without premeasured velocity, without initial value, and without square root operations (from [37])

1.3 Iterative Localization Methods

1.3.1 Linear Iterative Methods

1.3.1.1 Geiger Method

The current most used localization methods are usually based on the classic method proposed by Geiger in 1910 [13, 14], which until the 1970s, with the rapid development of the computer technology, was then applied to the MS/AE locating field [38]. The main idea of the Geiger method is to linearize the problem, in order to make the residual arrivals reach the minimum or a certain level of accuracy (by using iterative computations), and finally obtain the coordinates of the source. The specific steps are in detail in Chap. 2.

1.3.1.2 Optimized Algorithms Based on the Geiger Method

The Geiger method of seismic source localization is an iterative method. The iterative process usually converges rapidly unless the data is badly configured or the initial approximate is very far from the true solutions. However, if it also happens that the solution converges to a local minimum it would be difficult to detect in the output unless the residuals are very bad. Many scholars have proposed optimized algorithms based on Geiger method. Buland [39] combined QR algorithm with Geiger method to solve the solution. Aki and Lee [40] proposed a series of improved schemes based on Geiger method, including parameter separation, joint inversion of the 3-D velocity structure, and the source and coupling velocity with source resolution to solve the problem. Thurder [41] expanded the residuals formula with the second-order Tailor expansions and solved the coordinates of the epicenter. The second-order Tailor expansions were used to improve the convergence and stability of the algorithm, but it also increased the amount of computations required.

In 1975, Lee [15] released a series of computer programmers for determining hypocenter, magnitude, and first motion pattern of local earthquakes named HYPO71 and HYPO78-81 written in FORTRAN. Instead of carrying out the traditional procedure, a stepwise multiple regression was used in the HYPO71. A statistical analysis is first performed to see which independent variable should be included in the regression, and the normal equations are then set up for only those significant variables. Therefore, the adjustment vector is obtained by solving a matrix which is never ill-conditioned. Furthermore, convergence to a final hypocenter solution is also more rapid. The HYPO71 program also requires considerable efforts: accurate station coordinates, reasonable crustal structure model, as well as reliable P and S arrivals. Herrmann et al. [42] relocated the Denver earthquakes of 1967–1968 using HYPO71.

Byerlee [43] applied the Geiger method to the fluid injection test of rocks where six sensors were used to receive the AE signals and established the 3-D location equations with multi-parameters.

Klein [16] developed the HYPOINVERSE program to obtain the coordinates and the magnitude of the hypocenter. Lienert et al. [17] combined features of the two well-known methods HYPO71 and HYPOINVERSE, with a new technique—adaptive damping—and proposed a localization method—HYPOCENTER. Each column of the linearized condition matrix **T**, which relates changes in arrival time to changes in the hypocenter position, is centered and scaled to have zero mean and a norm of one, solved iteratively by adding a variable damping factor, θ^2, to their diagonal terms before inversion. If the residual sum of squares increases, return to the previous iteration, increase θ^2, and then try again. This procedure, which we called adaptive damping, always results in residuals which are less than or equal to the HYPO71 and the HYPOINVERSE residuals. The HYPOINVERSE fails to converge when the true depth is greater than about 40 km, while the HYPO71 fails at depths of greater than 50 km. The HYPOCENTER has a better performance on the difference in true versus calculated coordinates, especially depth.

Nelson and Vidale [44] improved the HYPOCENTER method and presented a new method for localization in a region with arbitrarily complex 3-D velocity structure called QUAKE3D. The method is a grid search method which searches all possible hypocenters and occurance times and finds the global minimum travel-time residual location within the volume. For a grid search method, the criterion is very important. QUAKE3D employs the L1 criterion (absolute value) instead of the commonly used L2 criterion (least squares method) based on the fact that L1 criterion is more robust than L2 criterion while the station coverage is sparse [45]. The HYPOINVERSE and the QUAKE3D were used to locate the source of earthquakes which occurred in Bear Valley from 1977 to 1986. The results proved that the QUAKE3D has more accuracy in location results than HYPOINVERSE.

1.3.1.3 Joint Inversion Methods

All the MS/AE source localization methods follow the basic distance-time constraints which can be expressed as

$$t - t_0 = \int \frac{1}{v(s)} \mathrm{d}s, \tag{1.1}$$

where t is the moment when the sensor receives the signal, t_0 is the time when the event happens, s is the wave propagation path from the source to sensor, and $v(s)$ is the wave velocity on the propagation path which is a path-dependent function.

It is vital to note that the simplification of the wave velocity structure is an important factor affecting the location accuracy. The effect of arrivals can be corrected by adding a station correction parameter to Eq. (1.1). The error caused by the simplified wave velocity structure can be solved by joint inversion of the source parameters and the wave velocity structure. The joint inversion problem is no longer centered on the location of a single source, but the location of the earthquake area. Two main methods of the joint inversion, joint inversion of the source location and the station correction, are described below.

Cleary and Hales [46] argued that Eq. (1.1) should consider a term for the correction of the station to describe the difference between the observed travel time and the travel time recorded in travel time table. A joint hypocenter location method-joint epicenter determination (JED)—was first proposed by Douglas in 1967 [18]. The method suggested that the travel time residual could be obtained by applying the station correction Δt_0^{s} to the arrivals,

$$r_i = \Delta t_0 + \frac{\partial T_i}{\partial x_0}\Delta x_0 + \frac{\partial T_i}{\partial y_0}\Delta y_0 + \frac{\partial T_i}{\partial z_0}\Delta z_0 + \Delta t_0^{\mathrm{s}}. \tag{1.2}$$

The coordinates of the events and the correction of the stations could also be obtained by applying standard least squares techniques to Eq. (1.2). Dewey [19] proposed a modification of Douglas' method called joint hypocenter determination (JHD) and applied it to relocate the earthquakes that occur in western Venezuela [47]. For a locating system, $m \times n$ equations can be obtained where m is the number of events and n is the number of stations. It should be noted that the number of equations is always so large that the JHD is usually inefficient. To solve the problem of oversized matrices due to the large number of AE events and stations, the Progressive multiple event location (PMLE) method using parameter separation based on the JHD was proposed by Pavlis and Booker in 1983 [48]. The PMLE requires less computation and still has strong stability. Pujol [49] discussed the relationship between the JHD method and the rough solution of Frohlich [50] to optimize the PMLE.

The JHD method and parameter separation method were applied to the monitoring data of the Kunming seismic network to obtain the corrected P-wave arrivals [51]. The result shows that an accurate locating result can be obtained by using the corrected arrivals. Ratchkovsky et al. [52] applied the JHD to relocate 1604 earthquakes occurred from 1988 to 1993 in the Wadati-Beinioff zone. Guo et al. [53] used Geiger method and the JHD to locate the earthquakes of the Jiashi area in Xinjiang Region, which suggests that the JHD has a much higher locating accuracy than Geiger method.

1.3.1.4 Relative Localization Methods

The master event localization method was developed from the JED. The principle of the method is that when the distance between two events is far less than the distance between the event and the station, it can be considered that the time difference between the two events and the stations is determined by the relative distance and the wave speed between them. Therefore, it can eliminate the influence of complicated wave velocity structure between source and station. Spence [54] put forward the master event localization method based on multi-event joint location method. In this method, one specially well-located event called master event is selected and the location of a group of earthquakes around it is calculated to determine the source location of these events. The accuracy of the method depends on the selection of the master event.

Ma [55] used the master event localization method to relocate the focal position of the Huoshan earthquake swarm in 1973 in Anhui province. Poupinet et al. [56] used the method to locate the earthquake of the Calaveras Fault in California. Hu et al. [57] developed a relative localization method for AE in the laboratory. This method has less dependence on velocity structure and is of great significance when applying to the AE locating in rock tests which is non-integral and anisotropy media.

If the focal mechanisms of the two earthquakes are similar and closely spaced, the propagation paths and the waveforms recorded on the same station are similar. Through the use of waveform cross correlation, the time difference can be accurate to the millisecond travel time and the relative errors between the two earthquakes can be reduced to tens of meters [58]. Waldhauser and Ellsworth [21] proposed DD method and developed a localization program—hypoDD [59]—based on this method in 2000. The method was applied to the localization of earthquakes in the Northern Hayward Fault of California.

The double differences between the station k and event i can be defined as

$$\mathrm{d}r_k^{ij} = \left(t_k^i - t_k^j\right)^{\mathrm{obs}} - \left(t_k^i - t_k^j\right)^{\mathrm{cal}}. \quad (1.3)$$

The following equation can be obtained by applying the first order of Taylor expansion:

$$\frac{\partial t_k^i}{\partial m}\Delta m^i - \frac{\partial t_k^j}{\partial m}\Delta m^j = \mathrm{d}r_k^{ij}. \quad (1.4)$$

Equation (1.4) can be written as

$$\begin{aligned} &\frac{\partial t_k^i}{\partial x}\Delta x^i + \frac{\partial t_k^i}{\partial y}\Delta y^i + \frac{\partial t_k^i}{\partial z}\Delta z^i + \Delta\tau^i - \frac{\partial t_k^j}{\partial x}\Delta x^j \\ &-\frac{\partial t_k^j}{\partial y}\Delta y^j - \frac{\partial t_k^j}{\partial z}\Delta z^j - \Delta\tau^j = \mathrm{d}r_k^{ij}. \end{aligned} \quad (1.5)$$

For all stations and events, the following matrix can be obtained:

$$\boldsymbol{WGm} = \boldsymbol{Wd}, \quad (1.6)$$

where **G** is a matrix of size $p \times 4q$ (p is the number of double-difference observations and q is the number of events) containing the partial derivatives, **d** is the data vector containing the double differences, **m** is a vector of length $4q$, $[\Delta x_i, \Delta y_i, \Delta z_i, \Delta t_i, \ldots]^{\mathrm{T}}$ containing the changes in hypocenter parameters which are to determined, and **W** is a diagonal matrix to weight each equation. The location of the events can be obtained by solving Eq. (1.6).

The difference between the DD and other methods is that no station correction term is needed to eliminate the effect of the velocity structure. The outstanding advantage of the DD is that it can use the cross-correlation analysis of waveform to pick the arrival time of the event and greatly improve the accuracy of the data. The DD inverts the relative position of each earthquake in a cluster of earthquakes relative to the centroid of the cluster, which is much different from the master event localization method and this gives it better applicability. In addition, the anti-interference and robustness of the DD are also strong and the method has been widely used in the localization of earthquakes [60]. Another multi-event localization method is PMEL [61], which is a grid search method. The JHD, HDC, DD, and PMEL were applied to the localization of the earthquakes in Izmit/Duzce, and the results were compared [62]. Recently, Feng et al. [63] has proposed the reverse double-difference time imaging method which can locate multiple seismic sources with high-precision when the velocity model has errors.

1.3.2 Nonlinear Iterative Methods

The above localization methods are all based on linear methods. In recent years, the nonlinear method theory has become the frontier in the fields of natural science. Since most of the geophysical problems are nonlinear problems, nonlinear methods tend to be more realistic than the linear methods while solving the geophysical problems. Various nonlinear optimization methods have been developed rapidly, including the methods based on derivatives such as the steepest descent method, Newton method, and conjugate gradient method and those based on no derivative including Monte Carlo method, genetic algorithm (GA), simulated annealing (SA), and random search and simplex search algorithm. Moreover, the waveform-based source localization method based on the wave equation is also a nonlinear localization method, which can obtain both the source location and velocity inversion information [64]. Due to the rapid development of computer technology, nonlinear methods have played an important role in geophysical inversion.

The essence of nonlinear localization methods is solving the least squares problem based on Eq. (1.1). The principles of some nonlinear localization methods are listed below.

1.3.2.1 Localization Methods Based on Nonlinear Optimization

In addition to linearizing the nonlinear formula and correcting them with iterations, many scholars has also proposed the localization methods based on the least squares method. For the arrivals data received by the AE sensors, the following function is constructed:

$$\Phi(t_0, x_0, y_0, z_0)=\sum_{i=1}^{n} r_i^2 \tag{1.7}$$

where r_i is the residual between the observed arrivals t_i and calculated arrivals $t_0 + T_i(x_0, y_0, z_0)$. $\Phi(t_0, x_0, y_0, z_0)$ is the quadratic sum of the arrivals residual r_i which represents the fitting degree between the hypothetical source and true source.

Twenty-two significant AE events during rock failure were located using the least squares method basing on the time difference between the S-wave arrivals and the six sensors [65]. The location results were used to analyze the propagation of micro-cracks during rock loading. Fedorov [66] described the least squares method, introduced a consistent estimator with normal distribution and solved it iteratively, and applied this method to AE localization. Since $\Phi(t_0, x_0, y_0, z_0)$ is a nonlinear function, some scholars proposed localization methods using the nonlinear optimization theory.

Powell's method [67] is a direct method of searching for the minimum value of the objective function. The method does not require partial derivatives or inverse matrices, has low requirements on the initial iteration value, and has good adaptability. The basic principle is to divide the whole calculation process into several stages. Each stage (one iteration) consists of $n + 1$ 1-D search. At each stage, we search first along the known n directions to get the best point and then search along the line connecting the initial point and the best point of this stage to find the best point. After this, the last search direction is used to replace one of the first n directions to start the next stage until the calculated residual value is less than the given allowable error or the number of iterations has reached the restraint. Many scholars have applied Powell's method for the localization of earthquake [68–70].

A method that uses Newton's method to solve the nonlinear equations was proposed in [71]. After removing the abnormal results basing on mean and standard deviation of all results, the average of remaining set is regarded as the source position. Han et al. [72] introduced modified factors, which could be calculated previously, to nonlinear equations. The factors modify the effects of the inhomogeneity or propagation delay but are determined by complicated tests. Hekmati [73] proposed a heuristic algorithm for source location considering indirect acoustic path. To improve the overall accuracy of the localization and the stationary of the objective function, Wu et al. [74] combined the L2-norm and variance function.

1.3.2.2 Localization Methods Based on Probabilistic Method

Monte Carlo method was first proposed by Metropolis and Ulam [75]. The method does not search thoroughly in the model space but searches randomly compared to the exhaustion method. Practice shows that if we randomly select the model in the model space and find the global minimum of the objective function. We will save a lot of time compared with planning the space of model space to calculate the global minimum of the model, but the workload is still huge. Therefore, there is no guarantee that the minimum value we found is the global minimum value. Meanwhile, the inherent randomness of the Monte Carlo method can lead to the failure of the calculation. Billings et al. [76] applied the Monte Carlo method to investigate the effect of picking errors.

Besides the Monte Carlo method, there are some probabilistic theories that can be applied to AE source localization. Niri et al. [77] proposed a method to estimate the mean and covariance of the AE source location distribution using unscented transformation. The method takes into account uncertainty in time of flight measurements and wave velocity and eventually estimates the AE source locations together with quantitative measures of confidence associated with those estimates. Tang et al. [78] proposed a probabilistic approach to estimate the AE source position and wave velocity simultaneously. The uncertainty of results can be also obtained. Kundu et al. [79] proposed a method to identify the generation cracks and predicted the position of damage based on the Bayesian inference. Zhou et al. [80] proposed an AE location method using tri-variate kernel density estimator, which has the higher location accuracy and tolerance for outliers comparing with traditional method. Jones et al. [81] proposed a Bayesian source localization strategy that is robust to composite materials and structures that contain non-trivial geometrical features.

1.3.2.3 Localization Methods Based on Simplex Method

In 1962, Spendley et al. [22] proposed the SM which is widely used as a mathematical tool in a variety of engineering fields. As a geometric search method, the SM ensures that each iteration is better than the previous one, so the optimal solution can be obtained only by repeated iterations. Meanwhile, if there is no optimal solution, the SM can also be used as an evaluation method. Nelder and Mead [82] proposed a search iterative method based on the work of Spendley in 1965.

Prugger and Gendzwill [83] introduced the SM into the seismic location and got a satisfactory localization result. This method avoids the derivative operation and the matrix transpose operation which greatly reduces the computational complexity and adapts L1 norm to reduce the influence of residual on arrival. Ge [84] evaluated Prugger's method and refined it. Zhao et al. [85] used the SM to locate the earthquakes in Tibet.

1.3.2.4 Localization Methods Based on Genetic Algorithm

Genetic algorithm [86, 87] is also a nonlinear global optimization method. The basic idea from the algorithm is imitating the genetic process of the biological world. The first step in using a genetic algorithm is to determine the encoding of the problem parameters, usually encoding of the parameters in binary. For earthquake location, the parameters are (x, y, z, t_0). The upper and lower bounds of the parameters are coded, and a set of individuals are randomly generated, called population. The residual between the calculated time and the actual observed time is used as the fitness function. The smaller the residual is, the higher the survival probability of the individual is and so is the probability of being the parent. Offspring can be created by crossover, and a certain probability of mutation is introduced to enrich the diversity of the population. The above process is repeated for the offspring obtained until the stop rule is satisfied and the individual with the highest fitness function is obtained, which is the optimal hypocenter parameter. Shen et al. [88] developed a method using time difference extraction algorithm and genetic algorithm. Using this method, the sources outside the sensor network can be located as accurately as that inside the sensor network.

The advantage of genetic algorithm lies in that the process of solving is only related to the object. It only needs to carry out simple operations such as crossover and mutation without complicated mathematical operations like calculating derivatives and has good global optimization ability. The genetic algorithm has been widely used in the field of source location and geophysical inversion [89–95].

1.3.2.5 Localization Methods Without Pre-measured Velocity

Since the pre-measured wave velocity can result in a large error in source localization, the velocity-free methods can be developed to obtain a more accurate result. In 2011, Dong et al. [96] proposed the mathematical functions of velocity-free methods and found that method based on time difference (TD) produce better results than traditional methods and method based on trigger time. Mostafapour et al. [97] used wavelet transform and cross-time frequency spectrum to calculate time delay. As a result, the real-time determined wave velocity is obtained to replace constant velocity. Harley and Moura [98] proposed two methods (data-driven localization and calibration-free lamb wave source localization) to locate AE source without pre-measured velocity and the location errors of both methods are less than 1 cm in a 1.22 m by 1.22 m isotropic aluminum plate. Davis [99] proposed a method determining the arrival time and velocity by analyzing the time domain waveform and the corresponding frequency spectrum. Then, the source position is estimated. Wang et al. [100] proposed a 1-D localization method without velocity calculation and insensitive to the threshold of arrival time estimation. Dong et al. [101] presented a MS/AE source localization method without the need for a pre-measured wave velocity. The location accuracy of the proposed method is significantly improved, which is superior to that of the traditional method using pre-measured wave velocity.

Based on the localization function with the model of arrival-time difference, Dong et al. [102] proposed a multi-step localization method without pre-measured velocity. The method can improve the location accuracy and computation efficiency in the complex environment compared to both the traditional localization method and TD method. A weighted linear least squares location method for an AE source without measuring wave velocity was presented in [103]. Zhou et al. [104] exploited the Cramer-Rao lower bound principle of unknown wave velocity, and the Newton's iterative method was combined to give the optimal solution.

1.4 Emerging Methods

The location accuracy would be affected by many factors, such as temperature [144] and abnormal arrival [145]. To locate the source position accurately and timely under complex structures and environment, scholars have developed a number of methods. These methods are classified into three categories and are introduced below.

1.4.1 Combination Methods

Considering the complex environment in practical application, using only an analytical method or iterative method for localization may get a poor result. Therefore, some combination methods that use different methods to solve corresponding problems and combine their advantages have been developed by many scholars.

The abnormal arrival of a certain sensor is another main factor that affects the localization accuracy. Note that a stable solution with high precision can be obtained using the analytical localization methods when the accurate data is provided. Dong et al. [105] proposed a collaborative localization method using analytical and iterative solutions to solve the source coordination. The location accuracy of the method proved to be better than that of the localization method with pre-measured velocity and TD method. Wang et al. [106] proposed a combination localization method based on TDOA method and beam forming method, which does not only improve the accuracy of the TDOA method, but also significantly reduce the localization calculation cost of the beam forming method. Zhou et al. [107] increased the locating speed of the virtual field optimization method by combining heuristic algorithms, and suggested that the particle swarm optimization algorithm can provide better location accuracy and computational efficiency for the virtual field optimization method than those obtained with the genetic algorithm. Liu et al. [108] combined the analytical solution and data field for source localization. The former is helpful for rapid AE location on a cylindrical shell structure, and the latter can overcome the local optimal problem. Yang et al. [109] applied the finite element method to generate training data and used delta-T mapping for localization on a complex plate, which reduces the location time and effort required for manually collecting and processing the training data.

Due to the complex environment, arrivals and velocity structure would contain some errors. A method that can accurately quarantine the bad sensor and localize sources with extreme outliers in an anisotropic medium was investigated in [110]. Considering real-time localization, some scholars [111, 112] proposed the methods to reduce calculation time. These methods can be a part of combination methods to make them more practical.

1.4.2 Localization Methods Based on Non-straight Wave Travel Paths

In most methods, the propagation paths of acoustic wave are assumed as the straight segments from sources to sensors. However, the assumption can hardly be satisfied in complex structure. Therefore, many scholars developed the localization methods considering non-straight wave travel paths in complex structures, which are introduced below.

Gollob et al. [113] proposed an algorithm, called FastWay, considering non-straight multi-segment wave travel paths. The FastWay has a better location accuracy than traditional method in heterogeneous media. Hu and Dong [146] then use ant colony algorithm to search fastest way path from the source to the sensor in complex structure. They [114] also proposed a velocity-free localization method (called ALM) for 2-D irregular complex structures. The method introduced A* search algorithm and the idea of velocity-free method, which avoids manual repetitive training by searching for paths between equidistant grid nodes and sensors. It also takes advantage of the localization method without requiring pre-measured velocity. The application and promotion of the autonomous driverless vehicles in underground intelligent mines are in great demand. Therefore, some tests about autonomous driverless vehicles localization are conducted in [115]. Furthermore, Dong et al. [116] proposed the velocity-free localization (called VFH) for the 3-D hole-containing structures, which can effectively locate sources in the complex 3-D structure. Xiao et al. [117] proposed a similar method for quasi-cylindrical structure with complex hole. However, all ALM, VFH and NBLM need to previously know the empty areas in the structure. To overcome the limitation, the localization methods with empty area detection and identification in 2-D structures were developed in [118] and [119]. To achieve a high-accuracy location in a complex cavern-containing structure, Jiang et al. [120] developed localization method using the fast marching method (FMM) with a second-order difference approach which offers a more accurate rock fracture location and facilitate the delineation of damage zones inside the surrounding rock mass. Because the travel time computation of microseismic waves in different directions in 3-D space has been found to be inaccurate, Li et al. [121] developed a method of calculating the P-wave travel time based on a 3-D high-order FMM. The shortest path method was proposed to compute the travel time for seismic event location in heterogeneous velocity models [122]. To improve location efficiency, only the boundaries of each geological unit is discretized to nodes.

Considering transversely isotropic medium, Wu et al. [123] proposed a Microseismic location algorithm for inclined strata using grid searching method. Zhou et al. [124] proposed a method that considers multi-layered media. However, the two methods need to measure the velocity of each layer in advance. Guo et al. [125] proposed a method based on the database of the arrival time differences which were calculated by fast sweeping method. Jiang et al. [126] used the FMM and the linear travel-time interpolation ray tracing to obtain relatively accurate first-arrival travel time, applied Runge-Kutta method to calculate the ray propagation path, and obtained source position by the least squares method. Huang et al. [127] proposed an inversion algorithm for simultaneous inversion of the model and event location.

These methods are based on a more practical assumption, so the location accuracy can be improved. Although considering the complex structure would lead to an inevitable increase in location time, the problem can be solved by the development of computer technology and the improved algorithms.

1.4.3 Localization Methods Based on Machine Learning

The development of machine learning provides a promising approach for localization in the complex environment and composite structure. Fu et al. [128] proposed a method based on back-propagation artificial neural network with the time difference of arrivals and the coordinates of sensors. Ebrahimkhanlou and Salamone [129] investigated the zonal AE source localization with only a single sensor using two deep learning approaches, among which the stacked auto-encoder approach achieved 100% accuracy. Based on the method, Ebrahimkhanlou et al. [130] proposed a generalizable deep learning framework for localizing and characterizing AE sources in riveted metallic panels. Huang et al. [131] proposed a method using convolution neural network and deep learning techniques to determine the source position in underground mines. Liu et al. [132] proposed generalized regression neural network based on time difference mapping which performed a better result than traditional time difference mapping. A framework based on stacked de-noising auto-encoders was proposed to localize AE sources in common and complex metallic panels [133]. The root mean squared location error of the method is smaller than that of support vector machine and artificial neural network. Using the relative P-wave arrival times as the input and AE source positions as the output, machine learning methods can provide effective and accurate approaches for relocating seismic events in a medium with unknown velocity structures [134]. Feng et al. [135] proposed a method for locating microseismic sources using deep reinforcement learning to improve the accuracy and efficiency of localization. Ma et al. [136] proposed a fully convolutional neural network for source localization overcoming the inaccuracy of velocity model and arrival picking which is a challenge in traditional method. A deep learning model was proposed which does not require prior information (wave velocity structure in the medium and arrival-time picking) to locate the AE source in rocks [137].

The application of machine learning improves the location accuracy compared to traditional methods. In fact, machine learning has been applied in many aspects of MS/AE technology, for example source discrimination [147]. However, there are some limitations that need to be overcome, such as, time-consuming training process and also the fact that the training data is hard to obtain.

References

1. Codeglia D, Dixon N, Fowmes GJ, Marcato G (2017) Analysis of acoustic emission patterns for monitoring of rock slope deformation mechanisms. Eng Geol 219:21–31. https://doi.org/10.1016/j.enggeo.2016.11.021
2. Warpinski N (2009) Microseismic monitoring: inside and out. J Petrol Technol 61(11):80–85. https://doi.org/10.2118/118537-JPT
3. Ge M (2005) Efficient mine microseismic monitoring. Int J Coal Geol 64(1):44–56. https://doi.org/10.1016/j.coal.2005.03.004
4. Wang H, Ge M (2008) Acoustic emission/microseismic source location analysis for a limestone mine exhibiting high horizontal stresses. Int J Rock Mech Min Sci 45(5):720–728. https://doi.org/10.1016/j.ijrmms.2007.08.009
5. Dong L, Sun D, Li X, Ma J, Zhang L, Tong X (2018) Interval non-probabilistic reliability of surrounding jointed rockmass considering microseismic loads in mining tunnels. Tunn Undergr Space Technol 81:326–335. https://doi.org/10.1016/j.tust.2018.06.034
6. Cheng JL, Song GD, Sun XY, Wen LF, Li F (2018) Research developments and prospects on microseismic source location in mines. Engineering 4(5):653–660
7. Baig A, Urbancic T (2010) Microseismic moment tensors: a path to understanding frac growth. Leading Edge 29(3):320–324. https://doi.org/10.1190/1.3353729
8. Moradian ZA, Ballivy G, Rivard P, Gravel C, Rousseau B (2010) Evaluating damage during shear tests of rock joints using acoustic emissions. Int J Rock Mech Min Sci 47(4):590–598. https://doi.org/10.1016/j.ijrmms.2010.01.004
9. Kundu T (2014) Acoustic source localization. Ultrasonics 54(1):25–38. https://doi.org/10.1016/j.ultras.2013.06.009
10. Novikov E, Oshkin R, Shkuratnik V, Epshtein S, Dobryakova N (2018) Application of thermally stimulated acoustic emission method to assess the thermal resistance and related properties of coals. Int J Min Sci Technol 28(2):243–249
11. Kaplan A, Gomez MP, Pumarega MIL, Nieva N (2015) Characterization of metal/ceramic interfaces in dental materials by acoustic emission. Proc Mater Sci 8:683–692

12. Dong L (2021) Localization and discrimination of microseismic/ae sources in mining: from data to information. In: Shen G, Zhang J, Wu Z (eds) Advances in acoustic emission technology. Springer, Singapore, pp 3–16
13. Geiger L (1910) Herdbestimmung bei Erdbeben aus den Ankunftszeiten. Nachrichten von der Gesellschaft der Wissenschaften zu Göttingen. Mathematisch-Physikalische Klasse 1910:331–349
14. Geiger L (1912) Probability method for the determination of earthquake epicenters from the arrival time only. Bull Saint Louis Univ 8(1):56–71
15. Lee W (1975) A computer program for determining hypocenter, magnitude, and first motion pattern of local earthquakes. US Geol Surv 75(311):1–116
16. Klein FW (1978) Hypocenter location program HYPOINVERSE: Part I. Users guide to Versions 1, 2, 3, and 4. Part II. Source listings and notes. US Geol Surv 78–694. https://doi.org/10.3133/ofr78694
17. Lienert BR, Berg E, Frazer LN (1986) HYPOCENTER: an earthquake location method using centered, scaled, and adaptively damped least squares. Bull Seismol Soc Am 76(3):771–783
18. Douglas A (1967) Joint epicentre determination. Nature 215(5096):47–48. https://doi.org/10.1038/215047a0
19. Dewey JW (1971) Seismicity studies with the method of joint hypocenter determination. University of California, Berkeley
20. Crosson RS (1976) Crustal structure modeling of earthquake data: 1. Simultaneous least squares estimation of hypocenter and velocity parameters. J Geophys Res (1896–1977) 81(17):3036–3046. https://doi.org/10.1029/JB081i017p03036
21. Waldhauser F, Ellsworth WL (2000) A double-difference earthquake location algorithm: method and application to the Northern Hayward Fault, California. Bull Seismol Soc Am 90(6):1353–1368. https://doi.org/10.1785/0120000006
22. Spendley W, Hext GR, Himsworth FR (1962) Sequential application of simplex designs in optimisation and evolutionary operation. Technometrics 4(4):441–461
23. Inglada V (1928) The calculation of the stove coordinates of a Nahbebens. Gerlands Beitrage Zur Geophys 19:73–98
24. Christy JJ (1982) A comparative study of the 'miner' and least squares location techniques as used for the seismic location of trapped coal miners. Pennsylvania State University
25. Mogi K (1968) Source locations of elastic shocks in the fracturing process in rocks. Bull Earthq Res Inst Univ Tokyo 46(4):1103–1125
26. Leighton F, Blake W (1970) Rock noise source location techniques, vol 7432. U.S. Department of Interior, Bureau of Mines
27. Blake W, Leighton F, Duvall WI (1974) Microseismic techniques for monitoring the behavior of rock structures. University of North Texas Libraries, Washington, D.C, USA
28. Godson R, MC B, BM M (1980) A 32-channel rock noise source location system. In: Proceedings of the 2nd Acoustic Emission Activity Conference, Pennsylvania State University, PA, USA, November 1978
29. Strang G (1993) The fundamental theorem of linear algebra. Am Math Mon 100(9):848–855
30. Smith J, Abel J (1987) Closed-form least-squares source location estimation from range-difference measurements. IEEE Trans Acoust Speech Signal Process 35(12):1661–1669
31. Duraiswami R, Zotkin D, Davis L (1999) Exact solutions for the problem of source location from measured time differences of arrival. J Acoust Soc Am 106(4):2277–2277
32. Brandstein MS, Adcock JE, Silverman HF (1997) A closed-form location estimator for use with room environment microphone arrays. IEEE Trans Speech Audio Process 5(1):45–50
33. Zhou Z, Rui Y, Zhou J, Dong L, Chen L, Cai X, Cheng R (2018) A new closed-form solution for acoustic emission source location in the presence of outliers. Appl Sci Basel 8(6). https://doi.org/10.3390/app8060949
34. Dong L, Li X, Ma J, Tang L (2017) Three-dimensional analytical comprehensive solutions for acoustic emission/microseismic sources of unknown velocity system. Chin J Rock Mechan Eng 36(1):186–197

35. Zhou Z, Rui Y, Cai X, Lan R, Cheng R (2020) A closed-form method of acoustic emission source location for velocity-free system using complete TDOA measurements. Sensors-Basel 20(12). https://doi.org/10.3390/s20123553
36. Rui Y, Zhou Z, Cai X, Dong L (2022) A novel robust method for acoustic emission source location using DBSCAN principle. Measurement 191. https://doi.org/10.1016/j.measurement.2022.110812
37. Dong L-J, Zou W, Sun D-Y, Tong X-J, Li X-B, Shu W-W (2019) Some developments and new insights for microseismic/acoustic emission source localization. Shock and Vibration 2019
38. Mowrey GL (1975) Computer processing and analysis of microscismic data. In: Proceedings of the 1st conference on AEIMA in geologia structures and materials, Pennsylvania State University, PA, USA, June 1975
39. Buland R (1976) The mechanics of locating earthquakes. Bull Seismol Soc Am 66(1):173–187
40. Aki K, Lee W (1976) Determination of three-dimensional velocity anomalies under a seismic array using first P arrival times from local earthquakes: 1. A homogeneous initial model. J Geophys Res 81(23):4381–4399
41. Thurber CH (1985) Nonlinear earthquake location: theory and examples. Bull Seismol Soc Am 75(3):779–790
42. Herrmann RB, Park S-K, Wang C-Y (1981) The Denver earthquakes of 1967–1968. Bull Seismol Soc Am 71(3):731–745
43. Byerlee JD (1977) Acoustic emission during fluid injection into rock. In: Proceedings of the 1st conference on acoustic emission/microseismic activity in geologic structures and materials, pp 87–98
44. Nelson GD, Vidale JE (1990) Earthquake locations by 3-D finite-difference travel times. Bull Seismol Soc Am 80(2):395–410
45. Draper NR, Smith H (2014) Applied regression analysis. John Wiley & Sons
46. Cleary J, Hales AL (1966) An analysis of the travel times of P waves to North American stations, in the distance range 32 to 100. Bull Seismol Soc Am 56(2):467–489
47. Dewey JW (1972) Seismicity and tectonics of western Venezuela. Bull Seismol Soc Am 62(6):1711–1751
48. Pavlis GL, Booker JR (1983) Progressive multiple event location (PMEL). Bull Seismol Soc Am 73(6A):1753–1777
49. Pujol J (1988) Comments on the joint determination of hypocenters and station corrections. Bull Seismol Soc Am 78(3):1179–1189
50. Frohlich C (1979) An efficient method for joint hypocenter determination for large groups of earthquakes. Comput Geosci 5(3–4):387–389
51. Wang C, Wang X, Yan Q (1993) Preliminary study on problems of multi-events location in Kunming seismic networks. Acta Seismol Sin 15(2):136–145
52. Ratchkovsky NA, Pujol J, Biswas NN (1997) Relocation of earthquakes in the Cook Inlet area, south central Alaska, using the joint hypocenter determination method. Bull Seismol Soc Am 87(3):620–636
53. Guo B, Liu Q-Y, Chen J-H, Li S-C (2002) Relocation of earthquake sequences using joint hypocenter determination method: portable seismic array study in Jiashi region, xinjiang. Seismol Egol 24(2):199
54. Spence W (1980) Relative epicenter determination using P-wave arrival-time differences. Bull Seismol Soc Am 70(1):171–183
55. Ma X-F (1982) Locating the epicenters of the 1973 Huoshan swarm, Anhui province, by the relative locating method. J Seismol Res 5(1):99–113
56. Poupinet G, Ellsworth W, Frechet J (1984) Monitoring velocity variations in the crust using earthquake doublets: an application to the Calaveras Fault, California. J Geophys Res Solid Earth 89(B7):5719–5731
57. Hu X, Ma S, Gao J, Liu L-q, Diao G-l, Song F-x, Liu S-g (2004) Location of acoustic emissions in non-integral rock using relative locating method. Chin J Rock Mechan Eng 34(12):277–283
58. Got JL, Fréchet J, Klein FW (1994) Deep fault plane geometry inferred from multiplet relative relocation beneath the south flank of Kilauea. J Geophys Res Solid Earth 99(B8):15375–15386

59. Waldhauser F (2001) hypoDD—a program to compute double-difference hypocenter locations Open-File Report, Version 1.0 edn. https://doi.org/10.3133/ofr01113
60. Miller SA, Collettini C, Chiaraluce L, Cocco M, Barchi M, Kaus BJ (2004) Aftershocks driven by a high-pressure CO_2 source at depth. Nature 427(6976):724–727
61. Rodi W, Toksoz M (2000) Grid-search techniques for seismic event location. massachusetts inst of tech cambridge earth resources lab
62. Rodi W, Engdahl ER, Bergman EA, Waldhauser F, Pavlis GL, Israelsson H, Dewey JW, Toksöz MN (2002) A new grid-search multiple-event location algorithm and a comparison of methods. In: The 24th seismic research review, 2002, pp 403–411
63. Feng Q, Pan B-Z, Han L-G, Zhang P (2021) Microseismic source location estimation using reverse double-difference time imaging. IEEE Access 9:66032–66042. https://doi.org/10.1109/access.2021.3076874
64. Zheng Y, Wang Y, Chang X (2016) Wave equation based microseismic source location and velocity inversion. Phys Earth Planet Inter 261:46–53
65. Scholz C (1968) Experimental study of the fracturing process in brittle rock. J Geophys Res 73(4):1447–1454
66. Fedorov V (1974) Regression problems with controllable variables subject to error. Biometrika 61(1):49–56
67. Powell MJ (1964) An efficient method for finding the minimum of a function of several variables without calculating derivatives. Comput J 7(2):155–162
68. Tang G-X (1979) A general method for determination of earthquake parameters by computer. Acta Seismol Sin 1(2):186–196
69. Yan Z, Xue J (1987) Redeterminations of weak earthquakes occurring in the Gorge region of the Yangtze river. Earthq Res China 3(1):52–59
70. Wang S-Y, Gao A-J, Xu Z-H, Zhang X-D, Guo Y (2000) Relocation of earthquakes in north-eastern region of Qinghai-Xizang plateau and characteristics of earthquake activity. Acta Seismol Sin 13(3):257–264
71. Antony D, Punekar GS (2015) Improvements in AEPD location identification by removing outliers and post processing. In: International conference on condition assessment techniques in electrical systems (CATCON), Bengaluru, India, December 2015, pp 66–69
72. Han Q, Xu J, Carpinteri A, Lacidogna G (2015) Localization of acoustic emission sources in structural health monitoring of masonry bridge. Struct Control Health Monit 22(2):314–329. https://doi.org/10.1002/stc.1675
73. Hekmati A (2016) A novel acoustic method of partial discharge allocation considering structure-borne waves. Int J Electr Power Energy Syst 77:250–255. https://doi.org/10.1016/j.ijepes.2015.11.083
74. Wu LZ, Li SH, Huang RQ, Wang SY (2020) Micro-seismic source location determined by a modified objective function. Eng Comput 36(4):1849–1856. https://doi.org/10.1007/s00366-019-00800-6
75. Metropolis N, Ulam S (1949) The monte carlo method. J Am Stat Assoc 44(247):335–341
76. Billings S, Sambridge M, Kennett B (1994) Errors in hypocenter location: picking, model, and magnitude dependence. Bull Seismol Soc Am 84(6):1978–1990
77. Niri ED, Farhidzadeh A, Salamone S (2015) Determination of the probability zone for acoustic emission source location in cylindrical shell structures. Mech Syst Signal Process 60–61:971–985. https://doi.org/10.1016/j.ymssp.2015.02.004
78. Tang J, Yan G, Cai C (2016) A particle filter-based method for acoustic emission source localization. Int J Appl Electromagnet Mech 52(3–4):975–981. https://doi.org/10.3233/jae-162188
79. Kundu A, Eaton MJ, Al-Jumali S, Sikdar S, Pullin R, Iop (2017) Acoustic emission based damage localization in composites structures using Bayesian identification. In: 12th International conference on damage assessment of structures (DAMAS), Kyushu Inst Technol, Kitakyushu, Japan, July 2017. J Phys Conf Ser. https://doi.org/10.1088/1742-6596/842/1/012081

80. Zhou Z, Lan R, Rui Y, Zhou J, Dong L, Cheng R (2019) A new acoustic emission source location method using tri-variate kernel density estimator. IEEE Access 7:158379–158388. https://doi.org/10.1109/access.2019.2950225
81. Jones MR, Rogers TJ, Worden K, Cross EJ (2022) A Bayesian methodology for localising acoustic emission sources in complex structures. Mech Syst Signal Process 163:14. https://doi.org/10.1016/j.ymssp.2021.108143
82. Nelder JA, Mead R (1965) A simplex method for function minimization. Comput J 7(4):308–313
83. Prugger AF, Gendzwill DJ (1988) Microearthquake location: A nonlinear approach that makes use of a simplex stepping procedure. Bull Seismol Soc Am 78(2):799–815
84. Ge M (1995) Comment on "microearthquake location: a nonlinear approach that makes use of a Simplex stepping procedure" by A. Prugger and D. Gendzwill. Bull Seismol Soc Am 85(1):375–377
85. Zhao Z, Ding ZF, Yi GX, Wang JG (1994) Hypocenter location for Tibet–a nonlinear location method optimized by Simplex algorithm. Acta Seismol Sin 16(2):212–219
86. Holland JH (1992) Adaptation in natural and artificial systems: an introductory analysis with applications to biology, control, and artificial intelligence. MIT press
87. Deb K, Pratap A, Agarwal S, Meyarivan T (2002) A fast and elitist multiobjective genetic algorithm: NSGA-II. IEEE Trans Evol Comput 6(2):182–197
88. Shen J-S, Zeng X-D, Li W, Jiang M-S (2017) An FBG acoustic emission source locating system based on PHAT and GA. Optoelectron Lett 13(5):330–334. https://doi.org/10.1007/s11801-017-7129-z
89. Wan Y, Li H (1995) The preliminary study on the seismic hypocenter location using genetic algorithms. Seismol Geomagnet Observ Res (in Chinese) 16(6):1–7
90. Han D-X, Wang G-Y (2009) Application of particle swarm optimization to seismic location. In: 2009 Third international conference on genetic and evolutionary computing. IEEE, pp 641–644
91. Sambridge M, Gallagher K (1993) Earthquake hypocenter location using genetic algorithms. Bull Seismol Soc Am 83(5):1467–1491
92. Nolte B, Frazer L (1994) Vertical seismic profile inversion with genetic algorithms. Geophys J Int 117(1):162–178
93. Sambridge M, Drijkoningen G (1992) Genetic algorithms in seismic waveform inversion. Geophys J Int 109(2):323–342
94. Billings S, Kennett BL, Sambridge MS (1994) Hypocentre location: genetic algorithms incorporating problem-specific information. Geophys J Int 118(3):693–706
95. Chen B, Feng X, Ding X, Xu P (2005) Back analysis on rheological parameters based on pattern-genetic-neural network. Chin J Rock Mechan Eng 24(4):553–558
96. Dong L-J, Li X-B, Tang L-Z, Gong F-Q (2011) Mathematical functions and parameters for microseismic source location without pre-measuring speed. Chin J Rock Mechan Eng 30(10):2057–2067
97. Mostafapour A, Davoodi S, Ghareaghaji M (2014) Acoustic emission source location in plates using wavelet analysis and cross time frequency spectrum. Ultrasonics 54(8):2055–2062. https://doi.org/10.1016/j.ultras.2014.06.022
98. Harley JB, Moura JMF (2015) Data-driven and calibration-free lamb wave source localization with sparse sensor arrays. IEEE Trans Ultrason Ferroelectr Freq Control 62(8):1516–1529. https://doi.org/10.1109/tuffc.2014.006860
99. Davis MIC, Rosalie C, Norman P, Rajic N (2017) Acoustic emission detection and characterisation using networked FBG sensors. In: 6th Asia-Pacific workshop on structural health monitoring (APWSHM), Hobart, Australia, December 2017. Proc Eng 440–447. https://doi.org/10.1016/j.proeng.2017.04.506
100. Wang C, Au YHJ, Li L, Cheng K (2016) Guided wave mode dispersion of transient acoustic emission on copper pipes-Its visualisation and application to source location. Mech Syst Signal Process 70–71:881–890. https://doi.org/10.1016/j.ymssp.2015.09.013

101. Dong L, Sun D, Li X, Du K (2017) Theoretical and experimental studies of localization methodology for AE and microseismic sources without pre-measured wave velocity in mines. IEEE Access 5:16818–16828
102. Dong L, Shu W, Han G, Li X, Wang J (2017) A multi-step source localization method with narrowing velocity interval of cyber-physical systems in buildings. IEEE Access 5:20207–20219
103. Zhou Z, Rui Y, Cai X, Cheng R, Du X, Lu J (2020) A weighted linear least squares location method of an acoustic emission source without measuring wave velocity. Sensors-Basel 20(11). https://doi.org/10.3390/s20113191
104. Zhou ZL, Rui YC, Cai X, Lu JY (2021) Constrained total least squares method using TDOA measurements for jointly estimating acoustic emission source and wave velocity. Measurement 182:12. https://doi.org/10.1016/j.measurement.2021.109758
105. Dong L, Zou W, Li X, Shu W, Wang Z (2019) Collaborative localization method using analytical and iterative solutions for microseismic/acoustic emission sources in the rockmass structure for underground mining. Eng Fract Mech 210:95–112. https://doi.org/10.1016/j.engfracmech.2018.01.032
106. Wang X, Liu X, He T, Tai J, Shan Y (2020) A novel joint localization method for acoustic emission source based on time difference of arrival and beamforming. Appl Sci-Basel 10(22). https://doi.org/10.3390/app10228045
107. Zhou J, Shen XJ, Qiu YG, Li EM, Rao DJ, Shi XZ (2021) Improving the efficiency of microseismic source locating using a heuristic algorithm-based virtual field optimization method. Geomech Geophys Geo-Energy Geo-Resour 7(3):18. https://doi.org/10.1007/s40948-021-00285-y
108. Liu C, Shang X, Miao R (2022) Acoustic emission source location on a cylindrical shell structure through grouped sensors based analytical solution and data field theory. Appl Acoust 192. https://doi.org/10.1016/j.apacoust.2022.108747
109. Yang H, Wang B, Grigg S, Zhu L, Liu DD, Marks R (2022) Acoustic emission source location using finite element generated delta-T mapping. Sensors-Basel 22(7):17. https://doi.org/10.3390/s22072493
110. Das AK, Lai TT, Chan CW, Leung CKY (2019) A new non-linear framework for localization of acoustic sources. Struct Health Monit 18(2):590–601. https://doi.org/10.1177/1475921718762154
111. Li F, Qin Y, Song W (2019) Waveform inversion-assisted distributed reverse time migration for microseismic location. IEEE J Sel Top Appl Earth Observ Rem Sens 12(4):1327–1332. https://doi.org/10.1109/jstars.2019.2904206
112. Tai J, He T, Pan Q, Zhang D, Wang X (2019) A fast beamforming method to localize an acoustic emission source under unknown wave speed. Materials 12(5). https://doi.org/10.3390/ma12050735
113. Gollob S, Kocur GK, Schumacher T, Mhamdi L, Vogel T (2017) A novel multi-segment path analysis based on a heterogeneous velocity model for the localization of acoustic emission sources in complex propagation media. Ultrasonics 74:48–61. https://doi.org/10.1016/j.ultras.2016.09.024
114. Hu Q, Dong L (2020) Acoustic emission source location and experimental verification for two-dimensional irregular complex structure. IEEE Sens J 20(5):2679–2691. https://doi.org/10.1109/jsen.2019.2954200
115. Dong L, Sun D, Han G, Li X, Hu Q, Shu L (2020) Velocity-free localization of autonomous driverless vehicles in underground intelligent mines. IEEE Trans Veh Technol 69(9):9292–9303. https://doi.org/10.1109/tvt.2020.2970842
116. Dong L, Hu Q, Tong X, Liu Y (2020) Velocity-free MS/AE source location method for three-dimensional hole-containing structures. Eng-Prc 6(7):827–834. https://doi.org/10.1016/j.eng.2019.12.016
117. Xiao P, Hu Q, Tao Q, Dong L, Yang Z, Zhang W (2020) Acoustic emission location method for quasi-cylindrical structure with complex hole. IEEE Access 8:35263–35275. https://doi.org/10.1109/access.2020.2972411

118. Dong L, Tong X, Hu Q, Tao Q (2021) Empty region identification method and experimental verification for the two-dimensional complex structure. Int J Rock Mech Min Sci 147:104885. https://doi.org/10.1016/j.ijrmms.2021.104885
119. Dong L, Tao Q, Hu Q, Deng S, Chen Y, Luo Q, Zhang X (2022) Acoustic emission source location method and experimental verification for structures containing unknown empty areas. Int J Min Sci Technol 32(3):487–497. https://doi.org/10.1016/j.ijmst.2022.01.002
120. Jiang RC, Dai F, Liu Y, Li A (2021) Fast marching method for microseismic source location in cavern-containing rockmass: performance analysis and engineering application. Eng-Prc 7(7):1023–1034. https://doi.org/10.1016/j.eng.2020.10.019
121. Li YJ, Wang J, Wang ZF, Sui QM, Xiong ZM (2021) Microseismic P-wave travel time computation and 3D localization based on a 3D high-order fast marching method. Sensors-Basel 21(17):24. https://doi.org/10.3390/s21175815
122. Duan Y, Luo X, Si GY, Canbulat I (2022) Seismic source location using the shortest path method based on boundary discretisation scheme for microseismic monitoring in underground mines. Int J Rock Mech Min Sci 149:13. https://doi.org/10.1016/j.ijrmms.2021.104982
123. Wu S, Chen Z, Zhang S, Xu M, Bai T (2018) Microseismic location algorithm for gently inclined strata and its numerical verification. Rock Soil Mech 39(1):297–307
124. Zhou Z, Rui Y, Zhou J, Dong L, Cai X (2018) Locating an acoustic emission source in multilayered media based on the refraction path method. IEEE Access 6:25090–25099. https://doi.org/10.1109/access.2018.2805384
125. Guo C, Gao Y, Wu S, Cheng Z, Zhang S, Han L (2019) Research of micro-seismic source location method in layered velocity medium based on 3D fast sweeping algorithm and arrival time differences database technique. Rock Soil Mech 40(3):1229–1238
126. Jiang R, Xu N, Dai F, Zhou J (2019) Research on microseismic location based on fast marching upwind linear interpolation method. Rock Soil Mech 40(9):3697–3708
127. Huang G, Ba J, Du Q, Carcione JM (2019) Simultaneous inversion for velocity model and microseismic sources in layered anisotropic media. J Petrol Sci Eng 173:1453–1463. https://doi.org/10.1016/j.petrol.2018.10.071
128. Fu T, Zhang Z, Liu Y, Leng J (2015) Development of an artificial neural network for source localization using a fiber optic acoustic emission sensor array. Struct Health Monit 14(2):168–177. https://doi.org/10.1177/1475921714568406
129. Ebrahimkhanlou A, Salamone S (2018) Single-sensor acoustic emission source localization in plate-like structures using deep learning. Aerospace 5(2). https://doi.org/10.3390/aerospace5020050
130. Ebrahimkhanlou A, Dubuc B, Salamone S (2019) A generalizable deep learning framework for localizing and characterizing acoustic emission sources in riveted metallic panels. Mech Syst Signal Process 130:248–272. https://doi.org/10.1016/j.ymssp.2019.04.050
131. Huang L, Li J, Hao H, Li X (2018) Micro-seismic event detection and location in underground mines by using Convolutional Neural Networks (CNN) and deep learning. Tunn Undergr Sp Tech 81:265–276. https://doi.org/10.1016/j.tust.2018.07.006
132. Liu ZH, Peng QL, Li X, He CF, Wu B (2020) Acoustic emission source localization with generalized regression neural network based on time difference mapping method. Exp Mech 60(5):679–694. https://doi.org/10.1007/s11340-020-00591-8
133. Yang L, Xu F (2020) A novel acoustic emission sources localization and identification method in metallic plates based on stacked denoising autoencoders. IEEE Access 8:141123–141142. https://doi.org/10.1109/access.2020.3012521
134. Zhao Q, Glaser SD (2020) Relocating acoustic emission in rocks with unknown velocity structure with machine learning. Rock Mech Rock Eng 53(5):2053–2061. https://doi.org/10.1007/s00603-019-02028-8
135. Feng Q, Han LG, Pan BZ, Zhao BH (2022) Microseismic source location using deep reinforcement learning. IEEE Trans Geosci Remote Sens 60:9. https://doi.org/10.1109/tgrs.2022.3182991

136. Ma K, Sun X-Y, Zhang Z-H, Hu J, Wang ZR (2022) Intelligent location of microseismic events based on a fully convolutional neural network (FCNN). Rock Mech Rock Eng 17. https://doi.org/10.1007/s00603-022-02911-x
137. Pu Y-Y, Chen J, Jiang D-Y, Apel DB (2022) Improved method for acoustic emission source location in rocks without prior information. Rock Mech Rock Eng 15. https://doi.org/10.1007/s00603-022-02909-5
138. Dong L, Chen Y, Sun D, Zhang Y, Deng S (2022) Implications for identification of principal stress directions from acoustic emission characteristics of granite under biaxial compression experiments. J Rock Mech Geotech Eng
139. Dong L, Yang L, Chen Y (2022) Acoustic emission location accuracy and spatial evolution characteristics of granite fracture in complex stress conditions. Rock Mech Rock Eng
140. Dong L, Chen Y, Sun D, Zhang Y (2021) Implications for rock instability precursors and principal stress direction from rock acoustic experiments. Int J Min Sci Technol 31(5):789-798
141. Dong L, Pei Z, Xie X, Zhang Y, Yan X (2022) Early identification of abnormal regions in rock-mass using traveltime tomography. Engineering
142. Dong L, Tong X, Ma J (2021) Quantitative investigation of tomographic effects in abnormal regions of complex structures. Engineering 7(7):1011-1022
143. Wang Z, Li X, Zhao D, Shang X, Dong L (2018) Time-lapse seismic tomography of an underground mining zone. Int J Rock Mech Min 107:136-149
144. Dong L-j, Tao Q, Hu Q-c (2021) Influence of temperature on acoustic emission source location accuracy in underground structure. Trans Nonferrous Met Soc China 31(8):2468-2478
145. Dong L, Shu W, Li X, Han G, Zou W (2017) Three dimensional comprehensive analytical solutions for locating sources of sensor networks in unknown velocity mining system. IEEE Access 5:11337-11351
146. Hu Q, Dong L An acoustic emission source localization method based ant colony without premeasured velocity. In: Singapore, 2021. Advances in Acoustic Emission Technology. Springer, Singapore, pp 71-78
147. Dong L-j, Tang Z, Li X-b, Chen Y-c, Xue J-c (2020) Discrimination of mining microseismic events and blasts using convolutional neural networks and original waveform. J Cent South Univ 27(10):3078-3089

Chapter 2
The Basic Theory of Source Localization

Geiger proposed a seismic source localization, named Geiger method, which becomes the beginning of source localization research. With the rapid development of computers, Geiger's ideas became widely used in seismic/acoustic location [1–15]. There are many works on source localization. Therefore, to understand the limits and make improvements to them, it is imperative to know the basic theory and some traditional methods.

In this chapter, the source localization theory is introduced as a basic knowledge for subsequent chapters. The basic theory of microseismic and AE technology is introduced in Sect. 2.1. In Sect. 2.2, localization methods are classified into three dimensional cases according to different application scenarios. To make the theory more specific, three localization methods with known velocity are also introduced in Sect. 2.3, including traditional methods and novel methods [16, 17]. The location results of traditional method and a basic velocity-free method [18] are compared and analyzed in Sect. 2.4.

2.1 Introduction for Microseismic and AE Monitoring Technology

2.1.1 Acoustic Emission Monitoring Technology

Acoustic emission is a phenomenon that materials rapidly release energy and send out acoustic waves instantaneously during the damage process of objects (such as rock mass) under the action of external forces. Acoustic emission testing technology is a non-destructive testing technology that uses AE testing instruments to receive, record, process, analyze AE signals, and obtain the relevant information about AE sources. The acoustic signal is transformed into an electrical signal and then processed into

L. Dong and X. Li, *Velocity-Free Localization Methodology for Acoustic and Microseismic Sources*, https://doi.org/10.1007/978-981-19-8610-9_2

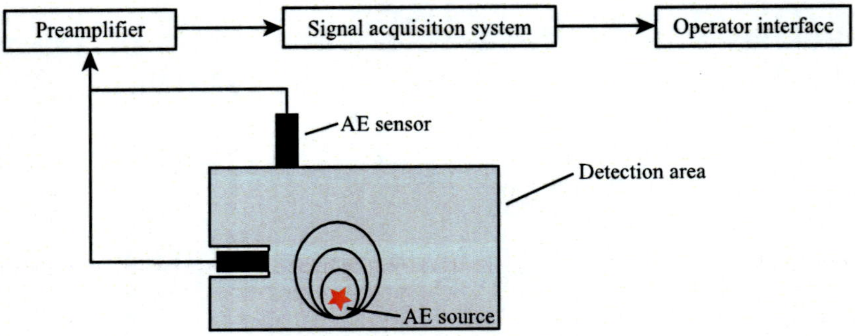

Fig. 2.1 Illustration of AE detection technology

a digital signal. The schematic diagram of AE detection technology is shown in Fig. 2.1.

The AE signals collected by the AE monitoring equipment can be characterized from many aspects, hence this results in a variety of analysis methods. With the improvement of computer ability to process data, various characteristic parameters can be obtained instantaneously. Some commonly of the main used AE parameters are shown below.

(1) Ring count
The threshold value should be set before the AE signal collection, and the value is related to the environment noise. Only the AE signal exceeding the threshold value will be recorded by the equipment. Ring count is the number of AE exceeding the threshold, as shown in Fig. 2.2. It is not only suitable for the analysis of burst AE signals, but also suitable for the analysis of continuous AE signals. It is widely used in the analysis of AE activities.

(2) Event count
The counting method is divided into ring counting and event counting. When the envelope of the pulse attenuation wave exceeds the threshold, a rectangular pulse will be generated. This count pulse is an event. An AE event is defined as a local transient change of material, which can be expressed by total count and count rate. The event count rate is defined as the events that occur per unit time. Event counting can evaluate the activity of AE source and the concentration of the source localization.

(3) Amplitude
The maximum value in the attenuation waveform of AE signal is called amplitude. It reflects the size of the event, and has nothing to do with the threshold value. The magnitude of the amplitude determines whether the event can be detected or not. It can be used to identify the type of AE source, evaluate the source intensity and the degree of attenuation.

(4) Energy count
Energy count is the area under the waveform envelope, which reflects the relative energy and intensity of AE events. Energy count can evaluate the activity of AE

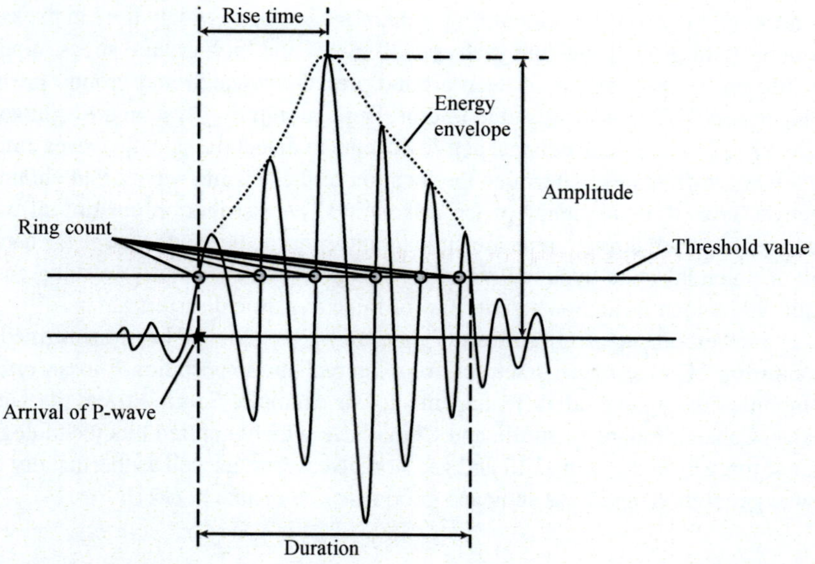

Fig. 2.2 Simplified waveform of burst AE signal

events, which is equivalent to the function of ring count. At the same time, energy count can be used to identify the type of wave source.

(5) Rise time
As shown in Fig. 2.2, the rise time is defined as the time interval between the first time the AE signal crosses the threshold and the maximum amplitude.

(6) Duration
The time interval between the AE signal crossing the threshold for the first time and finally falling to the threshold level is called the duration.

(7) Arrival of P wave
The time when the AE signal crosses the threshold for the first time. This is an important parameter since the arrival time of the P-wave is closely related to the setting of the threshold value. Setting the threshold too high will lead to delay.

(8) AE Sources
An acoustic emission source is generally defined as an energy released in a short period of time due to material deformation and fracture development. The elastic wave generated by AE sources contains information about the source, such as source position and characteristics of the source [19].

2.1.2 Microseismic Monitoring Technology

In recent years, it has been widely used in mine safety and underground engineering monitoring of hydropower. As an advanced and effective means of ground pressure monitoring, microseismic monitoring technology has been widely used in the safety monitoring of ground pressure in deep well mines and high ground stress mines. It has become a basic means of deep ground pressure research and ground pressure management [20]. Sensors used in microseismic monitoring system are arranged in the ore body where microseismic activity occurs to detect the seismic waves emitted by the micro-fractures, determine the location of the seismic waves, and obtain the strength as well as frequency of the seismicity. The distribution position of micro fractures obtained through microseismic monitoring can be used to judge the activity law of potential mine dynamic disasters. The prediction and early warning can be achieved by identifying the activity law of mine dynamic disasters [21].

The development of microseismic monitoring technology has transformed the monitoring of mine micro fractures from "unrealistic expectations" to an organic component of mining safety management. For example, South Africa, the United States, Canada, Japan, Australia and China have widely applied this technology in mines, tunnels, underground oil and gas storage caverns, as well as thermal dry rock power generation, and have achieved good research results [22–27].

2.2 Three Application Cases of the Source Localization

2.2.1 One-Dimensional Case

When source localization is applied to a linear structure, it can be regarded as a 1-D case (Fig. 2.3). For instance, in practical projects, pipeline leakage as well as rock bust localization in roadways and tunnels in most can be considered as 1-D localization. Since the source position is an unknown variable, the residual between observed values and calculated values is generally used to judge whether the source coordinates obtained by localization method are accurate or not. A classical and simple residual is derived by difference between arrivals recorded by sensors and calculated arrivals. The method of calculating residuals by using the time difference of arrivals is a variant of that of calculating them by arrivals.

Traditional line localization (TLL) methods assume that wave velocity is a constant variable. Arrivals are calculated by:

$$t_i = t_0 + \frac{R_i}{v}, \tag{2.1}$$

where $R_i = |x_i - x|$.

Since the source position is between S_1 and S_2, it can be calculated by Eq. (2.1) with $i = 1, 2$, as follows:

$$x = \frac{x_1 + x_2 - v(t_2 - t_1)}{2}. \tag{2.2}$$

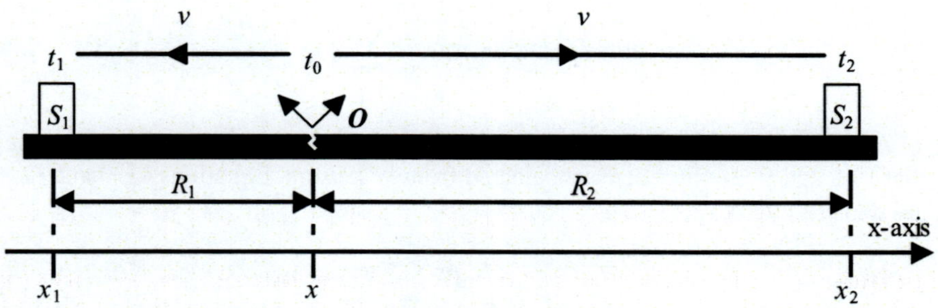

Fig. 2.3 Illustration of 1-D AE source localization. Variable v indicates wave velocity, t_0 the onset time of the event, S_i the i-th sensor, t_i the arrival time recorded by S_i, point O the source position, R_i the distance between O and S_i, x the coordinate of the rupture, as well as x_i the coordinate of S_i and i sensor number

If the number of sensors is greater than two, the optimal source position is derived by the least squares method so that all arrivals recorded by sensors can be fully utilized. For m sensors ($m > 2$), the source position can be derived by Eqs. (2.3) and (2.4). The localization based on Eqs. (2.3) and (2.4) are denoted by STT (speed and trigger time) and STD (speed and time difference), respectively.

$$(x, t_0) = \arg\min f(x, t_0) = \arg\min \sum_{i=1}^{m} \left(t_i - t_0 - \frac{R_i}{v} \right)^2 \tag{2.3}$$

$$x = \arg\min f(x) = \arg\min \sum_{i=2}^{m} \left(t_{i1} - \frac{R_i - R_1}{v} \right)^2 \tag{2.4}$$

where $t_{i1} = t_i - t_1$.

Equations (2.3) and (2.4) show different objective function represented by arrivals and the time difference of arrivals, respectively. In Eq. (2.3), onset time t_0 is an unknown variable and the independent variable is a vector containing x and t_0. By subtracting t_1 from t_i ($i = 2, \ldots, m$), t_0 is removed from the objective function, as shown in Eq. (2.4).

To circumvent the localization error caused by the difference between pre-measured wave velocity and actual wave velocity, the assumption that wave velocity is an unknown variable is made. Based on the assumption, at least three sensors are required to locate the source position. For three sensors, a solution is derived by Eq. (2.1) with $i = 1, 2, 3$. For m sensors ($m > 3$), Eqs. (2.3) and (2.4) can be rewritten as Eq. (2.5) and (2.6). The localization based on Eqs. (2.5) and (2.6) are denoted by TT (trigger times) and TD (time difference), respectively.

$$(x, v, t_0) = \arg\min f(x, v, t_0) = \arg\min \sum_{i=1}^{m} \left(t_i - t_0 - \frac{R_i}{v} \right)^2 \tag{2.5}$$

$$(x, v) = \arg\min f(x, v) = \arg\min \sum_{i=2}^{m} \left(t_{i1} - \frac{R_i - R_1}{v} \right)^2 \tag{2.6}$$

Another method is to eliminate the velocity v from objective function. Assuming that α, β and γ are three sensors selected from m sensors. Equations (2.7) and (2.8) are derived by Eq. (2.1).

$$t_\alpha - t_\beta = \frac{(R_\alpha - R_\beta)}{v} \tag{2.7}$$

$$t_\alpha - t_\gamma = \frac{(R_\alpha - R_\gamma)}{v} \tag{2.8}$$

Comparing Eqs. (2.7) and (2.8), we have

$$W = \frac{t_\alpha - t_\beta}{t_\alpha - t_\gamma} = \frac{R_\alpha - R_\beta}{R_\alpha - R_\gamma}. \tag{2.9}$$

The parameters to be solved in Eq. (2.9) are the source coordinates x. For each group of observation values $(x_{\alpha i}, x_{\beta i}, x_{\gamma i})$. Equation (2.9) can determine a regression value:

$$\hat{W} = \frac{R_\alpha - R_\beta}{R_\alpha - R_\gamma}. \tag{2.10}$$

The difference between the regression value $\hat{W}$ and the measured value W describes the degree of deviation between the regression value and the measured value. For $(x_{\alpha i}, x_{\beta i}, x_{\gamma i})$, the smaller the difference between $\hat{W}$ and W, the better the fitting degree between the fitting curve and observed values. The square sum of $\hat{W} - W$ can describe the deviation degree of all observed values from the fitted values. Therefore, x should make $f(x)$ minimum. So, the result becomes

$$x = \arg\min f(x) = \arg\min \sum_{\alpha>\beta>\gamma=1}^{m} \left(\hat{W} - \frac{R_\alpha - R_\beta}{R_\alpha - R_\gamma} \right)^2. \tag{2.11}$$

The localization based on Eq. (2.11) is denoted by TDQ (time difference quotient).

2.2.2 *Two-Dimensional Case*

When sources and sensors are in a plane, the source localization is considered as a 2-D case. In aerospace and construction industry, many plane materials and structures are employed. Therefore, 2-D localization methods are mostly applied to structural health monitoring and play an important role because a small crack or rupture may cause a big accident.

Figure 2.4 illustrates 2-D localization using arrivals and time difference of arrivals in geometry taking three sensors as an example. In a 2-D case, at least three sensors are used for source localization with pre-measured velocity and at least four sensors for velocity-free method. When there are redundant sensors, the objective functions of STT, STD, TT, TD and TDQ can be rewritten as Eqs. (2.12)–(2.16), respectively.

$$(x, y, t_0) = \arg\min f(x, y, t_0) = \arg\min \sum_{i=1}^{m} \left(t_i - t_0 - \frac{R_i}{v} \right)^2 \tag{2.12}$$

$$(x, y) = \arg\min f(x, y) = \arg\min \sum_{i=2}^{m} \left(t_{i1} - \frac{R_i - R_1}{v} \right)^2 \tag{2.13}$$

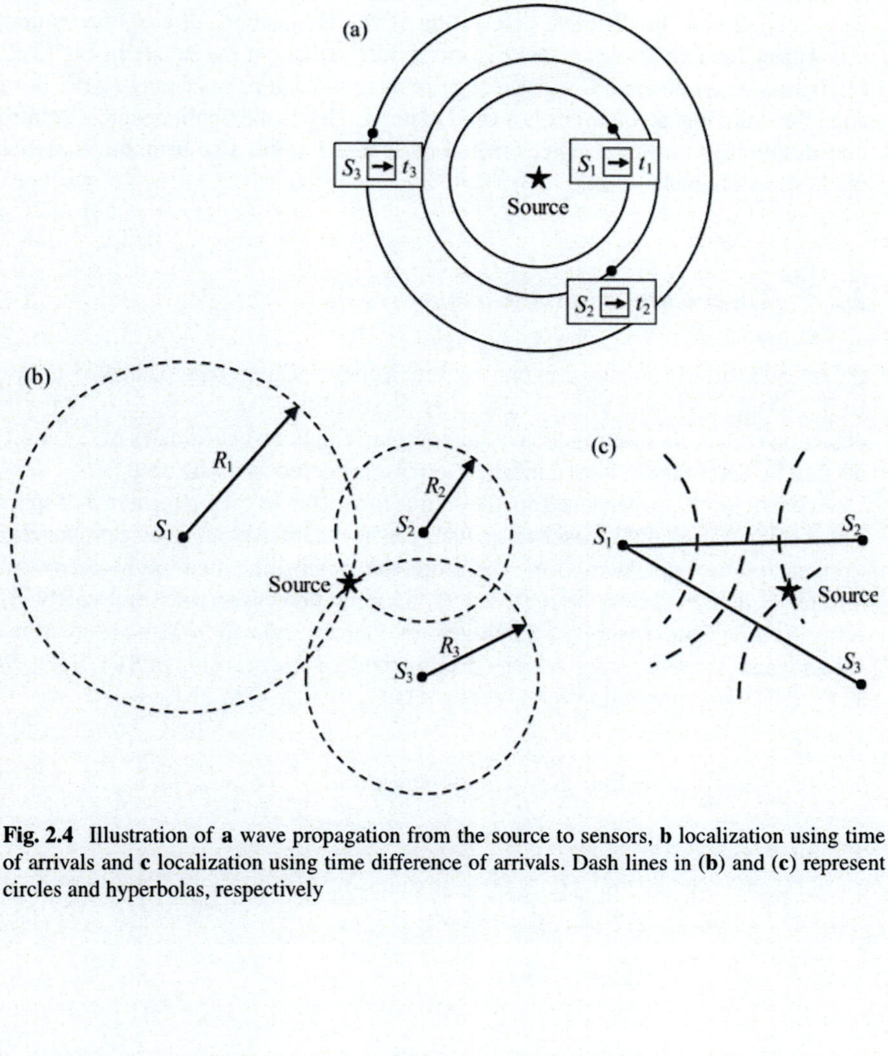

Fig. 2.4 Illustration of **a** wave propagation from the source to sensors, **b** localization using time of arrivals and **c** localization using time difference of arrivals. Dash lines in (**b**) and (**c**) represent circles and hyperbolas, respectively

$$(x, y, v, t_0) = \arg\min f(x, y, v, t_0) = \arg\min \sum_{i=1}^{m} \left(t_i - t_0 - \frac{R_i}{v} \right)^2 \tag{2.14}$$

$$(x, y, v) = \arg\min f(x, y, v) = \arg\min \sum_{i=2}^{m} \left(t_{i1} - \frac{R_i - R_1}{v} \right)^2 \tag{2.15}$$

$$(x, y) = \arg\min f(x, y) = \arg\min \sum_{\alpha>\beta>\gamma=1}^{m} \left(\hat{W} - \frac{R_\alpha - R_\beta}{R_\alpha - R_\gamma} \right)^2 \tag{2.16}$$

The distance between the source and ith sensor is derived by

$$R_i = \sqrt{(x - x_i)^2 + (y - y_i)^2}. \tag{2.17}$$

Equation (2.1) in 1-D case is a little different from that in 2-D case (and 3-D case), because of the different calculation of R_i. The symbol of absolute value in one-dimension can be eliminated by prior information, but the square in Eq. (2.17) can hardly be eliminated. Besides using optimization algorithm, changing nonlinear equations into linear equations is a valid method. This is the fundamental idea introduced not only from the Geiger method in Sect. 2.3.1, but also from the analytical methods in Chap. 4.

2.2.3 *Three-Dimensional Case*

In fact, all source localization in practical environment can be regarded as a 3-D case. Both 1-D and 2-D cases are 3-D cases that are simplified for convenience. When the detection area or monitoring area is stereoscopic and three coordinates of the source need to be determined, source localization is considered as a 3-D case.

Acoustic wave propagates from the source to sensors in sphere form in 3-D space. The localization methods based on arrivals can be regarded as an inverse propagation from each sensor in sphere form. Therefore, the localization methods based on time difference of arrivals can be regarded as the same process in rotating hyperboloid form. At least four sensors are employed for source localization with pre-measured velocity and five sensors for velocity-free methods. The expansion of STT, STD, TT, TD and TDQ to three-dimension is shown in Eqs. (2.18)–(2.22).

$$(x, y, z, t_0) = \arg\min f(x, y, z, t_0) = \arg\min \sum_{i=1}^{m} \left(t_i - t_0 - \frac{R_i}{v} \right)^2 \tag{2.18}$$

$$(x, y, z) = \arg\min f(x, y, z) = \arg\min \sum_{i=2}^{m} \left(t_{i1} - \frac{R_i - R_1}{v} \right)^2 \tag{2.19}$$

$$(x, y, z, v, t_0) = \arg\min f(x, y, z, v, t_0) = \arg\min \sum_{i=1}^{m} \left(t_i - t_0 - \frac{R_i}{v} \right)^2 \quad (2.20)$$

$$(x, y, z, v) = \arg\min f(x, y, z, v) = \arg\min \sum_{i=2}^{m} \left(t_{i1} - \frac{R_i - R_1}{v} \right)^2 \quad (2.21)$$

$$(x, y, z) = \arg\min f(x, y, z) = \arg\min \sum_{\alpha>\beta>\gamma=1}^{m} \left(\hat{W} - \frac{R_\alpha - R_\beta}{R_\alpha - R_\gamma} \right)^2 \quad (2.22)$$

The distance R_i is derived by

$$R_i = \sqrt{(x - x_i)^2 + (y - y_i)^2 + (z - z_i)^2}. \quad (2.23)$$

There are a kind of mathematical fitting forms without measuring the velocity in advance (i.e. TT, TD and TDQ) and two kinds of traditional mathematical fitting forms (i.e. STT and STD) are detailed above in 1-D, 2-D and 3-D cases. For velocity-free methods, the fitting parameters of TDQ are less than TT and TD, but the denominator term is $R_\alpha - R_\gamma$. If two sensors have the same distance to the source, the sensors will be invalid and the location accuracy will be affected. Therefore, TDQ needs more sensors than TT and TD, and this makes it expensive and complicated in practical engineering. For this reason, its specific simulation and application is not analyzed. Analyzing the mathematical fitting form of other two methods, TT needs at least five sensors to fit five unknown parameters through four known parameters, and TD needs at least four sensors to fit four unknown parameters through four known parameters. From the perspective of data fitting, TD is better than TT. It is verified and analyzed by the calculation examples and blasting tests in Sect. 2.4.

2.3 Source Localization Methods with Known Wave Velocity

2.3.1 Traditional Methods

Since the wave velocity of P wave is the fastest and the P-wave arrival time is easy to identify, source localization based on P-wave arrival time is generally adopted. Traditional methods assume that the medium is a uniform velocity model and the P-wave velocity is known. Then, at least four sensors are employed for source localization. Based on this assumption, the propagation velocity of P wave is considered a constant value. The objective functions in Eqs. (2.3) and (2.4) as well as their expansion to 2-D cases (as shown in Eqs. 2.12 and 2.13) and 3-D cases (as shown in Eqs. 2.18 and 2.19) are nonlinear. To solve the nonlinear objective function, some methods have been developed. In this section, Geiger method and SM are introduced.

2.3.1.1 Geiger Method

Geiger method [28] Taylor expands the arrival time around the initial values, and updates the values by iteration until the result achieves the convergence or the maximum iteration is achieved. Although a 3-D example is described here, Geiger method can be also applied to 2-D and 1-D cases. Arrival times can be calculated as

$$t_i = t_0 + \frac{\sqrt{(x - x_i)^2 + (y - y_i)^2 + (z - z_i)^2}}{v}. \tag{2.24}$$

The initial value is denoted by $\mathbf{x_0}$, where $\mathbf{x_0}$ represents (x_0, y_0, z_0, T_0) with T_0 being initial value of t_0. Equation (2.25) is deduced by first order Taylor expansion of Eq. (2.24) about $\mathbf{x_0}$.

$$t_i^{\mathrm{o}} = t_i^{\mathrm{c}} + \frac{\partial t_i}{\partial t_0}\Delta t_0 + \frac{\partial t_i}{\partial x}\Delta x + \frac{\partial t_i}{\partial y}\Delta y + \frac{\partial t_i}{\partial z}\Delta z \tag{2.25}$$

where t^{o} and t^{c} indicate observed and calculated values of arrival times. t^{c} is calculated by taking $\mathbf{x_0}$ into Eq. (2.24). The coefficients of Δx, Δy, Δz and Δt_0 are derived by

$$\begin{aligned}
\frac{\partial t_i}{\partial t_0} &= 1, \\
\frac{\partial t_i}{\partial x} &= \frac{x - x_i}{\sqrt{(x - x_i)^2 + (y - y_i)^2 + (z - z_i)^2}}, \\
\frac{\partial t_i}{\partial y} &= \frac{y - y_i}{\sqrt{(x - x_i)^2 + (y - y_i)^2 + (z - z_i)^2}}, \\
\frac{\partial t_i}{\partial z} &= \frac{z - z_i}{\sqrt{(x - x_i)^2 + (y - y_i)^2 + (z - z_i)^2}}.
\end{aligned} \tag{2.26}$$

The difference between initial value and the true value is estimated by solving linear Eq. (2.25). It is denoted by $\Delta\mathbf{x}$, where $\Delta\mathbf{x}$ is $(\Delta x, \Delta y, \Delta z, \Delta t_0)$. The initial value is then updated by $\mathbf{x_0} = \mathbf{x_0} + \Delta\mathbf{x}$ and the next iteration starts with new $\mathbf{x_0}$ until a condition is satisfied, for instance, $\|\Delta \boldsymbol{x}\|_2$ is less than 1×10^{-16}.

2.3.1.2 Simplex Method

Simplex method [29] is a geometric search method and the result will be closer to actual position after every iteration. The method constructs a simplex which is

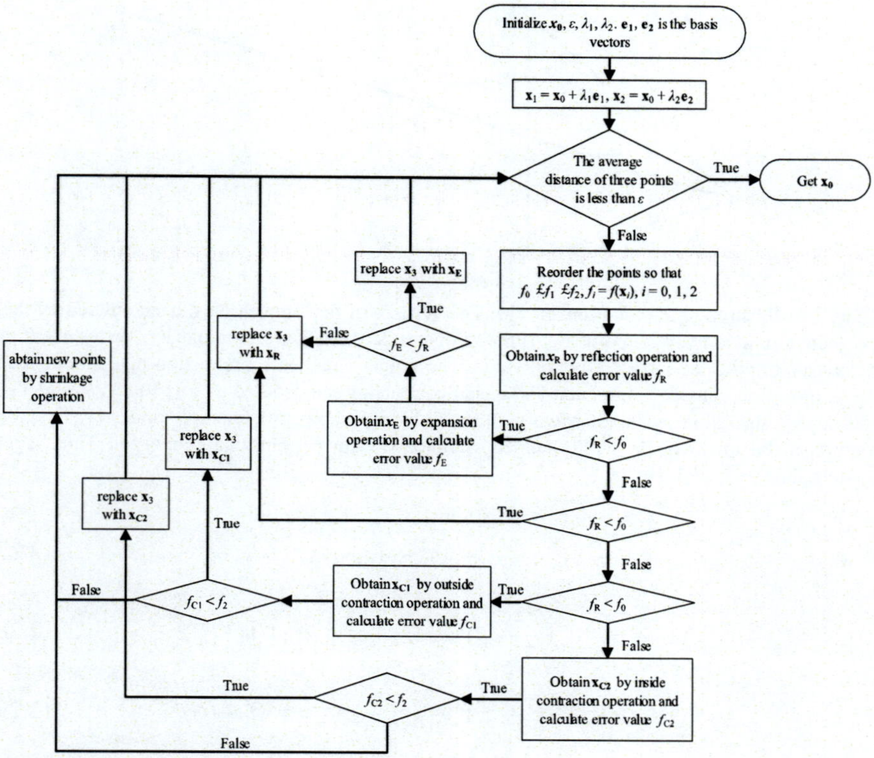

Fig. 2.5 Flow chart of SM method

a triangle in 2-D case. The vertices of the simplex become closer and closer to the minimum value of the objective function through iteration. Then the optimal solution is obtained. The simplex method is also suitable for 3-D case where the simplex is a tetrahedron. Figure 2.5 illustrates the flow chart of the algorithm. Some operations in SM are shown in Fig. 2.6. Considering divergence of location, a maximum iteration can be set.

2.3.2 Localization Method Based on Linearization of Nonlinear Equations

2.3.2.1 Methodology

An acoustic emission source is located at source (x, y, z) and the five geo-sensors are located at S_1 (x_1, y_1, z_1), S_2 (x_2, y_2, z_2), S_4 (x_3, y_3, z_3), S_4 (x_4, y_4, z_4) and S_5 (x_5, y_5, z_5). The equations governing the position of acoustic emission source and sensors are:

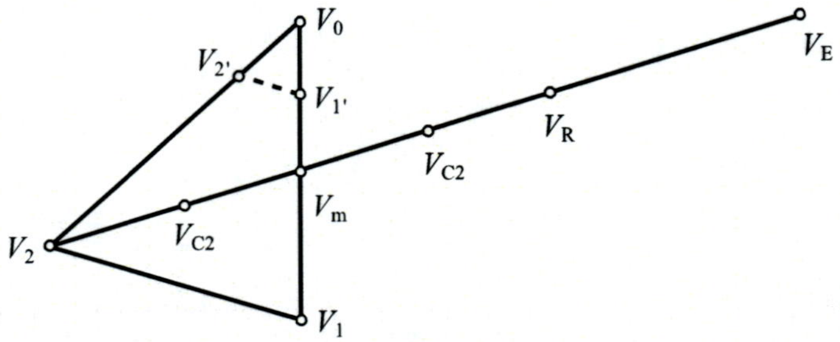

Fig. 2.6 Illustration of operations in SM. The vertices of the simplex have been ordered by their objective function values. Point V_m is the midpoint between Vertex V_0 and V_1. Vertex V_2 is the worst point with the largest function value. The points obtained by reflection operation, expansion operation, outside contraction and inside contraction are denoted as V_R, V_E, V_{C1}, and V_{C2}. Shrinkage operation means that two vertices with larger objective function values (V_1 and V_2) approach the point with the smallest objective function value (V_0) in equal proportion. Then, it gets a new simplex $\triangle\ V_0 V_{1'} V_{2'}$

$$(x_1 - x)^2 + (y_1 - y)^2 + (z_1 - z)^2 = v^2 t_0^2, \tag{2.27}$$

$$(x_2 - x)^2 + (y_2 - y)^2 + (z_2 - z)^2 = v^2 (t_0 + t_{12})^2, \tag{2.28}$$

$$(x_3 - x)^2 + (y_3 - y)^2 + (z_3 - z)^2 = v^2 (t_0 + t_{13})^2, \tag{2.29}$$

$$(x_4 - x)^2 + (y_4 - y)^2 + (z_4 - z)^2 = v^2 (t_0 + t_{14})^2, \tag{2.30}$$

$$(x_5 - x)^2 + (y_5 - y)^2 + (z_5 - z)^2 = v^2 (t_0 + t_{15})^2. \tag{2.31}$$

where t_0 is acoustic wave propagation time from source to the nearest sensor. The signal arrives first to the nearest sensor. After identification of the nearest sensor and calculation of propagation time delay between nearest sensor and other sensors, the sensors are numbered. The nearest sensor from source is numbered as sensor 1 (S_1). t_{12}, t_{13}, t_{14} and t_{15} are the propagation time delay between sensor 1 and sensors 2, 3, 4 and 5, respectively. The P wave velocity is indicated by v. Equations (2.27)–(2.31) represent five spheres (denoted by spheres (1)–(5)), with the center of respective sensor position, passing through the source. Any two of the spheres 1–5 intersect and the source is located on the intersecting circle. The equation of the intersecting plane for the spheres 1 and 2 can be obtained by taking difference of 1 and 2. This equation of intersecting plane is given in Eq. (2.32). In the following equations, l_i is a constant:

$$2(x_2 - x_1)x + 2(y_2 - y_1)y + 2(z_2 - z_1)z + 2t_{12}v^2 t_0 = l_1 \tag{2.32}$$

where $l_1 = x_2^2 - x_1^2 + y_2^2 - y_1^2 + z_2^2 - z_1^2 - v^2 t_{12}^2$.

Similarly, the equations for the intersecting planes for spheres (1) and (3), for spheres (1) and (4), and for spheres (1) and (5) are obtained by taking difference of (1) and (3) and (1) and (4), as well as (1) and (5), respectively. These equations are given in Eq. (2.33)–(2.35), respectively.

$$2(x_3 - x_1)x+2(y_3 - y_1)y+2(z_3 - z_1)z + 2v^2 t_{13}t_0 = l_2 \tag{2.33}$$

$$2(x_4 - x_1)x+2(y_4 - y_1)y+2(z_4 - z_1)z + 2v^2 t_{14}t_0 = l_3 \tag{2.34}$$

$$2(x_5 - x_1)x+2(y_5 - y_1)y+2(z_5 - z_1)z + 2v^2 t_{15}t_0 = l_4 \tag{2.35}$$

Equations (2.32)–(2.35) constitute a set of linear equations, which were rewritten as:

$$l_5x+l_6y+l_7z + l_8t_0 = l_1, \tag{2.36}$$

$$l_9x+l_{10}y+l_{11}z + l_{12}t_0 = l_2, \tag{2.37}$$

$$l_{13}x+l_{14}y+l_{15}z + l_{16}t_0 = l_3, \tag{2.38}$$

$$l_{17}x+l_{18}y+l_{19}z + l_{20}t_0 = l_4, \tag{2.39}$$

where $l_5 = 2(x_2 - x_1)$,$l_6 = 2(y_2 - y_1)$,$l_7 = 2(z_2 - z_1)$,$l_8 = 2t_{12}v^2$,$l_9 = 2(x_3 - x_1)$,$l_{10} = 2(y_3 - y_1)$,$l_{11} = 2(z_3 - z_1)$,$l_{12} = 2t_{13}v^2$,$l_{13} = 2(x_4 - x_1)$,$l_{14} = 2(y_4 - y_1)$,$l_{15} = 2(z_4 - z_1)$,$l_{16} = 2t_{14}v^2$,$l_{17} = 2(x_5 - x_1)$,$l_{18} = 2(y_3 - y_1)$,$l_{19} = 2(z_5 - z_1)$ and $l_{20} = 2t_{15}v^2$.

Therefore, the *x*, *y*, *z* which can be obtained by solving the linear equations, can be expressed and simplified as Eqs. (2.40)–(2.42).

$$\begin{aligned}
x =& [l_1 (l_{10}l_{16}l_{19} - l_{11}l_{16}l_{18} - l_{12}l_{14}l_{19} + l_{12}l_{15}l_{18} - l_{10}l_{15}l_{20} + l_{11}l_{14}l_{20}) \\
&+ l_6(l_{15}l_2l_{20} - l_{16}l_{19}l_2+l_{12}l_{19}l_3 - l_{11}l_{20}l_3 + l_{11}l_{16}l_4 - l_{12}l_{15}l_4) \\
&+ l_7(l_{16}l_{18}l_2 - l_{14}l_2l_{20} - l_{12}l_{18}l_3 - l_{10}l_{20}l_3 - l_{10}l_{16}l_4 + l_{12}l_{14}l_4) \\
&+ l_8 (l_{14}l_{19}l_2 - l_{15}l_{18}l_2 - l_{10}l_{19}l_{13} + l_{11}l_{18}l_3 + l_{10}l_{15}l_4 - l_{11}l_{14}l_4)]/[l_5(l_{10}l_{16}l_{19} \\
&- l_{11}l_{16}l_{18} - l_{12}l_{14}l_{19} + l_{12}l_{15}l_{18} - l_{10}l_{15}l_{20} + l_{11}l_{14}l_{20}) \\
&+ l_6(l_{11}l_{16}l_{17} + l_{12}l_{13}l_{19} - l_{12}l_{15}l_{17} - l_{11}l_{13}l_{20}) \\
&+ l_7(l_{10}l_{13}l_{20} - l_{10}l_{16}l_{17} - l_{12}l_{13}l_{18} + l_{12}l_{14}l_{17}) \\
&+ l_8(l_{10}l_{15}l_{17} - l_{10}l_{13}l_{19} + l_{11}l_{13}l_{18} - l_{11}l_{14}l_{17}) \\
&+ l_9 (l_{15}l_6l_{20} - l_{19}l_{16}l_6 + l_7l_{16}l_{18} - l_{20}l_{14}l_{17} + l_{19}l_{14}l_8 - l_{15}l_{18}l_8)],
\end{aligned} \tag{2.40}$$

$$
\begin{aligned}
y = [&l_1 (l_{11}l_{16}l_{17} - l_{12}l_{13}l_{19} - l_{12}l_{15}l_{17} + l_{11}l_{13}l_{20} - l_{16}l_{19}l_9 + l_{15}l_{19}l_{20}) \\
&+ l_5(l_{16}l_2l_{19} - l_5l_{20}l_2 + l_{12}l_{19}l_3 - l_{11}l_{20}l_3 + l_{11}l_{16}l_4 - l_{12}l_{15}l_4) \\
&+ l_7(l_{16}l_{17}l_2 - l_{13}l_2l_{20} - l_{12}l_{17}l_3 - l_{12}l_{13}l_4) \\
&+ l_9(l_{20}l_3l_7 - l_{16}l_4l_7 - l_{19}l_8l_3 - l_{15}l_{14}l_8) \\
&+ l_8 (l_{13}l_{19}l_2 - l_{15}l_{17}l_2 - l_{11}l_{17}l_3 + l_{11}l_{13}l_4)]/[l_5(l_{10}l_{16}l_{19} \\
&- l_{11}l_{16}l_{18} - l_{12}l_{14}l_{19} + l_{12}l_{15}l_{18} - l_{10}l_{15}l_{20} + l_{11}l_{14}l_{20}) \\
&+ l_6(l_{11}l_{16}l_{17} + l_{12}l_{13}l_{19} - l_{12}l_{15}l_{17} - l_{11}l_{13}l_{20}) \\
&+ l_7(l_{10}l_{13}l_{20} - l_{10}l_{16}l_{17} - l_{12}l_{13}l_{18} + l_{12}l_{14}l_{17}) \\
&+ l_8(l_{10}l_{15}l_{17} - l_{10}l_{13}l_{19} + l_{11}l_{13}l_{18} - l_{11}l_{14}l_{17}) \\
&+ l_9 (l_{15}l_6l_{20} - l_{19}l_{16}l_6 + l_7l_{16}l_{18} - l_{20}l_{14}l_{17} + l_{19}l_{14}l_8 - l_{15}l_{18}l_8)],
\end{aligned}
\tag{2.41}
$$

$$
\begin{aligned}
z = [&l_5 (l_{18}l_{16}l_2 - l_{14}l_2l_{20} - l_{12}l_{18}l_3 + l_{10}l_3l_{20} - l_{16}l_{10}l_4 + l_{12}l_{14}l_4) \\
&+ l_6(l_{13}l_{20}l_2 - l_{16}l_2l_{17} + l_{12}l_{17}l_3 - l_{12}l_{13}l_4) \\
&+ l_8(l_{14}l_{17}l_2 - l_{13}l_2l_{18} - l_{10}l_{17}l_3 + l_{10}l_{13}l_4) \\
&+ l_9 (l_1l_{14}l_{20} - l_1l_{16}l_{18} - l_{20}l_6l_3 + l_{16}l_6l_4 + l_{18}l_3l_8 - l_{14}l_8l_4)]/[l_5(l_{10}l_{16}l_{19} \\
&- l_{11}l_{16}l_{18} - l_{12}l_{14}l_{19} + l_{12}l_{15}l_{18} - l_{10}l_{15}l_{20} + l_{11}l_{14}l_{20}) \\
&+ l_6(l_{11}l_{16}l_{17} + l_{12}l_{13}l_{19} - l_{12}l_{15}l_{17} - l_{11}l_{13}l_{20}) \\
&+ l_7(l_{10}l_{13}l_{20} - l_{10}l_{16}l_{17} - l_{12}l_{13}l_{18} + l_{12}l_{14}l_{17}) \\
&+ l_8(l_{10}l_{15}l_{17} - l_{10}l_{13}l_{19} + l_{11}l_{13}l_{18} - l_{11}l_{14}l_{17}) \\
&+ l_9 (l_{15}l_6l_{20} - l_{19}l_{16}l_6 + l_7l_{16}l_{18} - l_{20}l_{14}l_{17} + l_{19}l_{14}l_8 - l_{15}l_{18}l_8)].
\end{aligned}
\tag{2.42}
$$

Equations (2.36)–(2.39) can also rewrite as

$$
\boldsymbol{AS} = \boldsymbol{B} \tag{2.43}
$$

where $\boldsymbol{A}=\begin{bmatrix} l_5 & l_6 & l_7 & l_8 \\ l_9 & l_{10} & l_{11} & l_{12} \\ l_{13} & l_{14} & l_{15} & l_{16} \\ l_{17} & l_{18} & l_{19} & l_{20} \end{bmatrix}$, $\boldsymbol{S}=\begin{bmatrix} x \\ y \\ z \\ t_0 \end{bmatrix}$ and $\boldsymbol{B}=\begin{bmatrix} l_1 \\ l_2 \\ l_3 \\ l_4 \end{bmatrix}$

The x, y and z can be also obtained by solving the linear systems. The proposed method is suitable for known wave velocity system. The closed-form solution for unknown wave velocity system is reported in [17].

2.3.2.2 Validated Examples and Discussion

(1) Example 1: Simulations of Sound Sources
For example, consider five receivers and a speed of sound of 330 m/s. The Cartesian coordinates of five sensors are S_1 (250.25, 450.17, 150.28), S_2 (290.11,

180.13, 140.18), S_3 (150.35, 190.21, 30.33), S_4 (380.38, 250.28, 98.15), and S_5 (500.55, 350.25, 210.45). The two source locations with Cartesian coordinates of P_1 (310.75, 200.33, 125.11) and P_2 (210.88, 290.99, 175.89) and the original times of two sources are 0 s. The trigger times are obtained according to the distance formula between spatial two points as well as distance and velocity formula. The trigger times of five receivers (S_1–S_5) of P_1 are 0.782697, 0.098713, 0.565408, 0.272225, and 0.777218 s, respectively. The trigger times of five receivers (S_1–S_5) of P_2 are 0.502922, 0.426859, 0.566984, 0.578392, and 0.902068 s, respectively.

To validate the proposed method, the coordinates of receivers and their trigger time were used to calculate $l_i (i = 1, 2, \cdots, 20)$ according to Eq. (2.36)–(2.39). Then, the calculated $l_i (i = 1, 2, \cdots, 20)$ were substituted in Eq. (2.40)–(2.42) to resolve the source location. The calculated results of sources P_1 and P_2 are (310.7499, 200.3302, and 125.1101) and (210.8797, 290.9899, and 175.8904), respectively. Results show that the calculated coordinates of P_1 and P_2 are consistent with authentic results.

(2) Example 2: Simulations of a Mine Acoustic Emission Location System and Location Error Analysis
A numerical test for an acoustic emission/microseismic monitoring system of a mine was used to validate the proposed method and analyze location errors. The locations of five sensors are shown in Fig. 2.7, and the coordinates are (200, 500, 300), (300, 100, 100), (400, 350, 280), (600, 250, 250), and (700, 200, 180). The locations of acoustic emission/microseismic sources E_1, E_2, E_3, E_4 and E_5 are (250, 300, 120), (310, 450, 130), (450, 260, 110), (650, 180, 135), and (780, 250, 125), respectively, and the unit is m. P-wave velocity is 5,000 m/s. Errors of wave velocity and arrivals were discussed.

Firstly, the acoustic emission sources are located by the proposed method using true wave velocity. The results are shown in Fig. 2.8. It can be seen that the calculated results are consistent with the true locations, therefore, the proposed method is reasonable. To investigate the location errors induced by errors of velocities, the method STT was used for comparison and analysis. The locations of acoustic emissions by STT method without considering the errors of velocities and arrivals are consistent with true locations (Fig. 2.8). The errors of absolute distant are shown in Fig. 2.8. It also can be seen that the results of STT are consistent with true locations and results of the proposed analytical solution method. It is proved that the STT method is stable and reliable. The intervals of velocity errors are from 4,500 m/s to 5,500 m/s. True velocity is 5,000 m/s. The variable interval is 200 m/s. Location results with analytical solutions and numerical methods STT are shown in Fig. 2.7. The absolute distance errors between the real coordinates and calculated coordinates are shown in Fig. 2.8 under different levels of velocity error. The left vertical axis is the absolute distance error. The original time errors are also plotted in the right vertical axis of Fig. 2.8. It can be seen from location results of Figs. 2.7 and 2.8 that the location accuracy is affected by the velocity value. In view of the spatial distribution of positions, symmetrical changes of velocity will result in location results symmetry

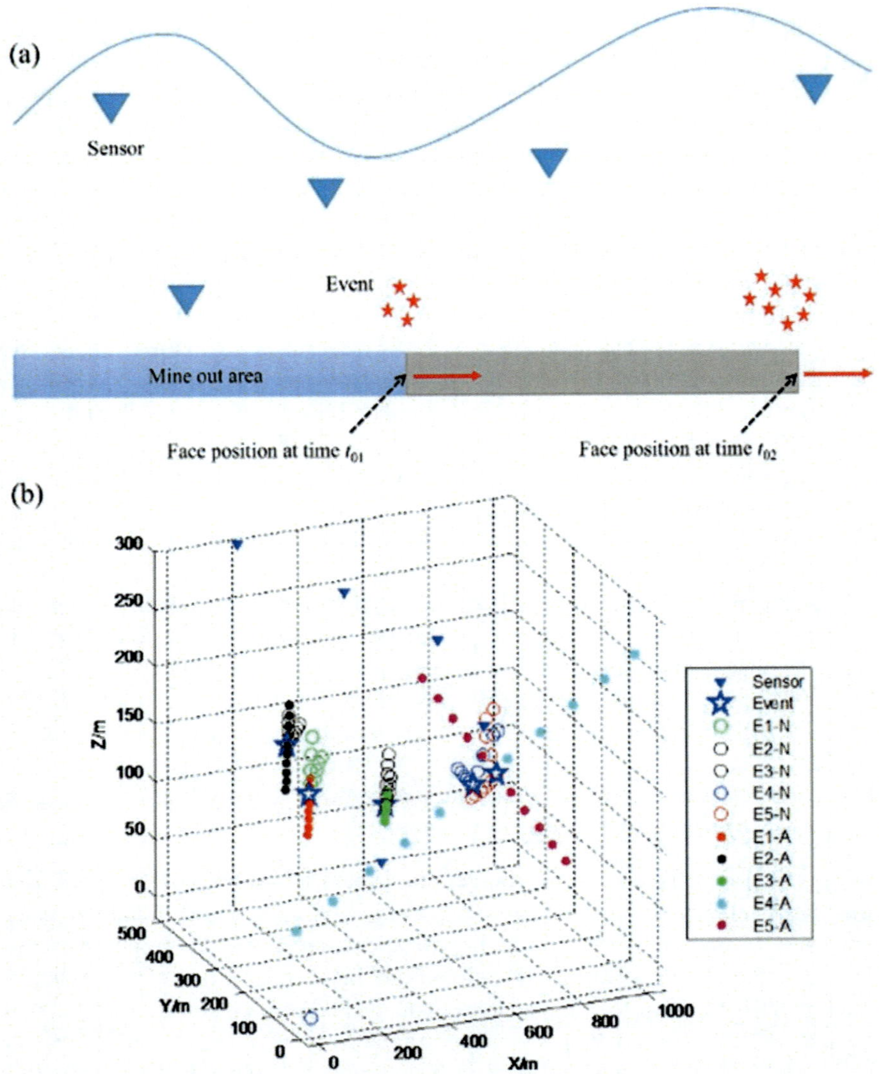

Fig. 2.7 Sketch map of acoustic emission/microseismic source location: A indicates analytical solution; N indicates numerical solution. **a** Locations of sensors and sources. **b** Spatial locations of sensors and sources (from [16])

spatial position using analytical solution, since analytical solutions are based on the results of each positioning operation parameters of the true value. The location errors are completely dependent on the operator error parameters which are offset from the true value. Symmetrical changes of velocity will induce irregular spatial position using the numerical iterative method, STT. If velocity error is large, it may cause serious positioning errors, which will locate a serious deviation from the true value,

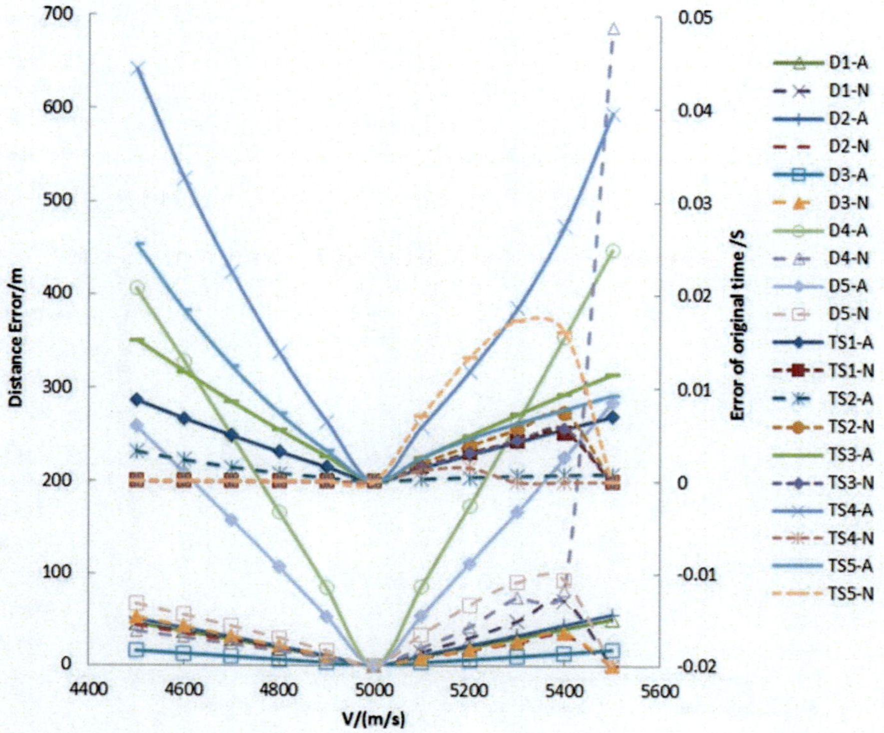

Fig. 2.8 Errors of distance error from located source to real source and original time induced by insufficient know velocity error: D indicates absolute distant error; TS indicates original time error (from [16])

such as the E4-N. It can be seen from Fig. 2.8 that higher positioning analytical location method requires measurement data with high quality, but when measurement data is reasonable, it can obtain accurate positioning results. Throughout all located microseismic sources, it can be easily found that the greatest absolute distance error is from the numerical solution of STT. A velocity error of 500 m/s can result in a big absolute distance error of 700 m using STT, while the positioning error of the analytical solution is 450 m.

Secondly, to investigate the location errors induced by errors of arrivals, the method STT was also used for comparison and analysis. The error of arrivals is limited between −1,000 s and +1,000 s. The 1,000 errors are generated by the random function of Microsoft Excel; then arrivals with errors were obtained by true arrivals plus generated errors. In this study, the minimum, average, and maximum of arrivals with errors were used to investigate the location errors induced by arrival errors. The calculated results are shown in Fig. 2.9. The absolute distance error is expressed in the left vertical axis. Original time error is expressed in the right vertical

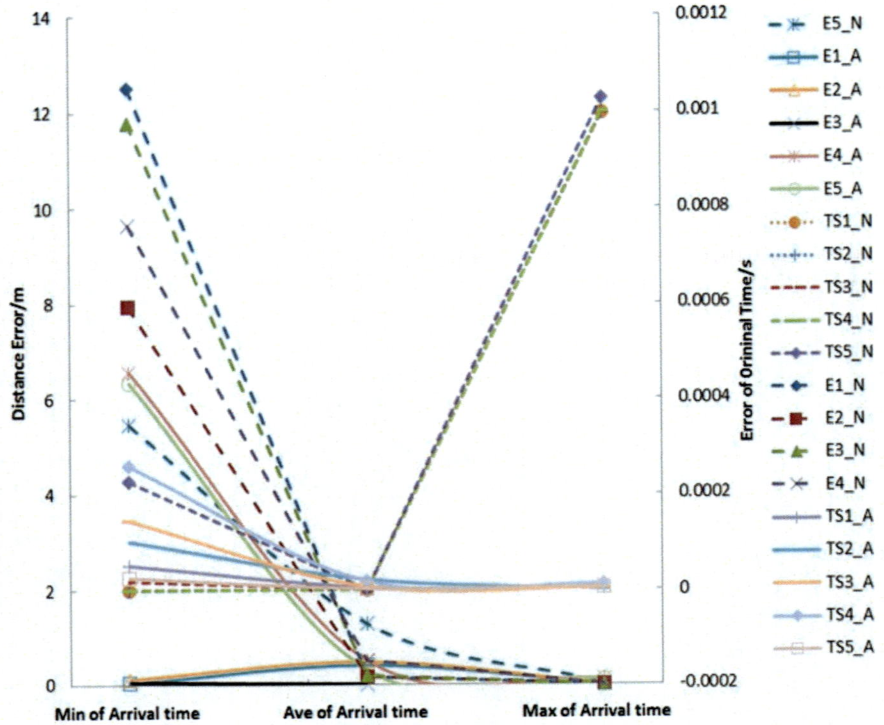

Fig. 2.9 Errors of Distance and original time induced by arrival time reading error: the left y axil expressing the distance error from located sources to real sources; and right y axil expressing the error of original time; E indicates absolute distant error; TS indicates original time errors; A indicates analytical solutions; and N indicates STT results (from [16])

axis. It can be clearly seen in Fig. 2.9 that arrival errors also significantly affect the location errors. The absolute distant errors and original errors of STT are always larger than those of analytical method. The maximums of absolute distant errors of STT and analytical method are 12.5 m and 6.5 m, respectively. The maximums of original time errors of STT and analytical method are 0.001 s and 0.00023 s, respectively.

Therefore, a unique three-dimensional analytical solution for acoustic emission source location can be obtained using TDOA measurements from N receivers, where $N \geq 5$ basing on solving simplified linear equations. No calculation of square root is required in the solution equations of the proposed method. The problems of the existence and multiplicity induced by calculations of square roots in existed close-form methods were solved successfully.

2.4 Comparison between the Traditional Method and Velocity-Free Method

2.4.1 Numerical Test and Location Results

Suppose that a localization system has eight sensors located at eight vertices of a cube, and the unit is m. The sensor coordinates are *A* (0, 0, 0), *B* (800, 0, 0), *C* (800, 800, 0), *D* (0, 800, 0), *E* (0, 0, 800), *F* (800, 0, 800), *G* (800, 800, 800) and *H* (0, 800, 800). The equivalent wave velocity in medium is 5,000 m/s. The onset time is set to 10 ms. Source coordinates are *O* (258, 336, 580), *P* (680, 290, 559), *Q* (190, 610, 380), *R* (789, 459, 280), *S* (308, 689, 1,200) and *T* (860, 910, 1,008). Sources *O*, *P*, *Q* and *R* are inside the sensor network, and sources *S* and *T* are outside the network. Figure 2.10 shows the spatial distribution of sources and sensors and the arrival times are shown in Table 2.1.

The main ideas of comparison are as follows.

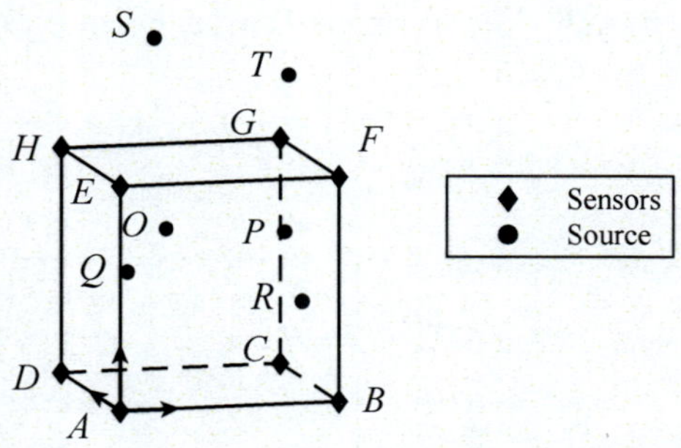

Fig. 2.10 Distribution of sources and sensors

Table 2.1 Trigger times of seismic sources recorded by sensors

Sensors	Trigger times (ms)					
	O	*P*	*Q*	*R*	*S*	*T*
A	153.6468	195.3625	158.6741	200.9557	293.5197	331.4818
B	182.4019	138.2156	198.5312	117.555	303.7199	281.865
C	193.8978	163.2294	158.6741	98.27276	270.3371	213.1516
D	167.259	213.468	103.0806	190.7985	258.7718	275.9146
E	105.4694	165.5096	162.9183	220.1049	180.8315	263.8475
F	144.9163	89.14063	201.8958	148.7374	197.2736	197.079
G	159.3265	125.3397	162.9183	134.3868	138.7455	58.56501
H	124.9365	186.701	109.7196	210.9181	113.3799	188.3215

1. Use two velocity-free methods (TT and TD) and two traditional methods (STT and STD) to locate the microseismic source with the actual velocity (i.e. error floating 0%)
2. Because it is difficult to measure the actual velocity in practical engineering, small errors of 1%, 2%, 3%, 4% and 5% are introduced to the velocity when using traditional method, that is, two traditional methods are used to locate the source with the velocity being 4,950, 4,900, 4,850, 4,800 and 4,750 m/s, respectively
3. Analyze the location results of ideas (1) and (2), calculate the error of each coordinate as well as the absolute distance error (ADE), and draw the ADEs of the traditional methods and the velocity-free methods in Fig. 2.11
4. To specifically compare the error size, Table 2.2 shows the location results and ADEs of TT and STT, and Table 2.3 shows the location results of TD and STD with three coordinates.

The results show that when the actual velocity is known, the location errors of STT, STD are small. However, a small error will lead to a large location error of source coordinates. From Fig. 2.11b, the location error increases significantly with the increase of velocity error. For the sources inside sensor network, when the velocity floats by 1%, the smallest absolute distance is 2.43 m, and the largest absolute distance

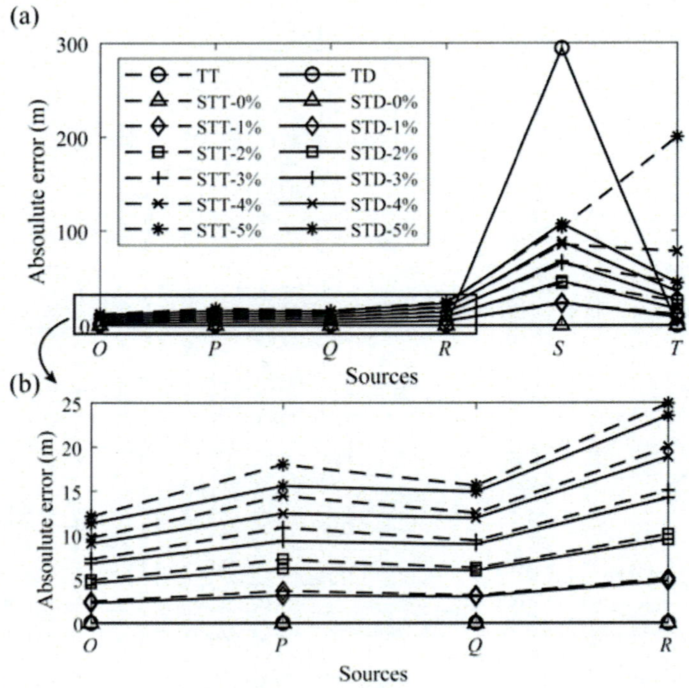

Fig. 2.11 Comparison curves of ADE by velocity-free methods and traditional methods. **a** Demonstrates the location results of all sources and **b** shows the location results of sources inside sensor network

Table 2.2 Result comparisons between velocity-free method TT and traditional method STT

Methods	Velocity error (%)	Velocity (m/s)	Source *O*				Source *P*			
			X (m)	*Y* (m)	*Z* (m)	D_{err} (m)	*X* (m)	*Y* (m)	*Z* (m)	D_{err} (m)
TT		1,000	258.00	336.00	580.00	0.00	680.00	290.00	559.00	0.00
STT	0	5,000	258.00	336.00	580.00	0.00	680.00	290.00	559.00	0.00
	1	4,950	259.43	336.62	578.14	2.43	676.93	291.05	557.40	3.62
	2	4,900	260.87	337.23	576.29	4.85	673.87	292.11	555.80	7.23
	3	4,850	262.30	337.85	574.43	7.27	670.82	293.16	554.20	10.83
	4	4,800	263.73	338.47	572.58	9.70	667.78	294.22	552.60	14.42
	5	4,750	265.17	339.09	570.73	12.12	664.75	295.28	551.01	18.01
Methods	Velocity error (%)	Velocity (m/s)	Source *Q*				Source *R*			
			X (m)	*Y* (m)	*Z* (m)	D_{err} (m)	*X* (m)	*Y* (m)	*Z* (m)	D_{err} (m)
TT		1,000	190.00	610.00	380.00	0.00	789.00	459.00	280.00	0.00
STT	0	5,000	190.00	610.00	380.00	0.00	789.00	459.00	280.00	0.00
	1	4,950	192.21	607.79	380.18	3.14	784.15	458.38	281.32	5.06
	2	4,900	194.42	605.58	380.37	6.27	779.35	457.76	282.63	10.08
	3	4,850	196.63	603.37	380.56	9.39	774.58	457.14	283.93	15.06
	4	4,800	198.83	601.17	380.74	12.51	769.86	456.53	285.23	20.00
	5	4,750	201.03	598.97	380.93	15.63	765.17	455.92	286.52	24.90

(continued)

Table 2.2 (continued)

Methods	Velocity error (%)	Velocity (m/s)	Source *S*				Source *T*			
			X (m)	*Y* (m)	*Z* (m)	D_{err} (m)	*X* (m)	*Y* (m)	*Z* (m)	D_{err} (m)
TT		1,000	339.67	589.51	924.60	294.53	860.00	910.00	1,008.00	0.00
STT	0	5,000	308.00	689.00	1,200.00	0.00	860.00	910.00	1,008.00	0.00
	1	4,950	310.77	680.67	1,178.41	23.30	852.57	903.02	1,002.29	11.68
	2	4,900	313.34	672.91	1,158.11	45.19	843.52	894.44	995.15	26.06
	3	4,850	315.72	665.66	1,138.95	65.82	831.53	882.95	985.29	45.36
	4	4,800	317.94	658.86	1,120.82	85.30	811.67	863.51	967.65	78.27
	5	4,750	320.01	652.48	1,103.63	103.76	746.94	794.29	889.87	200.31

Table 2.3 Result comparisons between velocity-free method TD and traditional method STD

Sources	TD						STD (velocity error: 0%, velocity: 5600 m/s)					
	X (m)	*Y* (m)	*Z* (m)	X_{err} (m)	Y_{err} (m)	Z_{err} (m)	*X* (m)	*Y* (m)	*Z* (m)	X_{err} (m)	Y_{err} (m)	Z_{err} (m)
O	258.00	336.00	580.00	0.00	0.00	0.00	258.00	336.00	580.00	0.00	0.00	0.00
P	680.00	290.00	559.00	0.00	0.00	0.00	680.00	290.00	559.00	0.00	0.00	0.00
Q	190.00	610.00	380.00	0.00	0.00	0.00	190.00	610.00	380.00	0.00	0.00	0.00
R	789.00	459.00	280.00	0.00	0.00	0.00	789.00	459.00	280.00	0.00	0.00	0.00
S	339.67	589.51	924.60	−31.67	99.49	275.40	308.00	689.00	1,200.00	0.00	0.00	0.00
T	258.00	336.00	580.00	0.00	0.00	0.00	860.00	910.00	1,008.00	0.00	0.00	0.00
Sources	STD (velocity error: 1%, velocity: 4950 m/s)						STD (velocity error: 2%, velocity: 4900 m/s)					
	X (m)	*Y* (m)	*Z* (m)	X_{err} (m)	Y_{err} (m)	Z_{err} (m)	*X* (m)	*Y* (m)	*Z* (m)	X_{err} (m)	Y_{err} (m)	Z_{err} (m)
O	259.35	336.57	578.28	−1.35	−0.57	1.72	260.71	337.14	576.56	−2.71	−1.14	3.44
P	677.47	291.01	557.53	2.53	−1.01	1.47	674.93	292.03	556.06	5.07	−2.03	2.94
Q	192.09	607.87	380.15	−2.09	2.13	−0.15	194.17	605.73	380.30	−4.17	4.27	−0.30
R	784.45	458.39	281.27	4.55	0.61	−1.27	779.92	457.79	282.54	9.08	1.21	−2.54
S	310.81	680.47	1,177.47	−2.81	8.53	22.53	313.42	672.51	1,156.31	−5.42	16.49	43.69
T	862.19	914.81	1,014.98	−2.19	−4.81	−6.98	864.42	919.71	1,022.08	−4.42	−9.71	−14.08

(continued)

Table 2.3 (continued)

Sources	STD (velocity error: 3%, velocity: 4850 m/s)						STD (velocity error: 4%, velocity: 4800 m/s)					
	X (m)	*Y* (m)	*Z* (m)	X_{err} (m)	Y_{err} (m)	Z_{err} (m)	*X* (m)	*Y* (m)	*Z* (m)	X_{err} (m)	Y_{err} (m)	Z_{err} (m)
O	262.07	337.71	574.84	−4.07	−1.71	5.16	263.43	338.28	573.12	−5.43	−2.28	6.88
P	672.38	293.04	554.59	7.62	−3.04	4.41	669.83	294.06	553.11	10.17	−4.06	5.89
Q	196.25	603.60	380.45	−6.25	6.40	−0.45	198.34	601.48	380.60	−8.34	8.52	−0.60
R	775.43	457.19	283.79	13.57	1.81	−3.79	770.96	456.59	285.05	18.04	2.41	−5.05
S	315.85	665.09	1,136.45	−7.85	23.91	63.55	318.10	658.17	1,117.78	−10.10	30.83	82.22
T	866.72	924.70	1,029.30	−6.72	−14.70	−21.30	869.07	929.80	1,036.66	−9.07	−19.80	−28.66
Sources	STD (velocity error: 5%, velocity: 4750 m/s)						Actual coordinates					
	X (m)	*Y* (m)	*Z* (m)	X_{err} (m)	Y_{err} (m)	Z_{err} (m)	*X* (m)	*Y* (m)	*Z* (m)			
O	264.79	338.86	571.39	−6.79	− 2.86	8.61	258.00	336.00	580.00			
P	667.26	295.09	551.63	12.74	− 5.09	7.37	680.00	290.00	559.00			
Q	200.42	599.35	380.76	−10.42	10.65	−0.76	190.00	610.00	380.00			
R	766.53	455.99	286.29	22.47	3.01	−6.29	789.00	459.00	280.00			
S	320.20	651.69	1,100.20	−12.20	37.31	99.80	308.00	689.00	1,200.00			
T	871.48	934.99	1,044.14	−11.48	− 24.99	−36.14	860.00	910.00	1,008.00			

is 5.06 m. When the velocity floats by 5%, the smallest absolute distance is 12.23 m, the largest absolute distance is 24.90 m. Therefore, if the location error is required to be less than 10 m, the velocity error must be within 2%. This requirement is very high. In the actual engineering monitoring, it can be difficult at times to achieve the velocity accuracy within 5%, because the mining environment is complex. Therefore, the traditional methods face a serious challenge on location accuracy in the actual environment. In addition, it can be seen from Fig. 2.11 as well as Tables 2.2 and 2.3 that when the source is outside the sensor network, the errors of four methods are large. Although velocity-free methods perform an excellent localization for sources inside the sensor network, they become unstable for sources localization outside the sensor network. This is shown by the accurate position that was obtained at source *T*, but poor position obtained at source *S*. The numerical test result shows that the arrangement of the sensor should try to make sources within its network.

2.4.2 Blasting Test and Location Results

The Dongguashan rockburst monitoring system has been operating since Aug. 25, 2005. It is composed of a seismic monitoring system and a conventional stress and deformation monitoring system. The seismic monitoring system was provided by ISS system from Integrated Seismic Systems International (South Africa). It has 24 channels and 16 sensors. All signals are transmitted by copper twisted cables to the monitoring control of underground, and then transmitted by an optical cable to the monitoring center on the ground surface as well as the safety and production management offices of the mine. Currently, the area of monitoring is the first mining area where there are four panels located between exploration lines 52 and 60 in the surrounding rock mass. The monitoring area will be extended to the entire mine later.

The measured velocity in the same area of Dongguashan Mine is 5,400–5,900 m/s, with an average floating rate of 3%–7%, and the velocity floating rate in different regions may be greater. Three blasting tests were conducted in Dongguashan Mine and trigger times were recorded by microseismic monitoring system. The sensor coordinates and trigger times are shown in Table 2.4. Some information about three blasting tests are shown in Table 2.5. The location results of four methods are shown in Table 2.6.

The location accuracy of TD method is the highest among the four methods. Without measuring the velocity in advance, the location error decreases comparing with traditional methods. The location error of TT is 21.26 m, which is relatively large. The main reason is that TT needs to use four known quantities to fit five unknown quantities, and the location accuracy is unstable. This proves that TD method is superior to STT and STD, and its prediction accuracy is higher than STT and STD. The reason is that TD algorithm can more accurately fit the relationship between the coordinate and time difference of each sensor. Although the basic idea also relies on

Table 2.4 Coordinates of sensors and triggered times by sensors (from [18])

No	Sensor coordinates (m)			Trigger times (s)		
	X	*Y*	*Z*	Event 1	Event 2	Event 3
1	84,345.73	22,474	−678.01	31.21414	0.563835	45.26793
2	84,157.08	22,717.2	−737.28			45.26493
3	84,256.71	22,587.9	−682.8	31.22597	0.574668	45.25826
4	84,493.74	22,395.4	−653.02	31.2103	0.567501	
5	84,299.94	22,861.7	−764.74			45.26118
6	84,377.81	22,755.5	−722.01	31.22294	0.566903	45.24801
7	84,487.86	22,612	−704.33	31.19561	0.54757	45.25868
8	84,580.14	22,489.6	−693.73	31.19694	0.55657	
9	84,591.12	22,453.2	−862.58	31.20644	0.556775	
11	84,429.88	22,332.3	−863.16	31.22661	0.573108	
12	84,509.8	22,391.8	−862.91	31.21328	0.561441	
13	84,076.11	22,705.4	−862.89			45.28031
14	84,182.39	22,775.1	−862.38			45.26864
15	84,259.16	22,840.2	−862.04			45.26714
16	84,307.19	22,943.1	−860.87			45.27964

Table 2.5 Coordinates and amount of dynamite of blasting

Event	Coordinates (m)			Amount of dynamite (kg)
	X	*Y*	*Z*	
1	84,528.4	22,556.2	−753.2	2.25
2	84,479.0	22,570.0	−814.4	2.40
3	84,359.0	22,673.0	−795.5	2.40

Table 2.6 Errors comparisons of position results using new methods TT and TD and traditional methods STT and STD

Event	TT				STT			
	X_{err} (m)	Y_{err} (m)	Z_{err} (m)	D_{err} (m)	X_{err} (m)	Y_{err} (m)	Z_{err} (m)	D_{err} (m)
1	3.51	7.65	7.07	10.99	8.55	8.93	3.68	12.90
2	13.81	3.32	14.53	20.32	9.96	2.02	12.58	16.17
3	6.88	6.00	31.15	32.46	7.04	3.20	23.67	24.90
Average	8.07	5.66	17.58	21.26	8.52	4.72	13.31	17.99
	TD				*STD*			
	X_{err} (m)	Y_{err} (m)	Z_{err} (m)	D_{err} (m)	X_{err} (m)	Y_{err} (m)	Z_{err} (m)	D_{err} (m)
1	3.51	7.65	6.97	10.93	8.55	8.76	4.15	12.93
2	1.14	3.10	7.72	8.40	9.32	0.72	11.52	14.84
3	8.02	5.63	5.30	11.14	7.05	3.20	23.67	24.90
Average	4.22	5.46	6.66	10.16	8.31	4.23	13.11	17.55

the average velocity, the average velocity is dynamically adjusted. The best velocity value of an event is sought in continuous iterations to meet the nonlinear relationship between the coordinate of each sensor and the time difference.

2.5 Conclusions

The velocity-free source localization methods can save the personnel, time and economic cost caused by measuring the velocity through many blasting tests in the early stage of microseismic location. In addition, the methods avoid the location error caused by the difference between the measured velocity area and the actual area. Therefor the velocity-free methods are more convenient and practical than the traditional methods. Three velocity-free methods (TT, TD and TDQ) are introduced and their rationality is demonstrated. Through numerical examples, the location results of the velocity methods TT, TD and the traditional methods STT, STD with six levels of velocity error are analyzed and compared. The results show that the location results of TT and TD are not affected by the velocity error, which is consistent with that of STT and STD without velocity error. However, the traditional methods will lead to a large location error when the velocity has error of 1–5%. Finally, the methods are verified in Dongguashan Copper Mine where a microseismic monitoring system is built. Through three blasting tests, it is found that TD has the highest location accuracy. The mean absolute distance errors of TT, TD, STT and STD are 21.25861 m, 10.15664 m, 17.98860 m and 17.55453 m, respectively. The location results of numerical test and field blasting tests further prove that TD has higher location accuracy than TT.

From the aspects of mathematical significance, numerical simulation test and field blasting test, TD shows better advantages than TT, STT and STD. It overcomes the defect that it is difficult to accurately determine the speed in the traditional positioning method, improves the microseismic positioning method, and makes the field application more convenient than the traditional methods. It only needs to modify the data processing module in the existing localization system and can be popularized and used in the actual source location.

References

1. Lienert BR, Berg E, Frazer LN (1986) HYPOCENTER: an earthquake location method using centered, scaled, and adaptively damped least squares. Bull Seismol Soc Am 76(3):771–783
2. Nelson GD, Vidale JE (1990) Earthquake locations by 3-D finite-difference travel times. Bull Seismol Soc Am 80(2):395–410
3. Pujol J (1988) Comments on the joint determination of hypocenters and station corrections. Bull Seismol Soc Am 78(3):1179–1189
4. Pujol J (2000) Joint event location—the JHD technique and applications to data from local seismic networks. In: Advances in seismic event location. Springer, pp 163–204

5. Crosson RS (1976) Crustal structure modeling of earthquake data: 1. Simultaneous least squares estimation of hypocenter and velocity parameters. J Geophys Res 81(17):3036–3046
6. Aki K, Lee W (1976) Determination of three-dimensional velocity anomalies under a seismic array using first P arrival times from local earthquakes: 1. A homogeneous initial model. J Geophys Res 81(23):4381–4399
7. Aki K, Christoffersson A, Husebye ES (1977) Determination of the three-dimensional seismic structure of the lithosphere. J Geophys Res 82(2):277–296
8. Pavlis GL, Booker JR (1980) The mixed discrete-continuous inverse problem: application to the simultaneous determination of earthquake hypocenters and velocity structure. J Geophys Res Solid Earth 85(B9):4801–4810
9. Spencer C, Gubbins D (1980) Travel-time inversion for simultaneous earthquake location and velocity structure determination in laterally varying media. Geophys J Int 63(1):95–116. https://doi.org/10.1111/j.1365-246X.1980.tb02612.x
10. Spence W (1980) Relative epicenter determination using P-wave arrival-time differences. Bull Seismol Soc Am 70(1):171–183
11. Lomnitz C (1977) A fast epicenter location program. Bull Seismol Soc Am 67(2):425–431
12. Garza T, Lomnitz C, de Velasco CR (1979) An interactive epicenter location procedure for the RESMAC seismic array: II. Bull Seismol Soc Am 69(4):1215–1236
13. Romney C (1957) Seismic waves from the dixie valley-fairview peak earthquakes. Bull Seismol Soc Am 47(4):301–319
14. Tarantola A, Valette B (1982) Inverse problems=quest for information. J Geophys 50(1):159–170
15. Matsu'ura M (1984) Bayesian estimation of hypocenter with origin time eliminated. J Phys Earth 32(6):469–483
16. Dong L-j, Li X-b, Zhou Z-l, Chen G-h, Ma J (2015) Three-dimensional analytical solution of acoustic emission source location for cuboid monitoring network without pre-measured wave velocity. Trans Nonferrous Met Soc China 25(1):293–302
17. Li X, Dong L (2014) An efficient closed-form solution for acoustic emission source location in three-dimensional structures. AIP Adv 4(2):027110–027111–027118. https://doi.org/10.1063/1.4866170
18. Dong L-J, Li X-B, Tang L-Z, Gong F-Q (2011) Mathematical functions and parameters for microseismic source location without pre-measuring speed. Chin J Rock Mechan Eng 30(10):2057–2067
19. Eitzen DG (1977) Wadley HNG (1984) acoustic emission: establishing the fundamentals. J Res Natl Bur Stand 89(1):75–100. https://doi.org/10.6028/jres.089.008
20. Li S-L, Yin X-G, Zheng W-D, Trifu C (2005) Research of multichannel microseismic monitoring system and its application to Fankou Lead-zinc mine. Chin J Rock Mechan Eng 24(12):2048–2053
21. Zhao X-D, Tang C-A, Li Y-H, Yuan R-F Prediction method of rock burst based on microseismic monitoring and stress field analysis. In: The 9th national conference on rock dynamics, Wuchang, Wuhan, Hubei, China, 2005. pp 149–153
22. McCreary R, McGaughey J, Potvin Y, Ecobichon D, Hudyma M, Kanduth H, Coulombe A (1992) Results from microseismic monitoring, conventional instrumentation, and tomography surveys in the creation and thinning of a burst-prone sill pillar. Pure Appl Geophys 139(3):349–373
23. Milev A, Spottiswoode SM, Rorke AJ, Finnie G (2001) Seismic monitoring of a simulated rockburst on a wall of an underground tunnel. J South Afr Inst Min Metall 101(5):253–260
24. Urbancic TI, Trifu C-I (2000) Recent advances in seismic monitoring technology at Canadian mines. J Appl Geophys 45(4):225–237
25. Wang H, Ge M (2006) Acoustic emission/microseismic source location analysis for a limestone mine exhibiting high horizontal stresses. In: Golden Rocks 2006, The 41st US Symposium on Rock Mechanics (USRMS)
26. Ge M (2005) Efficient mine microseismic monitoring. Int J Coal Geol 64(1–2):44–56

27. Hirata A, Kameoka Y, Hirano T (2007) Safety management based on detection of possible rock bursts by AE monitoring during tunnel excavation. Rock Mech Rock Eng 40(6):563–576
28. Geiger L (1912) Probability method for the determination of earthquake epicenters from the arrival time only. Bull. St. Louis. Univ. 8(1):56–71
29. Chong KP, Żak HS (2013) An introduction to optimization. John Wiley & Sons Inc., Hoboken

Chapter 3
Factors Affecting the Accuracy of Acoustic Emission Sources Localization

Four experiments were carried out in this chapter and the influence of temperature, velocity, sensor position, stress stages, and optimization algorithms on AE source location accuracy was investigated in this chapter.

To explore the influence of temperature on the source location accuracy, an experiment was designed in Sect. 3.1. A hollow hemispherical specimen was heated with the flame to simulate the fire in the underground structure. During the heating process, the AE source was simulated by a constant frequency pulse. The Node block localization method (NBLM) which is detailed in [1], was then used for localization. Improvements on AE location accuracy in the changing temperature field is explored and some suggestions for the AE source localization is put forward.

In Sect. 3.2, the numerical tests were given three dimensional cases and were then designed to show the influence of velocity error on location accuracy of traditional methods and a velocity-free methods. In addition, the influence of sensor arrangement on location accuracy is discussed. A comparison of the location error between the traditional methods (TLL, SM, and Geiger method) described in Chap. 2 and the microseismic source localization method without the pre-measured wave velocity (MSLM-MV) which will be introduced in detail in Chap. 5 was done. The results showed that the location accuracy of MSLM-MV with low initial velocity is not affected by velocity error, however the closer the source is to the sensors position, the worse the location accuracy is.

The influence of the stress stages on location accuracy is investigated in Sect. 3.3. The uniaxial and biaxial compression tests were conducted and the velocity-free method which was introduced in Chap. 2 is used for localization so that the location error caused by velocity error is avoided. The variation in MS/AE location accuracy under complex stress conditions was then analyzed and some suggestions as well as qualitative evaluation of the location accuracy are discussed.

The influence of the different optimization algorithms on location accuracy is descripted in Sect. 3.4. Since the objective function of velocity methods is nonlinear, using different optimization algorithm to solve the nonlinear problem will result in difference location accuracy. Therefore, it is significant to choose a reasonable

L. Dong and X. Li, *Velocity-Free Localization Methodology for Acoustic and Microseismic Sources*, https://doi.org/10.1007/978-981-19-8610-9_3

algorithm. The famous optimization algorithms, such as the Levenberg-Marquardt Method (LM), the Simplex Method (SM), the Quasi-Newton Method (QN), the Self-Organizing Migrating Algorithms (SOMA), Particle Swarm Optimization (PSO), as well as the Global Optimization (using PSO) coupled with LM, SM, and QN, were used to locate and compare the seismic source coordinates.

3.1 The Influence of Temperature on Location Accuracy

3.1.1 Experimental Materials and Procedures

An experiment was designed to explore the effect of temperature on the source location accuracy. The test selected a hollow hemisphere specimen to simulate underground structures (such as large underground warehouses). Generally, underground structures are constructed of concrete and rock. The AE performance of these materials is consistent with the gypsum [2, 3].

Due to its excellent plasticity, gypsum material was used to make the model. The inner radius of the hollow hemisphere gypsum specimen was 200 cm, and the outer radius is 300 cm. To facilitate the analysis of the experimental results, a point heat source was used to heat the specimen at the center of the sphere. The temperature of the inner and outer walls of the structure was measured easily during the heating process. However, the internal temperature of the structure could not be directly obtained. For the sake of measuring the temperature inside the gypsum, a flat section was taken at 260 cm from the bottom (see Fig. 3.1a), so that the temperature of the gypsum at different radius could be measured. Eight and seven spring-loaded K-type thermocouples were uniformly adhered to the inner and outer walls of the specimen with inorganic high-temperature resistant adhesive, respectively (see Fig. 3.1b, c).

The thermocouple wide temperature range of 0-600 °C and its fast response speed makes it advantageous. The UT terminals of the thermocouple were connected to the K-type thermocouple temperature collection system, which could be used to collect the real-time temperature during the heating process. AE monitoring equipment adopted multi-channel equipment (composed of parallel measurement channels). It has 32 channels for collecting AE signals. The sampling frequency was 10 MHz, and the gain of the preamplifier was 34 dB (select the type of amplifier). The sensor model is VS45-H, and the sensor coordinates are shown in Table 3.1.

Thirty-two AE sensors were symmetrically arranged in the specimen for localization, which were divided into three layers according to the height of the specimen: the first layer arranged 12 sensors, the second layer arranged 12 sensors, and the third layer arranged eight sensors. The relative position of each sensor is shown in Fig. 3.1a. The sensors were fixed on the surface of the specimen with custom-made supporting fixtures. Vaseline was used as a coupling agent on the joint surface of

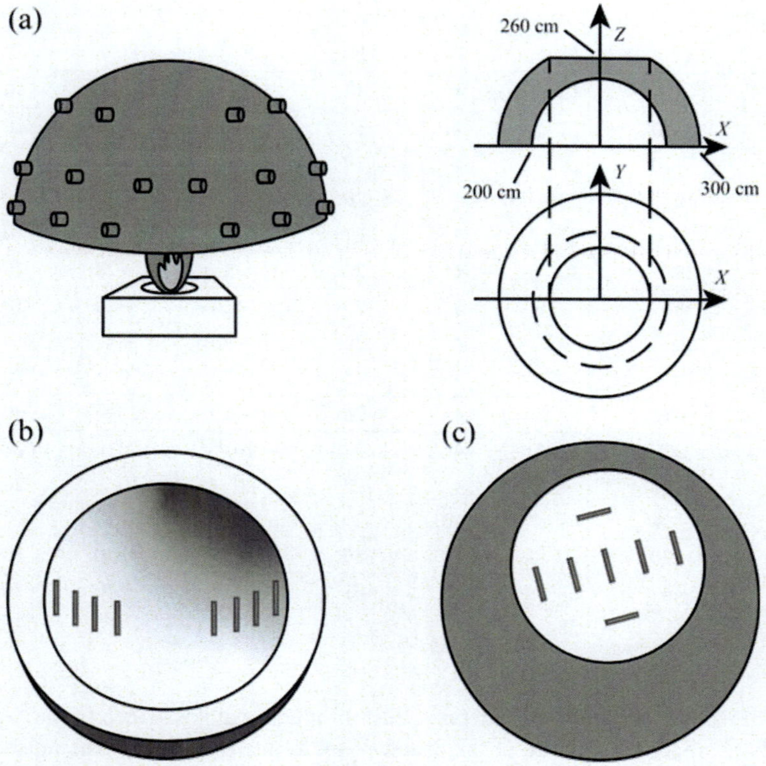

Fig. 3.1 Experimental diagram. **a** Device and sensors arrangement. **b** K-type thermocouple layout top view. **c** K-type thermocouple layout bottom view

the sensors and the specimen to reduce signal attenuation, and then following by the connection of sensors, preamplifier, and AE acquisition system.

Before the formal experiment, lead-break tests were conducted near each sensor to check if the amplitude of the lead-break signal received by each sensor was above 95 dB to ensure the coupling quality of the sensor and the specimen [4]. During the lead-break process, it was found that the amplitudes of the 30th and 31st sensors were lower than 90 dB, while the signals of other sensors were greater than 98 dB. The intensity of the signal received by these two sensors would become weak and could not exceed the threshold value. It would also delay the arrival times, which would in-turn affect the location accuracy of the AE source. Therefore, it can be concluded that the coupling effect of the 30th and 31st sensors was poor, so these two sensors were excluded during the test data acquisition and localization. Before continuous monitoring, environmental noise was collected, and the amplitude of the environmental noise received by the sensor was mostly concentrated below 40 dB. Considering the two aspects of reducing the impact of environmental noise and improving the reliability of signal acquisition, the threshold value was set to be 40 dB.

Table 3.1 Coordinates of sensors (cm) (from [5])

No.	*X*	*Y*	*Z*	No.	*X*	*Y*	*Z*
1	−28	10	4	17	20	20	10
2	−20	22	4	18	28	4	10
3	−10	28	4	19	28	−4	10
4	10	28	4	20	20	−20	10
5	20	22	4	21	4	−28	10
6	28	10	4	22	−4	−28	10
7	28	−10	4	23	−20	−20	10
8	20	−22	4	24	−28	−4	10
9	10	−28	4	25	−20	10	20
10	−10	−28	4	26	−10	20	20
11	−20	−22	4	27	10	20	20
12	−28	−10	4	28	20	10	20
13	−28	4	10	29	20	−10	20
14	−20	20	10	30	10	−20	20
15	−4	28	10	31	−10	−20	20
16	4	28	10	32	−20	−10	20

The specimen was perched on a custom-made stainless-steel frame. A solid alcohol furnace was placed in the center of the frame (Fig. 3.1a). The pulse signal emitted by sensors was used as the AE source, and the interval of pulse emission was 1 s. One sensor channel emitted four pulses. Before heating, each channel transmitted four pulses in advance. Then, by lighting the alcohol lamp and adjusting the height of the alcohol furnace, the flame center was located at the geometric ball center of the specimen. After continuous heating, the temperature of the specimen rose continuously from 28 °C (ambient temperature). Thermocouples recorded the temperature in the whole process. Acoustic emission channels circularly emitted pulses and received the pulses signal at the same time. The monitoring process was stopped until the specimen was broken entirely. Limited by the site conditions, the whole test was conducted outdoors. Although the outdoor temperature may fluctuate slightly, the effect of temperature fluctuation can be ignored because the test time was short.

3.1.2 *Localization Method*

Node block localization method was used for the localization of the structures with excavation regions. The main steps are as follows.

(1) Meshing Entity
The entity and empty areas of the structure to be located have meshed into cubes of the same size. The size of the meshed cube is determined by the volume size and location accuracy requirements. A node matrix **M** is established, and the index tag (i, j, k) for the element in **M** corresponds one-to-one with the grid nodes N. If the grid node falls in the entity area, its corresponding element in the node matrix **M** is set to be 0, which means that it can pass; the others are set to be 1, which means that they cannot pass. A block matrix **m** is established, and the elements in the block matrix **m** correspond one-to-one with the cube. If the whole cube falls into the entity area, the corresponding element in the matrix **m** is set to be 0, which means that it can pass. And the others are set to be 1.

(2) Searching Shortest Wave Path
The converted coordinate value of the sensor is input, which corresponds to the grid node. Assuming that every grid node is a potential position of the AE source, the optimized A* search algorithm is used to search for the theoretical shortest distance L_{ijk}^{q} between the qth sensor and the grid node $N\ (i, j, k)$. If the grid node is located in the empty area, then $L_{ijk}^{q} = \infty$.

(3) Acquiring AE Signal
AE events are screened out and the actual arrival time of the AE sources is sort out. The difference between the actual arrival time of the AE source to the pth sensor (x_p, y_p, z_p) and to the qth sensor (x_q, y_q, z_q) is expressed by Δt_{pq}.

(4) Determining Location of AE Source
The theoretical arrival time difference between the AE source to the pth sensor (x_p, y_p, z_p) and to the qth sensor (x_q, y_q, z_q) is expressed by $\Delta t'_{pq}$, which is as:

$$\Delta t'_{pq} = \frac{L_{ijk}^{p}}{C} - \frac{L_{ijk}^{q}}{C}, \tag{3.1}$$

where L_{ijk}^{p} and L_{ijk}^{q} represent the shortest propagation path distance from the potential AE source to the pth and qth sensors, respectively and C represents the P-wave velocity. The deviation D_{ijk} represents the deviation between the actual AE source and the grid node $N\ (i, j, k)$, and is derived by

$$D_{ijk} = \sum \left(\Delta t_{pq} - \Delta t'_{pq}\right)^2. \tag{3.2}$$

When D_{ijk} takes $D_{\min}$, the corresponding coordinate can be regarded as the position of the AE source.

3.1.3 Results and Discussion

3.1.3.1 Location Results and Errors with Changing Temperature

During the heating process, it can be noted that there was no noticeable temperature change in the seven thermocouples on the outer wall of the test piece, while the temperature of the inner wall rises rapidly. Basing on the temperature collected by the K-type thermocouple temperature acquisition system, it is found that there is no noticeable temperature change in the seven thermo-couples on the outer wall of the test piece, while the temperature of the inner wall rises rapidly.

Therefore, the average value of the internal is taken as the temperature of the specimen at each moment. Taking all the pulse signals of AE channels as one round, the experiment process can be divided into the following five stages: room temperature (28 °C), 28–48 °C, 48–69 °C, 69–82 °C, and 82–94 °C. From the beginning of heating the specimen to the moment when the specimen cracked and the test could not be continued, the entire process only took nine minutes. Since an AE channel emits four pulses, each pulse point has four location results in every round (the size of the grid element is 1 cm for localization). It can be easily noticed that in the first and second round pulses, the locating results showed abnormal values when the third and fifth sensors launched for the first time, but the following data were within the normal range. It seems likely that the voltage of the third and fifth sensors is not stable when they start to transmit pulses. Therefore these two sets of abnormal data were excluded during the analysis. The location errors are shown in Table 3.2. To display the locating results more intuitively, the results of the five stages are visualized in Fig. 3.2. Holistically, it can be seen from the experimental results that with the increase of temperature, the overall trend of the average location errors of the pulse points increases, as well as the standard deviation of the errors.

It can also be seen from the results that when the specimen is not heated, the average location error of the source points is 1.7 cm, and the standard deviation of location errors is 1.3 cm. This location accuracy is used as the reference value of location accuracy at different temperatures during the test. In the process of heating the specimen, the temperature keeps rising. The temperature of the inner wall reached 48 °C after the AE pulse emitted for one round. In this stage of time, the influence of the temperature change on the location accuracy of the source can almost be ignored. In the third stage, the overall average location accuracy of AE pulses remains unchanged. However, after further comparison it can be noted that the location results of the 17 effective pulse points have 169 changed. The nodes with increased location accuracy and the nodes with decreased location accuracy account for about half, respectively. For every single pulse point, the temperature change of the specimen has affected its location accuracy in this stage.

As the temperature of the specimen continues to rise, and when the 28th sensor emits the pulse in the fourth round, the first crack appears on the specimen after a loud burst. When the fourth round of pulse emission is launched, the average temperature of eight thermocouples on the inner wall is 82 °C. It is obvious that

Table 3.2 Location errors of four pulses per sensor (cm)

No.	28 °C				28–48 °C				48–69 °C			
	1th	2nd	3rd	4th	1th	2nd	3rd	4th	1th	2nd	3rd	4th
1	1.4	1.4	1.4	1.4	1.4	1.4	1.4	1.4	2	2	2	1
2	2.2	2.2	2.2	2.2	2.2	2.2	2.2	2.2	2.2	2.2	2.2	2.2
3	5.8	1.4	1.4	1.4	5.8	1.4	1.4	1.4	2.2	1.4	1.4	1.4
4	2.4	2.4	2.4	2.4	2.4	2.4	2.4	2.4	2.4	2.4	2.4	2.4
5	6	2.4	2.4	2.4	6	2.4	2.4	2.4	2.4	2.4	2.4	2.4
6	1.4	0	0	0	1.4	0	0	0	0	0	0	0
7	1	1	1	1	1	1	1	1	0	1	0	0
8	2.4	2.4	3.3	2.4	2.4	2.4	3.3	2.4	2.4	2.4	2.4	2.4
9	3.6	2.2	2.2	2.2	3.6	2.2	2.2	2.2	2.2	2.2	2.2	2.2
10	4.2	4.2	4.2	4.2	4.2	4.2	4.2	4.2	4.2	4.2	4.2	4.2
11	2.4	2.4	2.4	2.4	2.4	2.4	2.4	2.4	2.4	2.4	2.4	2.4
12	1	1	1	1	1	1	1	1	2.8	1.4	1.4	2.8
13	0	1	1	1	0	1	1	1	1.4	1.4	1.4	1.4
14	3.6	3.6	3.6	3.6	3.6	3.6	3.6	3.6	3.6	3.6	3.6	3.6
15	2.2	2.2	2.2	2.2	0	2.2	2.2	2.2	2.2	2.2	2.2	2.2
16	2.2	2.2	1.4	2.2	2.2	2.2	1.4	2.2	2.2	2.2	2.2	2.2
17	0	0	0	0	0	0	0	0	0	0	0	0
18	0	0	0	0	0	0	0	0	0	0	0	0
19	1	1	1	1	1	1	1	1	1	1	1	1
20	0	0	0	1.4	0	0	0	1.4	1.4	1.4	1.4	1.4
21	5	3.6	5	5.8	0	3.6	5	5.8	3.6	3.6	3.6	4.5
22	3.6	3.6	3.6	3.6	3.6	3.6	3.6	3.6	0	0	0	0
23	0	0	0	0	0	0	0	0	0	0	0	0
24	3	1.4	1.4	1.4	3	1.4	1.4	1.4	0	0	1	0
25	2.4	2.4	2.4	2.4	2.4	2.4	2.4	2.4	3.7	3.7	3.7	3.7
26	2.2	1.4	1.7	1.7	2.2	1.4	1.7	1.7	1.7	1.7	1.7	1.7
27	2.2	1.4	1.4	1.4	2.2	1.4	1.4	1.4	1.4	1.4	1.4	1.4
28	0	0	0	0	0	0	0	0	0	0	0	0
29	0	0	0	0	0	0	0	0	0	0	0	0
32	0	0	0	0	0	0	0	0	0	0	0	0
Average error	1.7				1.6				1.6			
Deviation	1.3				1.2				1.3			

(continued)

Table 3.2 (continued)

No.	69–82 °C				82–94 °C							
	1th	2nd	3rd	4th	1th	2nd	3rd	4th				
1	1	2	2	2	9.3	9.3	10.1	10.1				
2	2.2	2.2	2.2	2.2	6.8	6.8	7.1	6.8				
3	1.4	1.4	1.4	1.4	6.1	7.1	6.1	7.1				
4	2.4	2.4	2.4	2.4	8.1	12.3	12.3	12.3				
5	2.4	2.4	2.4	2.4	7.8	7.1	7.1	7.8				
6	0	0	0	0	11.2	11.2	11.2	11.2				
7	1	0	0	1	49	11.7	11.7	49				
8	2.4	2.4	2.4	2.4	8.7	8.7	8.7	7.8				
9	2.2	2.2	2.2	2.2	10	10	10	10				
10	4.2	4.2	4.2	4.2	16.7	15.7	16.7	14.6				
11	2.4	2.4	2.4	2.4	5.9	5.9	5.9	5.9				
12	2.8	2.8	2.8	2.8	2.4	2.4	2.4	2.2				
13	1.4	1.4	1.4	1.4	9.7	9.7	9.7	9.7				
14	3.6	3.6	3.6	3.6	12	11.1	11.1	12				
15	2.2	2.2	2.2	2.2	3.6	3.6	3.6	3.6				
16	2.2	2.2	2.2	2.2	0	0	0	0				
17	0	0	0	0	10.8	10.8	10.8	10.8				
18	0	0	0	0	20.6	20.6	20.6	19				
19	1	1	1	1	2.4	2.4	2.4	2.4				
20	1.4	1.4	1.4	1.4	5.2	5.2	5.2	5.2				
21	4.5	4.5	4.5	4.5	11.4	11.9	11.9	11.9				
22	0	0	0	0	8.6	8.6	8.6	8.6				
23	0	0	0	0	1.7	1.7	6	6.5				
24	1	0	1	0	3.5	5.2	5.2	5.2				
25	3.7	2.4	2.4	2.4	3.7	6	3.7	3.7				
26	1.7	1.7	1.7	1.7	3.6	3.6	3.6	3.6				
27	1.4	1.4	1.4	1.4	16.1	16.1	15.3	16.1				
28	0	0	0	7.1	5.7	5.7	5.7	5.7				
29	7.3	7.3	6.2	6.2								
32	4.7	4.7	4.4	4.4								
Average error	2.0				8.9							
Deviation	1.5				6.1							

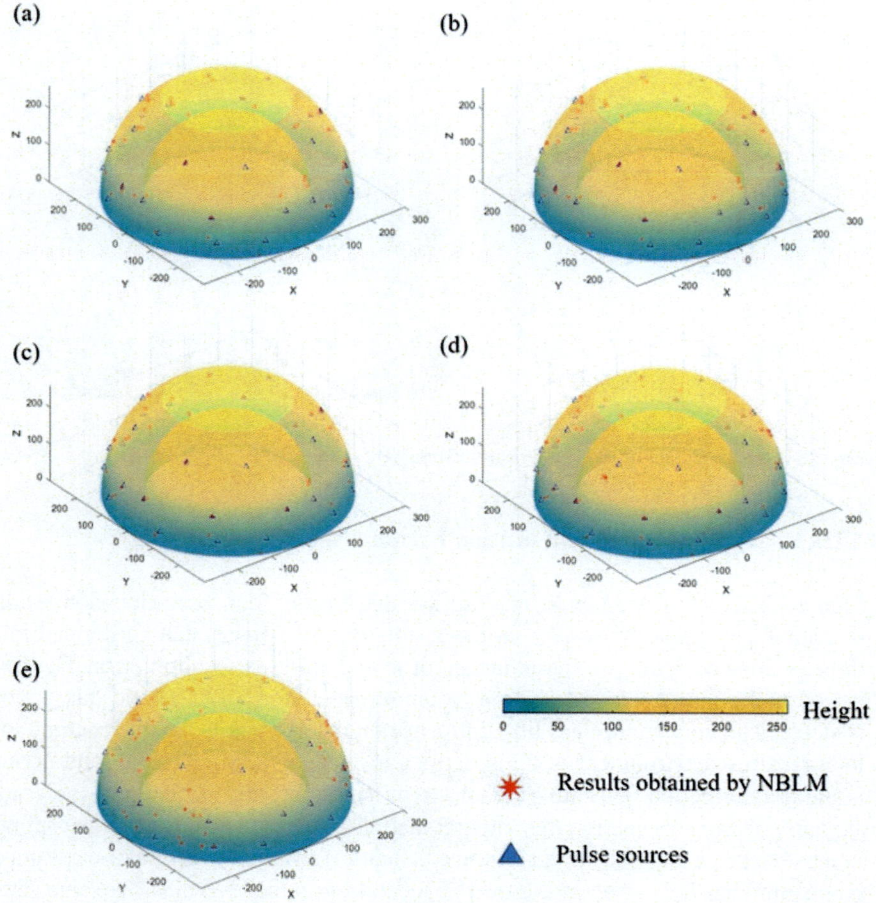

Fig. 3.2 Location results of AE sources obtained by NBLM: **a** 28 °C; **b** 28–48 °C; **c** 48–69 °C; **d** 69–82 °C; **e** 82–94 °C (from [5])

the temperature rise in this stage is slower than before. The possible reason is that when the temperature is higher than 70 °C, gypsum is easy to be dehydrated, and water vapor evaporation will take away a lot of heat [6]. In this stage, the location accuracy of the pulse point is the same as that of the third stage before the 28th sensor transmitted the pulse.

However, after the 28th sensor emitted pulse, the specimen cracked, and the location error obviously increased. After the 29th sensor emitted the pulse in the fifth stage, the specimen made two continuous burst sounds. The inner part of the specimen expands with heat and breaks into three pieces (Fig. 3.3), which makes it impossible to continue the test. In this stage, the location error increases obviously with the increase of temperature. The average location error reaches 8.9 cm, and its standard deviation reaches 6.1 cm.

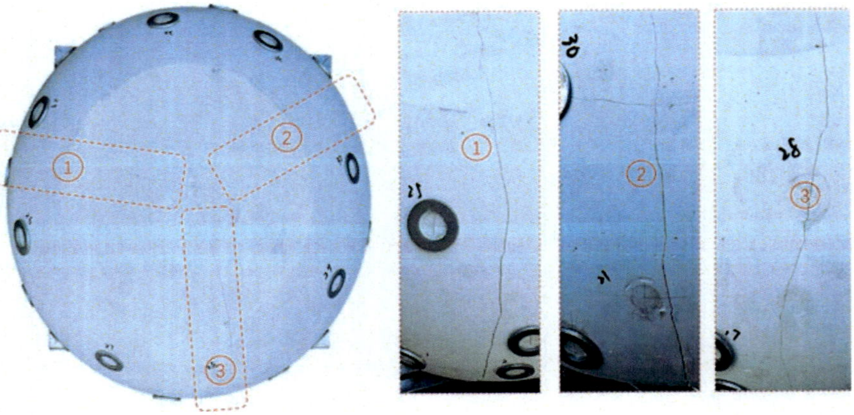

Fig. 3.3 Crack diagram of cracked gypsum (from [5])

3.1.3.2 Analysis Based on Location Errors

Figure 3.4 shows the visualization of the average error of four source location results in each stage. It can be seen from the results that five stages can be divided into three periods according to the influence of temperature on location error. The first period is the initial stage of heating (room temperature to 49 °C) and this period is stable. The temperature has little influence on the average location error, and the internal stress environment of the specimen is changed a little. The second period is the middle period of heating (from the third to the fourth stage until the crack appears). During this period, the internal thermal expansion of the specimen results in the change of internal stress and wave velocity. The location errors of most pulse points are changed, whereas the overall location accuracy remains the same. Due to the fact that most existing localization methods assume that the wave velocity of the whole specimen is homogeneous, the wave velocity inhomogeneity caused by material inhomogeneity is not taken into account. When the temperature inside the specimen increases, the stress inside the specimen will be redistributed. The change of the stress will lead to the change of the wave velocity of the specimen so that the location results of most pulse points in the middle period of heating are different from that in the room temperature state. But the results show that the overall location accuracy still remains unchanged. It reflects that the location error caused by the temperature is slight, even smaller than the location error caused by the materials in homogeneity. This is partly due to the selected NBLM, which takes the influence of wave velocity into account and sets the wave velocity value as an unknown. It can effectively reduce the locating error caused by the change of wave velocity.

The third period is the later period of heating (the specimen cracks). In this period, the location errors will increase significantly because the acoustic wave cannot pass through the crack and needs to bypass the crack (Fig. 3.5), which causes that the

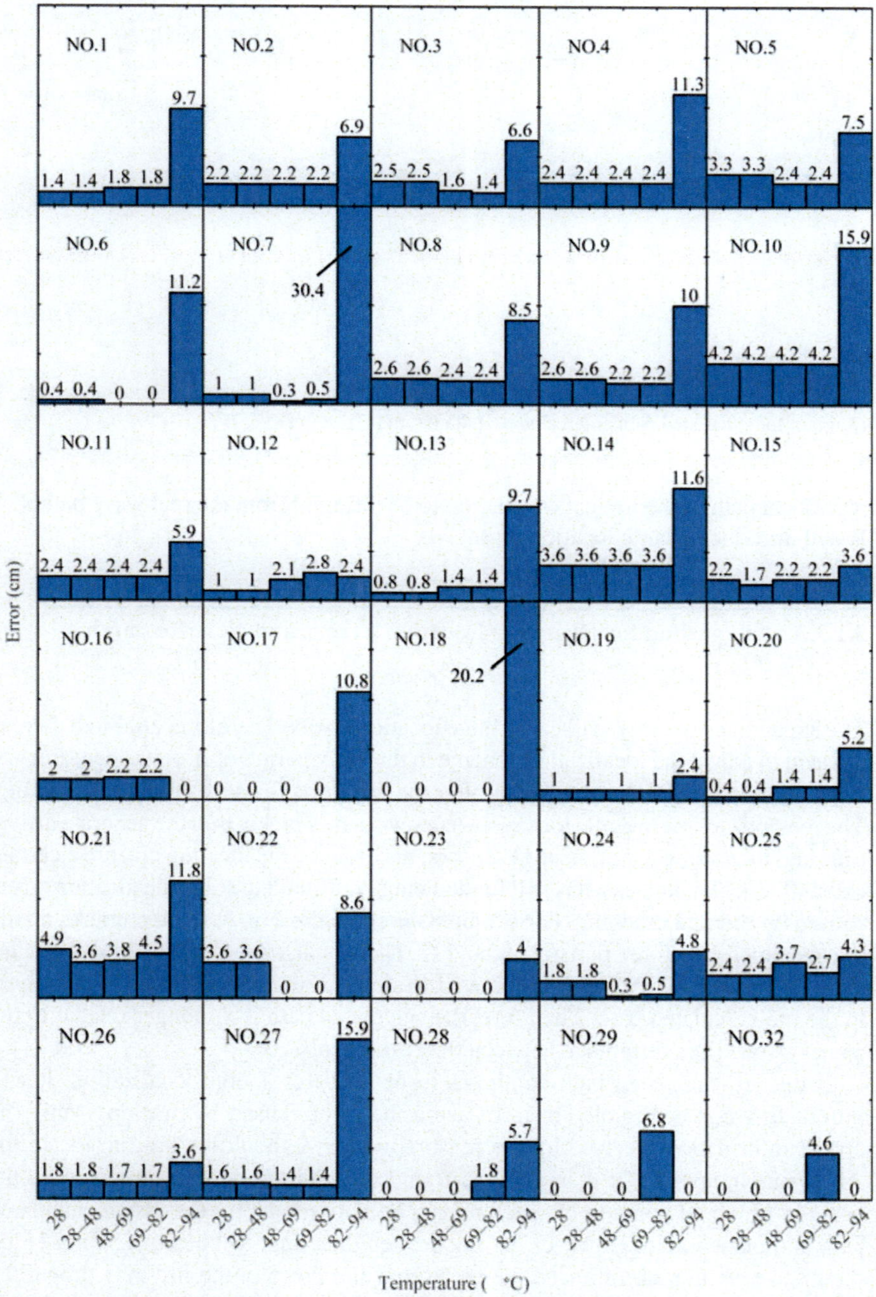

Fig. 3.4 Average error of four source location results in each stage

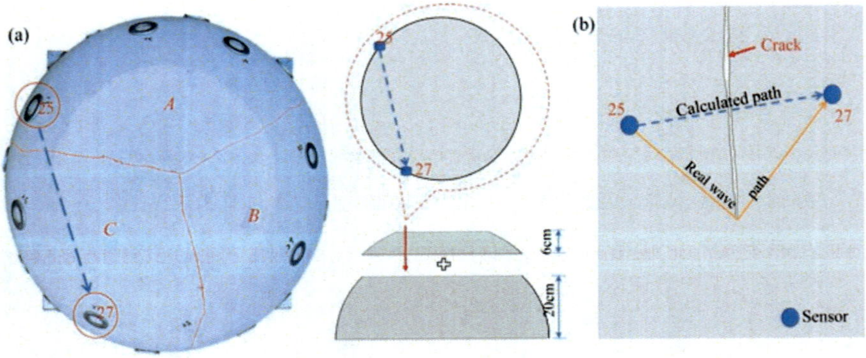

Fig. 3.5 Calculated path and real wave path for specimen with cracks: **a** Specimen with cracks; **b** Diagram for calculated path and real wave path (from [5])

waveform path of theoretical calculation to be different from the real wave path. [7]. It will introduce a large location error.

3.1.3.3 Suggestion for Reducing Location Errors and Next Research Plan

The appearance of crack will cause huge location errors. Therefore, although there is no need to adjust the localization scheme in the early period of fire, it is necessary to effectively reduce the location error after the fire starts to cause structural cracking. The methods to improve the location accuracy such as picking up the accurate arrival time and increasing the number of sensors, may reduce the location error to a certain extent [8, 9]. Nonetheless, it is still fundamentally difficult to solve the location errors caused by thermal cracking. For example, as shown in Fig. 3.5, three cracks divide the specimen into three parts: *A*, *B*, and *C*. The actual wave-form path of the pulse signal generated by 25th sensor in area *A* to signal 27th in area *C* is no longer equal to the pre-calculated path. Therefore, the calculated path is no longer related to the actual arrival time difference between the two sensors.

If the pre-calculated path continues to be used for source localization, it will introduce a non-negligible error. To avoid the error caused by cracking, only the arrival information received by the sensors in area *A* should be used to locate the AE events in area *A*. Of course, areas *B* and *C* should be in the same way. In the test, the time and position of cracking can be determined by the visual calculated path and burst sound. However, in practical engineering, technicians cannot be on the first scene in real-time, and the environmental noise in the fire may drown the crack sound of the structure. It is more difficult to find the crack position in a series of AE events. Therefore, it is of great safety significance to accurately determine the time and location of the structural cracking in the fire and select sensors with more suitable positions for localization in the actual project.

3.2 The Influence of Velocity Error and Sensor Position on Source Location Accuracy

Numerical tests were conducted to show the influence of the difference between pre-measured velocity and actual velocity on source location accuracy in 1-D, 2-D and 3-D cases.

3.2.1 *Design of Numerical Tests*

3.2.1.1 One-Dimensional Test

As shown in Fig. 3.6, two numerical tests were designed on a linear structure with a length of 2,000 mm. By comparing the location results of test a and test b, the influence of sensor position and the number of sensor employed in localization on location accuracy can be investigated. The Wave velocity is defined as 3,500 m/s in simulation. The source and sensor coordinates as well as arrival times are shown in Table 3.3.

3.2.1.2 Two-Dimensional Test

A 2-D numerical test is designed on a plane measuring 500 mm × 500 mm. Wave velocity is 3,500 m/s. Sources and sensors position are shown in Fig. 3.7. By comparing the location errors in each group, studies on whether the distance between the source and sensors affects the location errors can be conducted. By also comparing the location errors of the different groups the effect of the arrangement direction of the sources on the location errors can be investigated. Arrival times obtained by simulation are shown in Table 3.4. The source and sensor coordinates are given in Table 3.5.

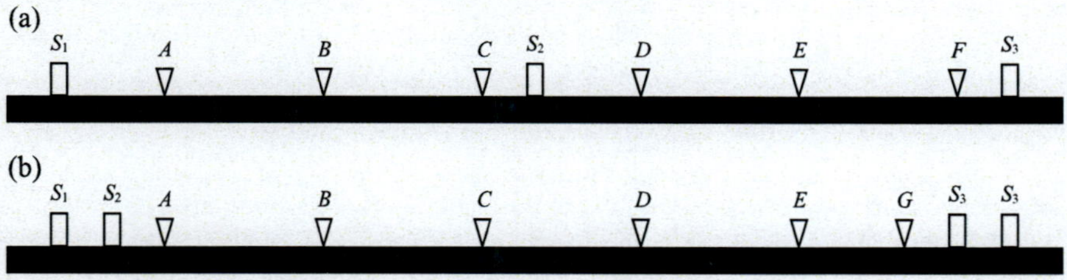

Fig. 3.6 Illustration of numerical **a** test a and **b** test b in the 1-D case. Three sensors are uniformly distributed on the structure in test a. Four sensors are symmetrically distributed at both ends of the structure in test b

Table 3.3 Source and sensor coordinates as well as arrival times obtained by numerical tests

Tests		Sensor coordinates (mm)	Arrival times (ms)					
Test a	Sources		*A*	*B*	*C*	*D*	*E*	*F*
	Source coordinates (mm)		300	600	900	1,200	1,500	1,800
	S_1	100	0.0571	0.1429	0.2286	0.3143	0.4000	0.4857
	S_2	1,000	0.2000	0.1143	0.0286	0.0571	0.1429	0.2286
	S_3	1,900	0.4571	0.3714	0.2857	0.2000	0.1143	0.0286
Test b	Sources		*A*	*B*	*C*	*D*	*E*	*G*
	Source coordinates (mm)		300	600	900	1,200	1,500	1,700
	S_1	100	0.0571	0.1429	0.2286	0.3143	0.4000	0.4571
	S_2	200	0.0286	0.1143	0.2000	0.2857	0.3714	0.4286
	S_3	1,800	0.4286	0.3429	0.2571	0.1714	0.0857	0.0286
	S_4	1,900	0.4571	0.3714	0.2857	0.2000	0.1143	0.0571

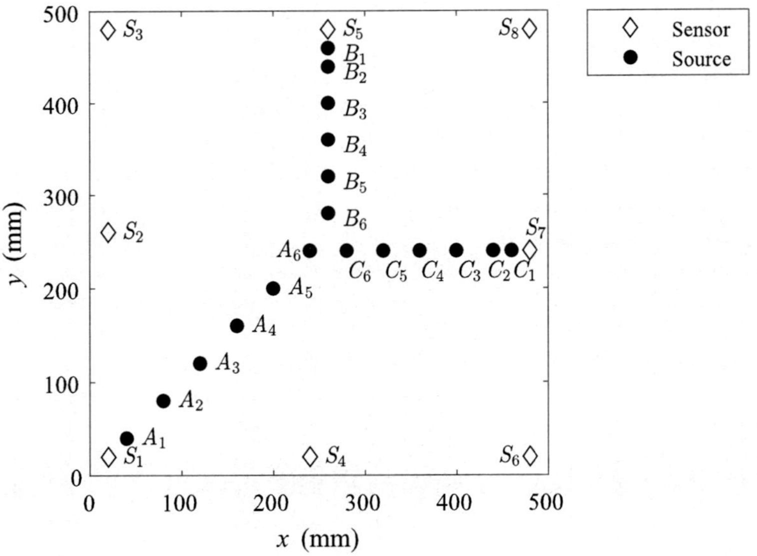

Fig. 3.7 Illustration of the 2-D numerical test. Sources are divided into three groups, group A, group B and group C. There are six sensors in each group, from subscript one to six away from S_1, S_5 or S_7

Table 3.4 Arrival times obtained by the numerical test

Sensors	Arrival times (ms)								
	A_1	A_2	A_3	A_4	A_5	A_6	B_1	B_2	B_3
S_1	0.0081	0.0242	0.0404	0.0566	0.0727	0.0889	0.1432	0.1382	0.1284
S_2	0.0631	0.0542	0.0492	0.0492	0.0542	0.0631	0.0893	0.0857	0.0794
S_3	0.1258	0.1156	0.1068	0.0998	0.0951	0.093	0.0688	0.0695	0.0723
S_4	0.0574	0.0488	0.0446	0.0461	0.0527	0.0629	0.1258	0.1201	0.1087
S_5	0.1406	0.1253	0.1104	0.0958	0.0818	0.0688	0.0057	0.0114	0.0229
S_6	0.1258	0.1156	0.1068	0.0998	0.0951	0.093	0.1406	0.1355	0.1255
S_7	0.1381	0.1231	0.1084	0.0942	0.0808	0.0686	0.0889	0.0849	0.0777
S_8	0.1778	0.1616	0.1455	0.1293	0.1131	0.097	0.0631	0.0639	0.0669
	B_4	B_5	B_6	C_1	C_2	C_3	C_4	C_5	C_6
S_1	0.1189	0.1098	0.1011	0.1406	0.1355	0.1255	0.1157	0.1063	0.0973
S_2	0.0743	0.0707	0.0688	0.1258	0.1201	0.1087	0.0973	0.0859	0.0745
S_3	0.0767	0.0824	0.0893	0.1432	0.1382	0.1284	0.1189	0.1098	0.1011
S_4	0.0973	0.0859	0.0745	0.0889	0.0849	0.0777	0.0716	0.0669	0.0639
S_5	0.0343	0.0457	0.0571	0.0893	0.0857	0.0794	0.0743	0.0707	0.0688
S_6	0.1157	0.1063	0.0973	0.0631	0.0639	0.0669	0.0716	0.0777	0.0849
S_7	0.0716	0.0669	0.0639	0.0057	0.0114	0.0229	0.0343	0.0457	0.0571
S_8	0.0716	0.0777	0.0849	0.0688	0.0695	0.0723	0.0767	0.0824	0.0893

Table 3.5 Source and sensor coordinates in the numerical test

Sources and sensors	Coordinates		Sources and sensors	Coordinates	
	x (mm)	y (mm)		x (mm)	y (mm)
S_1	20	20	A_6	240	240
S_2	20	260	B_1	260	460
S_3	20	480	B_2	260	440
S_4	240	20	B_3	260	400
S_5	260	480	B_4	260	360
S_6	480	20	B_5	260	320
S_7	480	240	B_6	260	280
S_8	480	480	C_1	460	240
A_1	40	40	C_2	440	240
A_2	80	80	C_3	400	240
A_3	120	120	C_4	360	240
A_4	160	160	C_5	320	240
A_5	200	200	C_6	280	240

3.2.1.3 Three-Dimensional Test

The test is designed on a cuboid measuring 400 mm × 200 mm × 200 mm and the wave velocity is 3,500 m/s. There are 45 sources in the simulation and they are divided into three groups according to difference height. The sources in the first group are P_1–P_{15} at the height 40 mm, the second are P_{16}–P_{30} at 100 mm and the third are P_{31}–P_{45} at 160 mm (see Fig. 3.8a). Figure 3.8b shows sensors arrangement. Arrival time obtained by simulation is shown in Table 3.6. Source and sensor coordinates are shown in Table 3.7.

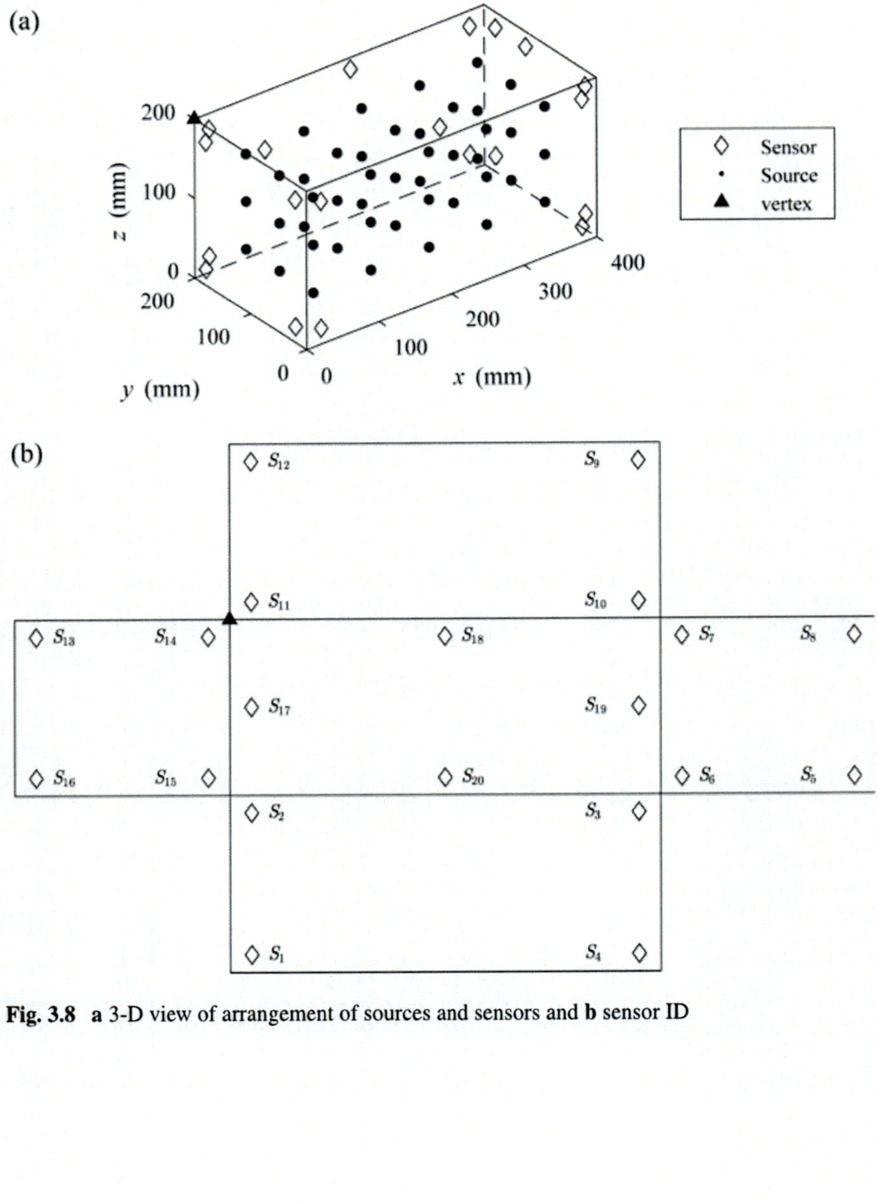

Fig. 3.8 a 3-D view of arrangement of sources and sensors and b sensor ID

Table 3.6 Arrival times obtained by simulation

Sensors	Arrival times (ms)								
	P_1	P_2	P_3	P_4	P_5	P_6	P_7	P_8	P_9
S_1	0.014	0.0297	0.0464	0.0313	0.0408	0.0542	0.053	0.0591	0.069
S_2	0.042	0.0495	0.061	0.0505	0.0569	0.0671	0.0661	0.0711	0.0796
S_3	0.1057	0.1089	0.1146	0.0851	0.0891	0.096	0.0661	0.0711	0.0796
S_4	0.098	0.1014	0.1075	0.0754	0.0798	0.0874	0.053	0.0591	0.069
S_5	0.1032	0.1055	0.1105	0.0804	0.0834	0.0896	0.0577	0.0618	0.07
S_6	0.1105	0.1127	0.1174	0.0896	0.0923	0.098	0.07	0.0734	0.0804
S_7	0.1174	0.1127	0.1105	0.098	0.0923	0.0896	0.0804	0.0734	0.07
S_8	0.1105	0.1055	0.1032	0.0896	0.0834	0.0804	0.07	0.0618	0.0577
S_9	0.1075	0.1014	0.098	0.0874	0.0798	0.0754	0.069	0.0591	0.053
S_{10}	0.1146	0.1089	0.1057	0.096	0.0891	0.0851	0.0796	0.0711	0.0661
S_{11}	0.061	0.0495	0.042	0.0671	0.0569	0.0505	0.0796	0.0711	0.0661
S_{12}	0.0464	0.0297	0.014	0.0542	0.0408	0.0313	0.069	0.0591	0.053
S_{13}	0.042	0.0262	0.014	0.053	0.0416	0.0352	0.07	0.0618	0.0577
S_{14}	0.0577	0.0475	0.042	0.0661	0.0574	0.053	0.0804	0.0734	0.07
S_{15}	0.042	0.0475	0.0577	0.053	0.0574	0.0661	0.07	0.0734	0.0804
S_{16}	0.014	0.0262	0.042	0.0352	0.0416	0.053	0.0577	0.0618	0.07
S_{17}	0.0492	0.0461	0.0492	0.0566	0.0539	0.0566	0.0709	0.0688	0.0709
S_{18}	0.076	0.0686	0.0649	0.0649	0.056	0.0514	0.0607	0.0511	0.0461
S_{19}	0.1087	0.1074	0.1087	0.0889	0.0872	0.0889	0.0709	0.0688	0.0709
S_{20}	0.0649	0.0686	0.076	0.0514	0.056	0.0649	0.0461	0.0511	0.0607
	P_{10}	P_{11}	P_{12}	P_{13}	P_{14}	P_{15}	P_{16}	P_{17}	P_{18}
S_1	0.0754	0.0798	0.0874	0.098	0.1014	0.1075	0.0262	0.037	0.0514
S_2	0.0851	0.0891	0.096	0.1057	0.1089	0.1146	0.0262	0.037	0.0514
S_3	0.0505	0.0569	0.0671	0.042	0.0495	0.061	0.1004	0.1038	0.1098
S_4	0.0313	0.0408	0.0542	0.014	0.0297	0.0464	0.1004	0.1038	0.1098
S_5	0.0352	0.0416	0.053	0.014	0.0262	0.042	0.1055	0.1078	0.1127
S_6	0.053	0.0574	0.0661	0.042	0.0475	0.0577	0.1055	0.1078	0.1127
S_7	0.0661	0.0574	0.053	0.0577	0.0475	0.042	0.1127	0.1078	0.1055
S_8	0.053	0.0416	0.0352	0.042	0.0262	0.014	0.1127	0.1078	0.1055
S_9	0.0542	0.0408	0.0313	0.0464	0.0297	0.014	0.1098	0.1038	0.1004
S_{10}	0.0671	0.0569	0.0505	0.061	0.0495	0.042	0.1098	0.1038	0.1004
S_{11}	0.096	0.0891	0.0851	0.1146	0.1089	0.1057	0.0514	0.037	0.0262
S_{12}	0.0874	0.0798	0.0754	0.1075	0.1014	0.098	0.0514	0.037	0.0262
S_{13}	0.0896	0.0834	0.0804	0.1105	0.1055	0.1032	0.0475	0.0343	0.0262
S_{14}	0.098	0.0923	0.0896	0.1174	0.1127	0.1105	0.0475	0.0343	0.0262

(continued)

Table 3.6 (continued)

Sensors	Arrival times (ms)								
S_{15}	0.0896	0.0923	0.098	0.1105	0.1127	0.1174	0.0262	0.0343	0.0475
S_{16}	0.0804	0.0834	0.0896	0.1032	0.1055	0.1105	0.0262	0.0343	0.0475
S_{17}	0.0889	0.0872	0.0889	0.1087	0.1074	0.1087	0.0338	0.0291	0.0338
S_{18}	0.0649	0.056	0.0514	0.076	0.0686	0.0649	0.0671	0.0586	0.0542
S_{19}	0.0566	0.0539	0.0566	0.0492	0.0461	0.0492	0.1027	0.1013	0.1027
S_{20}	0.0514	0.056	0.0649	0.0649	0.0686	0.076	0.0542	0.0586	0.0671
	P_{19}	P_{20}	P_{21}	P_{22}	P_{23}	P_{24}	P_{25}	P_{26}	P_{27}
S_1	0.0383	0.0464	0.0586	0.0574	0.0631	0.0725	0.0786	0.0828	0.0902
S_2	0.0383	0.0464	0.0586	0.0574	0.0631	0.0725	0.0786	0.0828	0.0902
S_3	0.0786	0.0828	0.0902	0.0574	0.0631	0.0725	0.0383	0.0464	0.0586
S_4	0.0786	0.0828	0.0902	0.0574	0.0631	0.0725	0.0383	0.0464	0.0586
S_5	0.0834	0.0863	0.0923	0.0618	0.0657	0.0734	0.0416	0.0471	0.0574
S_6	0.0834	0.0863	0.0923	0.0618	0.0657	0.0734	0.0416	0.0471	0.0574
S_7	0.0923	0.0863	0.0834	0.0734	0.0657	0.0618	0.0574	0.0471	0.0416
S_8	0.0923	0.0863	0.0834	0.0734	0.0657	0.0618	0.0574	0.0471	0.0416
S_9	0.0902	0.0828	0.0786	0.0725	0.0631	0.0574	0.0586	0.0464	0.0383
S_{10}	0.0902	0.0828	0.0786	0.0725	0.0631	0.0574	0.0586	0.0464	0.0383
S_{11}	0.0586	0.0464	0.0383	0.0725	0.0631	0.0574	0.0902	0.0828	0.0786
S_{12}	0.0586	0.0464	0.0383	0.0725	0.0631	0.0574	0.0902	0.0828	0.0786
S_{13}	0.0574	0.0471	0.0416	0.0734	0.0657	0.0618	0.0923	0.0863	0.0834
S_{14}	0.0574	0.0471	0.0416	0.0734	0.0657	0.0618	0.0923	0.0863	0.0834
S_{15}	0.0416	0.0471	0.0574	0.0618	0.0657	0.0734	0.0834	0.0863	0.0923
S_{16}	0.0416	0.0471	0.0574	0.0618	0.0657	0.0734	0.0834	0.0863	0.0923
S_{17}	0.0439	0.0404	0.0439	0.0613	0.0588	0.0613	0.0814	0.0796	0.0814
S_{18}	0.0542	0.0431	0.037	0.0492	0.0366	0.0291	0.0542	0.0431	0.037
S_{19}	0.0814	0.0796	0.0814	0.0613	0.0588	0.0613	0.0439	0.0404	0.0439
S_{20}	0.037	0.0431	0.0542	0.0291	0.0366	0.0492	0.037	0.0431	0.0542
	P_{28}	P_{29}	P_{30}	P_{31}	P_{32}	P_{33}	P_{34}	P_{35}	P_{36}
S_1	0.1004	0.1038	0.1098	0.042	0.0495	0.061	0.0505	0.0569	0.0671
S_2	0.1004	0.1038	0.1098	0.014	0.0297	0.0464	0.0313	0.0408	0.0542
S_3	0.0262	0.037	0.0514	0.098	0.1014	0.1075	0.0754	0.0798	0.0874
S_4	0.0262	0.037	0.0514	0.1057	0.1089	0.1146	0.0851	0.0891	0.096
S_5	0.0262	0.0343	0.0475	0.1105	0.1127	0.1174	0.0896	0.0923	0.098
S_6	0.0262	0.0343	0.0475	0.1032	0.1055	0.1105	0.0804	0.0834	0.0896
S_7	0.0475	0.0343	0.0262	0.1105	0.1055	0.1032	0.0896	0.0834	0.0804
S_8	0.0475	0.0343	0.0262	0.1174	0.1127	0.1105	0.098	0.0923	0.0896

(continued)

Table 3.6 (continued)

Sensors	Arrival times (ms)								
S_9	0.0514	0.037	0.0262	0.1146	0.1089	0.1057	0.096	0.0891	0.0851
S_{10}	0.0514	0.037	0.0262	0.1075	0.1014	0.098	0.0874	0.0798	0.0754
S_{11}	0.1098	0.1038	0.1004	0.0464	0.0297	0.014	0.0542	0.0408	0.0313
S_{12}	0.1098	0.1038	0.1004	0.061	0.0495	0.042	0.0671	0.0569	0.0505
S_{13}	0.1127	0.1078	0.1055	0.0577	0.0475	0.042	0.0661	0.0574	0.053
S_{14}	0.1127	0.1078	0.1055	0.042	0.0262	0.014	0.053	0.0416	0.0352
S_{15}	0.1055	0.1078	0.1127	0.014	0.0262	0.042	0.0352	0.0416	0.053
S_{16}	0.1055	0.1078	0.1127	0.042	0.0475	0.0577	0.053	0.0574	0.0661
S_{17}	0.1027	0.1013	0.1027	0.0214	0.0128	0.0214	0.0352	0.0308	0.0352
S_{18}	0.0671	0.0586	0.0542	0.0618	0.0524	0.0475	0.0475	0.0343	0.0262
S_{19}	0.0338	0.0291	0.0338	0.0993	0.0978	0.0993	0.0771	0.0752	0.0771
S_{20}	0.0542	0.0586	0.0671	0.0475	0.0524	0.0618	0.0262	0.0343	0.0475
	P_{37}	P_{38}	P_{39}	P_{40}	P_{41}	P_{42}	P_{43}	P_{44}	P_{45}
S_1	0.0661	0.0711	0.0796	0.0851	0.0891	0.096	0.1057	0.1089	0.1146
S_2	0.053	0.0591	0.069	0.0754	0.0798	0.0874	0.098	0.1014	0.1075
S_3	0.053	0.0591	0.069	0.0313	0.0408	0.0542	0.014	0.0297	0.0464
S_4	0.0661	0.0711	0.0796	0.0505	0.0569	0.0671	0.042	0.0495	0.061
S_5	0.07	0.0734	0.0804	0.053	0.0574	0.0661	0.042	0.0475	0.0577
S_6	0.0577	0.0618	0.07	0.0352	0.0416	0.053	0.014	0.0262	0.042
S_7	0.07	0.0618	0.0577	0.053	0.0416	0.0352	0.042	0.0262	0.014
S_8	0.0804	0.0734	0.07	0.0661	0.0574	0.053	0.0577	0.0475	0.042
S_9	0.0796	0.0711	0.0661	0.0671	0.0569	0.0505	0.061	0.0495	0.042
S_{10}	0.069	0.0591	0.053	0.0542	0.0408	0.0313	0.0464	0.0297	0.014
S_{11}	0.069	0.0591	0.053	0.0874	0.0798	0.0754	0.1075	0.1014	0.098
S_{12}	0.0796	0.0711	0.0661	0.096	0.0891	0.0851	0.1146	0.1089	0.1057
S_{13}	0.0804	0.0734	0.07	0.098	0.0923	0.0896	0.1174	0.1127	0.1105
S_{14}	0.07	0.0618	0.0577	0.0896	0.0834	0.0804	0.1105	0.1055	0.1032
S_{15}	0.0577	0.0618	0.07	0.0804	0.0834	0.0896	0.1032	0.1055	0.1105
S_{16}	0.07	0.0734	0.0804	0.0896	0.0923	0.098	0.1105	0.1127	0.1174
S_{17}	0.0554	0.0527	0.0554	0.0771	0.0752	0.0771	0.0993	0.0978	0.0993
S_{18}	0.0416	0.0256	0.0128	0.0475	0.0343	0.0262	0.0618	0.0524	0.0475
S_{19}	0.0554	0.0527	0.0554	0.0352	0.0308	0.0352	0.0214	0.0128	0.0214
S_{20}	0.0128	0.0256	0.0416	0.0262	0.0343	0.0475	0.0475	0.0524	0.0618

Table 3.7 Source and sensor coordinates

Sources and sensors	Coordinates			Sources and sensors	Coordinates		
	x (mm)	y (mm)	z (mm)		x (mm)	y (mm)	z (mm)
S_1	20	0	20	P_{14}	360	100	40
S_2	20	0	180	P_{15}	360	160	40
S_3	380	0	180	P_{16}	40	40	100
S_4	380	0	20	P_{17}	40	100	100
S_5	400	20	20	P_{18}	40	160	100
S_6	400	20	180	P_{19}	120	40	100
S_7	400	180	180	P_{20}	120	100	100
S_8	400	180	20	P_{21}	120	160	100
S_9	380	200	20	P_{22}	200	40	100
S_{10}	380	200	180	P_{23}	200	100	100
S_{11}	20	200	180	P_{24}	200	160	100
S_{12}	20	200	20	P_{25}	280	40	100
S_{13}	0	180	20	P_{26}	280	100	100
S_{14}	0	180	180	P_{27}	280	160	100
S_{15}	0	20	180	P_{28}	360	40	100
S_{16}	0	20	20	P_{29}	360	100	100
S_{17}	20	100	200	P_{30}	360	160	100
S_{18}	200	180	200	P_{31}	40	40	160
S_{19}	380	100	200	P_{32}	40	100	160
S_{20}	200	20	200	P_{33}	40	160	160
P_1	40	40	40	P_{34}	120	40	160
P_2	40	100	40	P_{35}	120	100	160
P_3	40	160	40	P_{36}	120	160	160
P_4	120	40	40	P_{37}	200	40	160
P_5	120	100	40	P_{38}	200	100	160
P_6	120	160	40	P_{39}	200	160	160
P_7	200	40	40	P_{40}	280	40	160
P_8	200	100	40	P_{41}	280	100	160
P_9	200	160	40	P_{42}	280	160	160
P_{10}	280	40	40	P_{43}	360	40	160
P_{11}	280	100	40	P_{44}	360	100	160
P_{12}	280	160	40	P_{45}	360	160	160
P_{13}	360	40	40				

3.2.2 Localization Methods

The traditional localization methods described in Chap. 2 are used for source localization to investigate the influence of velocity error on location accuracy of methods with pre-measured velocity. The TLL, SM and Geiger method are applied to 1-D, 2-D and 3-D source localization. To show the advantage of using velocity-free methods for localization, MSLM-WV which is introduced in Chap. 5 is also applied to localization.

Because the SM, Geiger method and MSLM-MV need initialization to start the first iteration, the reasonable initial values for three dimensional cases are as follows.

The initial value of source position for MSLM-MV in 1-D tests is 200 mm in test a and is 1,000 mm in test b. Parameters μ and ε are 0.01 and 1×10^{-16}, respectively.The two parameters in three cases remain the same. In the 2-D case, the initial source coordinate (x_0, y_0) in SM and MSLM-MV could be calculated by:

$$x_0 = \frac{1}{m}\sum_{i=1}^{m} S_{ix}, \quad y_0 = \frac{1}{m}\sum_{i=1}^{m} S_{iy}, \tag{3.3}$$

where S_{ix} indicates the x-coordinate of S_i, S_{iy} the y-coordinate of S_i, and m number of sensors. In SM, λ_1, λ_2 and ε are initialized to 10, 10 and 1×10^{-4} mm, respectively and the maximum iteration is 10,000. In the 3-D case, initial x-coordinate and y-coordinate of source are calculated by Eq. (3.3) and the initial z-coordinate is calculated by

$$z_0 = \frac{1}{m}\sum_{i=1}^{m} S_{iz}, \tag{3.4}$$

where S_{iz} indicates the z-coordinate of S_i. For Geiger method, the maximum iteration is 10,000 and the result is considered as the source position when $\|\Delta \boldsymbol{x}\|_2$ is less than 1×10^{-16}.

3.2.3 Results and Discussion

3.2.3.1 One-Dimensional Results and Discussion

The results of TLL and MSLM-MV are shown and discussed. When apply the TLL to source localization, only two sensors close to the ends of the structure are used. Given different initial values of velocity, location errors of different sources in test a and test b are shown in Fig. 3.9.

The location results of two tests show the linear relationship between velocity error and location error. The closer the source is to the middle of the structure, the

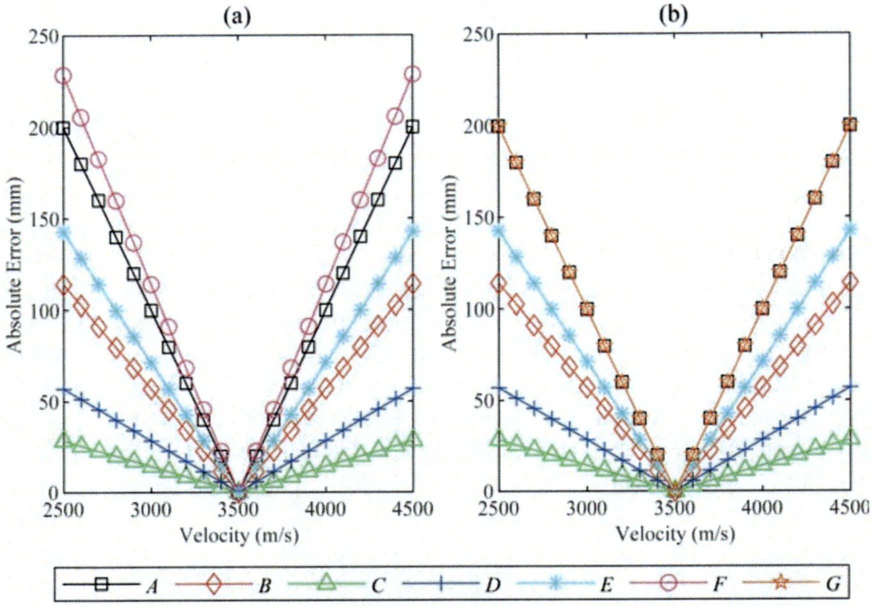

Fig. 3.9 Location results of **a** test a and **b** test b using TLL

less the location error is affected by velocity error. The same source of both tests has the same location error, and this indicates that the sensor array has no effect on the 1-D location error.

When using MSLM-MV for source localization, all sensors arranged on the structure are used and the location results of test a and test b are denoted by a-3 and b-4, respectively. To investigate the effect of sensors arrangement on location error, three sensors (S_1, S_2, and S_3) are also used for localization in test b and the location result is denoted by b-3. Given different initial velocity values, location errors of different sources in test a and test b are shown in Fig. 3.10. The Location error of each source shows a similar trend in the three location results (Fig. 3.10a, c, and e). When the initial velocity is less than a certain value, the location error is low and is not affected by velocity error. When the initial velocity is close to 3,500 m/s, the location error fluctuates greatly. Figure 3.10b, d, and f illustrate the phenomenon clearly after filtering large error. From Fig. 3.10a, c, and e, we can find that the location errors of source *A*, *C* and *F* in test a as well as source *A*, *E*, and *G* in test b increase significantly when the initial velocity is greater than a certain value and the location errors of others decrease gradually. Therefore, the location result of the source close to sensors are prone to divergence when initial velocity is large.

In the three location results, the location error of a-3 is the smallest, followed by b-3 and b-4 is the largest. So the uniform distribution of sensors could improve location accuracy. However, when the middle of the structure is not suitable for arranging sensors, symmetrical distribution at both ends can be adopted to reduce the location error caused by asymmetry and a low value of initial velocity is recommended.

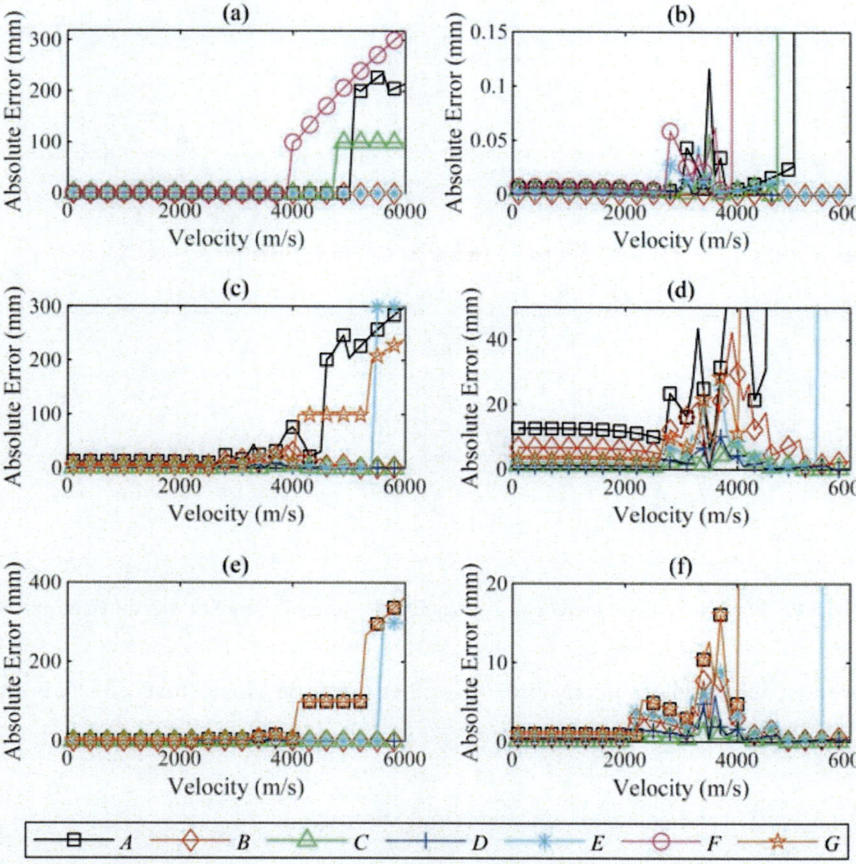

Fig. 3.10 Location result of MSLM-MV. **a** a-3, **b** a-3 with absolute error less than 0.15 mm, **c** b-3, **d** b-3 with absolute error less than 50 mm, **e** b-4, **f** b-4 with absolute error less than 20 mm

3.2.3.2 Two-Dimensional Results and Discussion

The Location result is shown in Fig. 3.11. The location error of sources close to the sensor (A_1, B_1, and C_1) is larger than others and is greatly affected by the velocity error. Interestingly, even though the distance between S_1 and the source in group A is larger than that between S_5 and corresponding source in group B, location error of group A is smaller than group B (comparing Fig. 3.11b, c). By comparing Fig. 3.11c, d, it is reasonable to note that their error is the same since the distance between sources and sensors is the same. The explanation to this is that the arrangement of sensors affects the location error and the effect is positive. Since the effect to group B and C is the same, it indicates that it has an axis of symmetry and sources in group A just on the axis.

Location result is shown in Fig. 3.12. It can be noted that the location result is similar to 1-D case. Comparing location error of different groups, group A is smaller than group B and C which have the same error. Concisely, the MSLM-MV can greatly

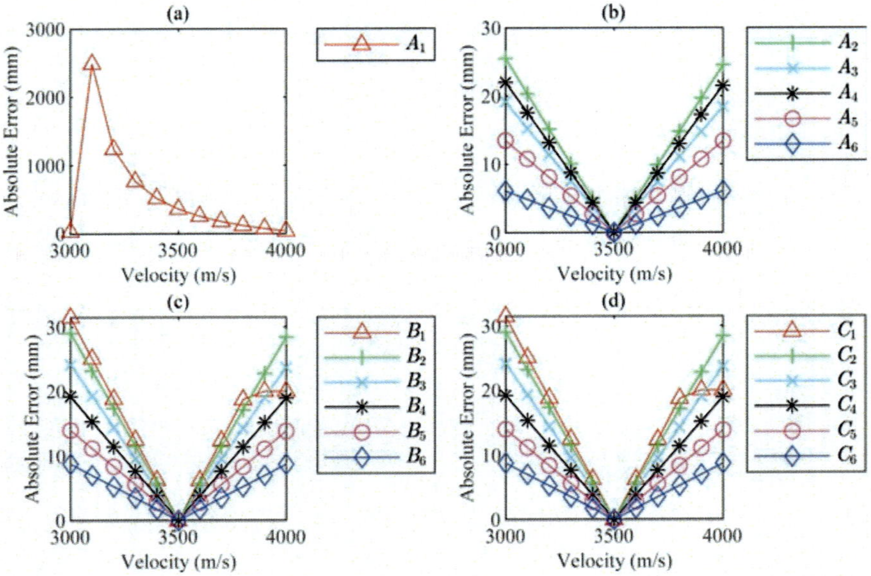

Fig. 3.11 Location result of **a** group A, **b** group B and **c** group C using SM with different velocity

decrease the effect of velocity error on location error and a low initial velocity is also recommended in 2-D case.

3.2.3.3 Three-Dimensional Results and Discussion

Figure 3.13 shows the location error of every source and its variation with velocity and spatial position using Geiger method. There is a strong linear relationship between velocity error and location error. Location errors of sources at the height of 160 mm are larger than that of 40 mm and 100 mm. On a horizontal plane, error of source at the edge is larger than that inside. Since the bottom of the cuboid does not arrange sensor, the number of sensors close to sources at a low height is less than that of sources at a high height. So the closer source is to the sensor, the larger its location error is.

Location result of MSLM-MV is shown in Fig. 3.14. The trend of curve is similar to that in 1-D and 2-D cases. Therefore, the recommendation that initial velocity in MSLM-MV should be lower than the true velocity, such as 1,000 m/s, is suitable for three cases. The MSLM-MV can achieve high accuracy that Geiger method need accurate pre-measured velocity to achieve.

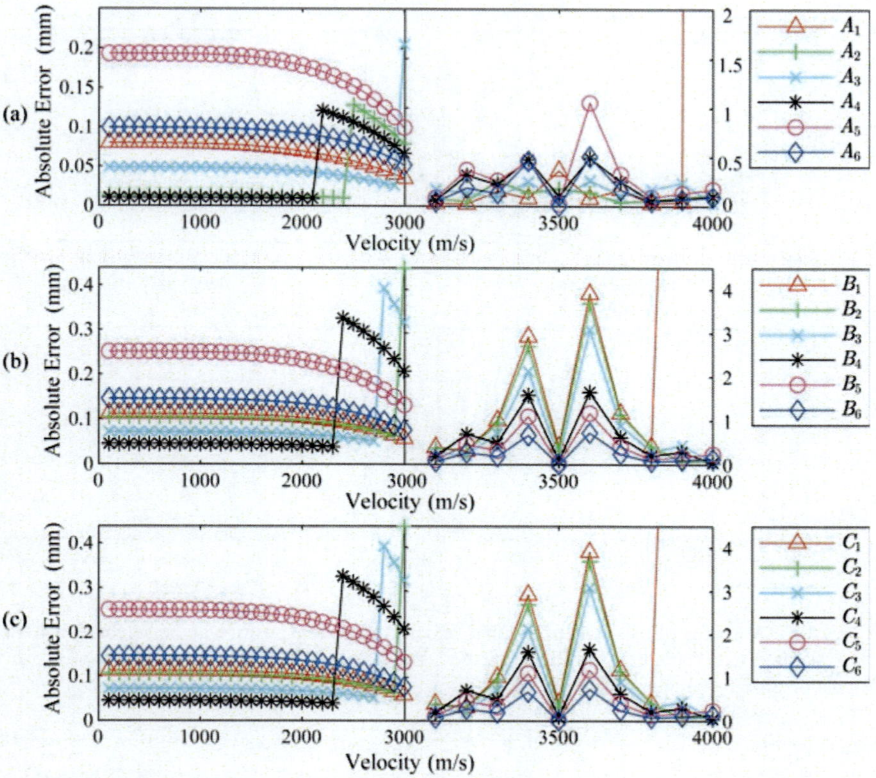

Fig. 3.12 Location result of **a** group A, **b** group B and **c** group C using MSLM-MV with different initial velocity

3.3 The Influence of Stress Stages on Source Location Accuracy

3.3.1 Experimental Materials and Procedure

In the experiments, coarse-grained granite was selected as the research object. In the uniaxial compression tests (UCT), the sample was prepared into a standard cube of 100 mm × 100 mm × 200 mm, which is denoted by S-UCT. Due to the size limitation of the mechanical loading equipment, the sample in biaxial compression tests (BCT) was prepared as standard cube of 100 mm × 100 mm × 100 mm, which is denoted by S-BCT. The surface of the samples is flat and smooth, without macroscopic joint fissures and weak structural surfaces. According to the ISRM rock mechanics test recommendations, the surface of the rock sample is polished to ensure that the flatness, parallelism, and verticality of each rock surface meet the test standards.

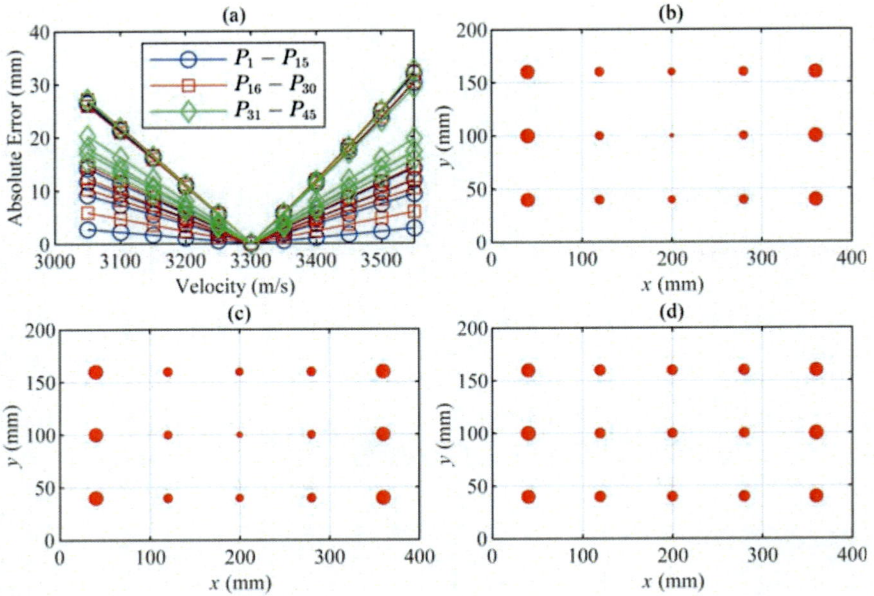

Fig. 3.13 Variation of location error with **a** velocity and spatial position at the height of **b** 40 mm, **c** 100 mm, and **d** 160 mm using Geiger method. The size of circles in **b–d** indicates relative size of location errors when the velocity is 4,000 m/s

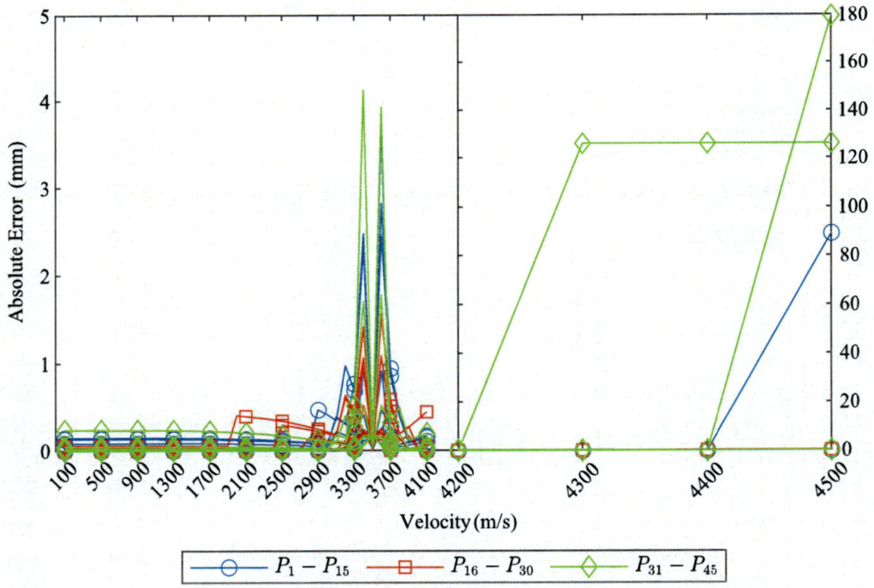

Fig. 3.14 Location result of MSLM-MV

The tests were carried out on the true triaxial electro-hydraulic servo mutagenesis experimental system (TRW-3000) of Central South University, China. The equipment can perform independent dynamic, static, and combined loading in three directions on samples of various sizes. In the UCT, only the vertical direction (Z direction) was used for loading and the other two directions remained unchanged. The axial stress rate was kept constant at 0.05 MPa/s until a clear cracking sound was heard to end the loading. In the BCT, the intermediate principal stress and the maximum principal stress were loaded in the Y and Z directions by the loading system, respectively. The loading method used in the tests is as follow: The Y and Z directions were loaded simultaneously to the preset value (10 MPa) at a loading rate of 0.05 MPa/s, and then the loading mode in Z direction is adjusted to displacement loading of 0.001 mm/s until the macroscopic instability of the rock occurs. Before the test, a prestress of 0.1 MPa was applied to the rock sample to ensure it was in full contact with the loader indenter. Parameters such as time, load, displacement, stress, and strain in three directions were collected in real-time during the test.

The experiments were conducted using an AMSY-6 multi-channel AE acquisition system from Vallen, Germany, to acquire AE characteristics and waveform data in real-time. The acquisition frequency of AE data was 10 MHz, the threshold value was 55 dB, and the amplifier gain was 34 dB. The sensors used in the experiment were VS45-H piezoelectric sensors with a diameter of 20 mm and the response frequency is 20–450 kHz. Special fixtures were used to fix the AE sensors on the surface of the rock sample to avoid accidental falling off during the loading process, and apply couplant on the surface of the sensor to obtain a good coupling effect. A total of 18 AE sensors were used in the UCT and the sensors were arranged at typical locations on the side of the rock sample in diamond and rectangular grids (Fig. 3.15a). A total of ten AE sensors were used in the BCT and arranged on the free surface of the rock sample in a rectangular grid (Fig. 3.15b).

In addition, identifiable AE pulse signals were cyclically emitted during the experiments. The pulse width and pulse amplitude of the Auto-Sensor-Test (AST) were 5.2 s and 200 Voltage-Peak-Peak (VPP), respectively. The sensor that emits the pulse signal in one pulse can be regarded as the active source, and the remaining AE sensors receive the pulse signal to realize the analysis of location accuracy. The pulse signal repeats four times for each channel in order to reduce random signal noise and thus improve the signal-to-noise ratio (SNR).

The time-difference location method without pre-measured velocity is used for source localization. In this method, the wave velocity is set as an unknown value, then the optimal real-time wave velocity and source coordinates in the target area are calculated iteratively through the algorithm to meet the nonlinear relationship between time difference and the coordinates of each sensor. This method breaks through the traditional technical bottleneck that pre-measured velocity affects the location accuracy and greatly improves the location accuracy which has great application values.

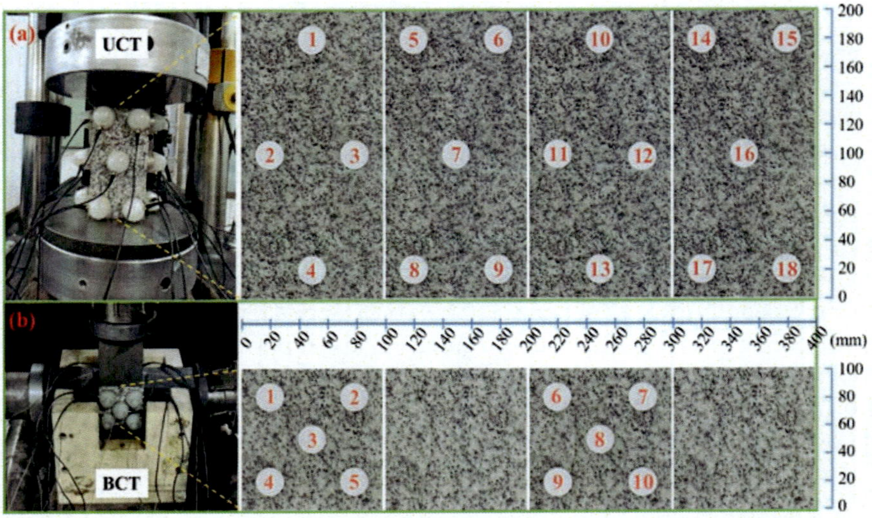

Fig. 3.15 Experimental test system and AE sensor distribution in the **a** UCT and **b** BCT

3.3.2 Rock Fracture Stage Division

There are certain stage characteristics in the process of rock fracture and instability. The properties, bearing capacity, and deformation characteristics of rocks in different stages are significantly different. The cumulative AE events curve change characteristic method proposed in [10] is used to divide the rock fracture stage. The closing stress point and the crack initiation stress point are determined from the starting point and ending point of the linear segment of the cumulative AE events curve, respectively. The maximum value of the second derivative of the segment curve determines the damage stress point. Fracture stage division of granite samples S-UCT and S-BCT are shown in Fig. 3.16.

In the micro-fracture compaction stage (stage I), the internal micro-fractures of the rock are compacted and closed under the action of external loads and a small amount of AE events are generated. The cumulative AE events curve is a convex curve with a gradually decreasing slope. In the elastic stage (stage II), the elastic deformation of rock causes some AE events with constant release rate and the cumulative AE events curve is close to linear. In the stage of micro-cracks initiation and stable expansion (stage III), the AE event rate begins to increase and the cumulative AE events curve is a concave curve with a gradually increasing slope, which indicates the damage inside the rock is increasing. In the stage of unstable propagation of micro-cracks (stage IV), the AE event rate and cumulative AE events curve in UCT exhibit a 'blowout' manner increase and exponential change characteristic, respectively. The AE event and cumulative AE events curve in BCT show the unstable release and 'step-like' rise characteristics, respectively.

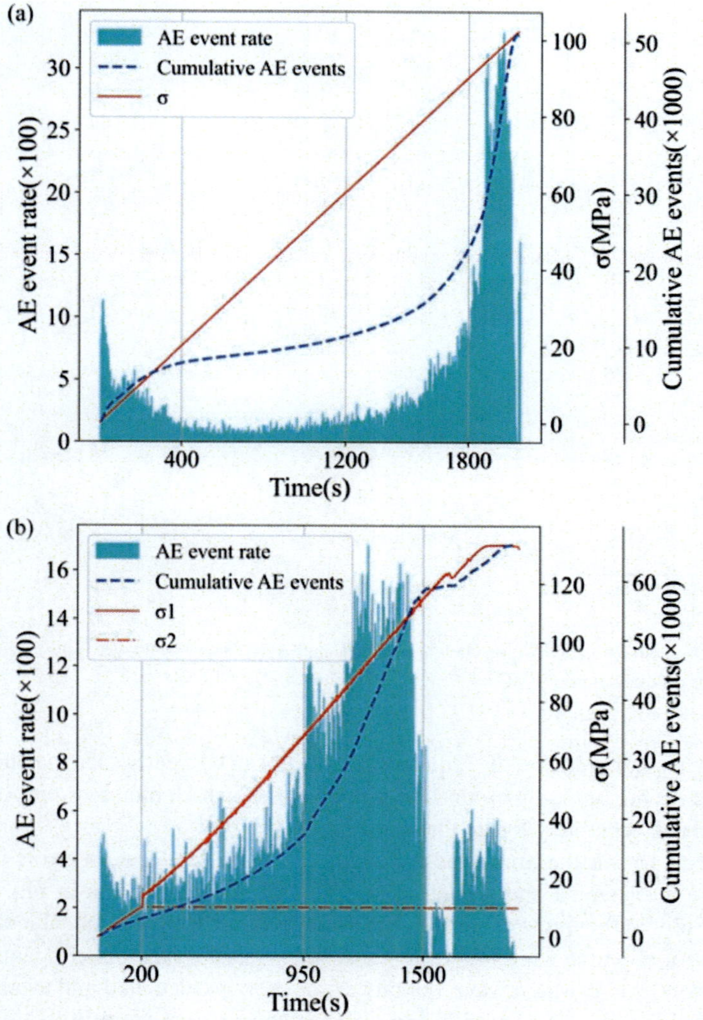

Fig. 3.16 Fracture stage division of **a** sample S-UCT and **b** S-BCT based on cumulative AE events curve

3.3.3 Results and Discussion

3.3.3.1 Analysis of AE Location Error

Figures 3.17 and 3.18 visually show the AE pulse source localization results for granite samples S-UCT and S-BCT, respectively. The red five-pointed star in the figures represent the AE sensors, the dots represent the location results, and the

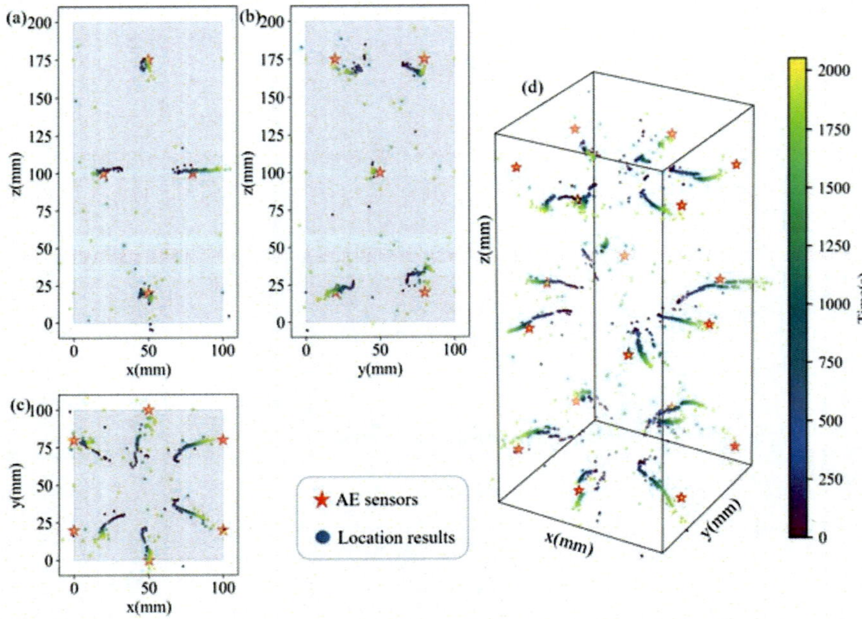

Fig. 3.17 **a** Front view, **b** right view, **c** top view and **d** 3-D view of the AE pulse source location result for granite sample S-UCT

change in color indicates the sequence of time. It can be clearly seen that the AE location results are basically around the pulse sources, which indicates the AE location method can accurately locate the sources.

Taking 15 s as the time window length, the AE pulse location errors in the time window are counted and plotted as boxplots of different colors in Fig. 3.19. The white point in the figure represents the average value of the statistical location error amount of this time window. At the same time, based on the pulse signals recorded by sensors, the real-time wave velocity of the rock is calculated and averaged every 15 s, and this is also plotted in Fig. 3.19. It can be seen from Fig. 3.19a that the location error shows a trend of first decreasing and then increasing during the whole UCT loading. The average location error decreases from 30 mm during static load to about 15 mm and then increases to 50 mm after the sample is broken. The discreteness of the location error also shows the characteristics of first decreasing and then increasing. The location error in the early and medium stages is about 25 mm and the discreteness is relatively small in BCT (Fig. 3.19b). After entering stage III, the location error begins to gradually increase and the discreteness also increases. The location error suddenly increases to about 50 mm in stage IV. In addition, the average AE pulse source location error and the real-time wave velocity show "mirror" variation characteristics both in UCT and BCT. The peak value of the wave velocity basically corresponds to the minimum value of the location error.

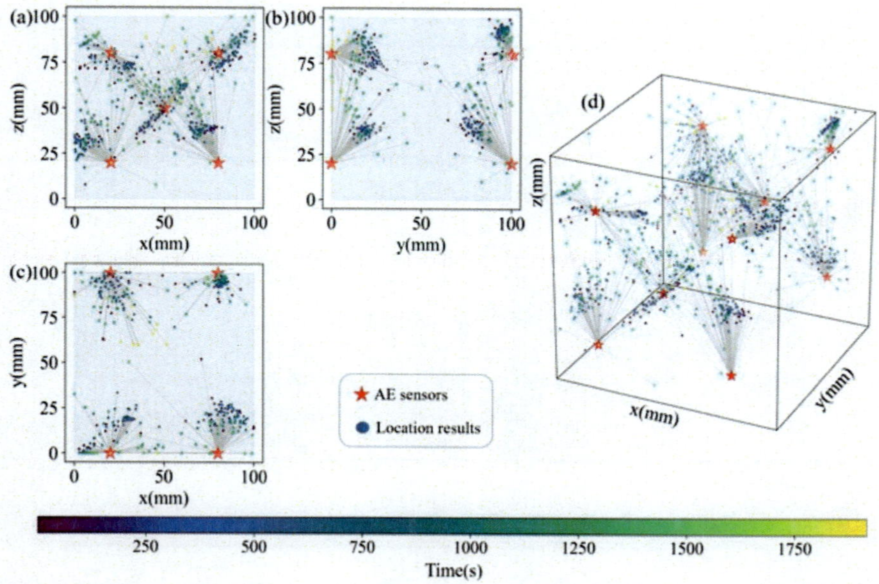

Fig. 3.18 **a** Front view, **b** right view, **c** top view and **d** 3-D view of the AE pulse source location result for granite sample S-BCT

The above trend can be explained by analyzing the stress state of the rock sample. In stage I, the initial micro-fractures in the rock gradually closes and the rock gradually gravitates to a compact and dense state. At this stage, the wave velocity increases significantly and the location error is reduced greatly with the increase of stress. Due to the existence of micro-fractures in the rock, there are many outliers in the location results. In stage II, the rock is further compacted and the wave velocity increases slightly and approaches the peak value. The increased rate of wave velocity is much slower than that in stage I. The macroscopic performance is that the location error maintains stability or decreases slowly with the increase of stress and wave velocity. In stage III, the wave velocity field is less homogeneous due to the initial generation of micro-cracks in the rock. The change in the wave velocity field leads to an inflection point of location error change, a greater dispersion of location error, and more outliers in the location results. In stage IV, with the massive expansion of micro-cracks and the formation of macroscopic fracture, the rock wave velocity greatly decreases and the location error increases significantly. The macroscopic performance is that the location error increases with increasing stress and decreasing wave velocity.

Distribution characteristics of AE location error at different stages and heights of the granite sample S-UCT are displayed in Fig. 3.20. Overall, the location errors in stage II and stage III are better than that in stage I and stage IV. It shows a very similar distribution law of location error at both ends of the sample ($z = 20$ mm and $z = 180$ mm). Specifically, sources in the middle of the sample ($z = 100$ mm) show lower location error than that at both ends of the sample in stage I, stage II, and stage

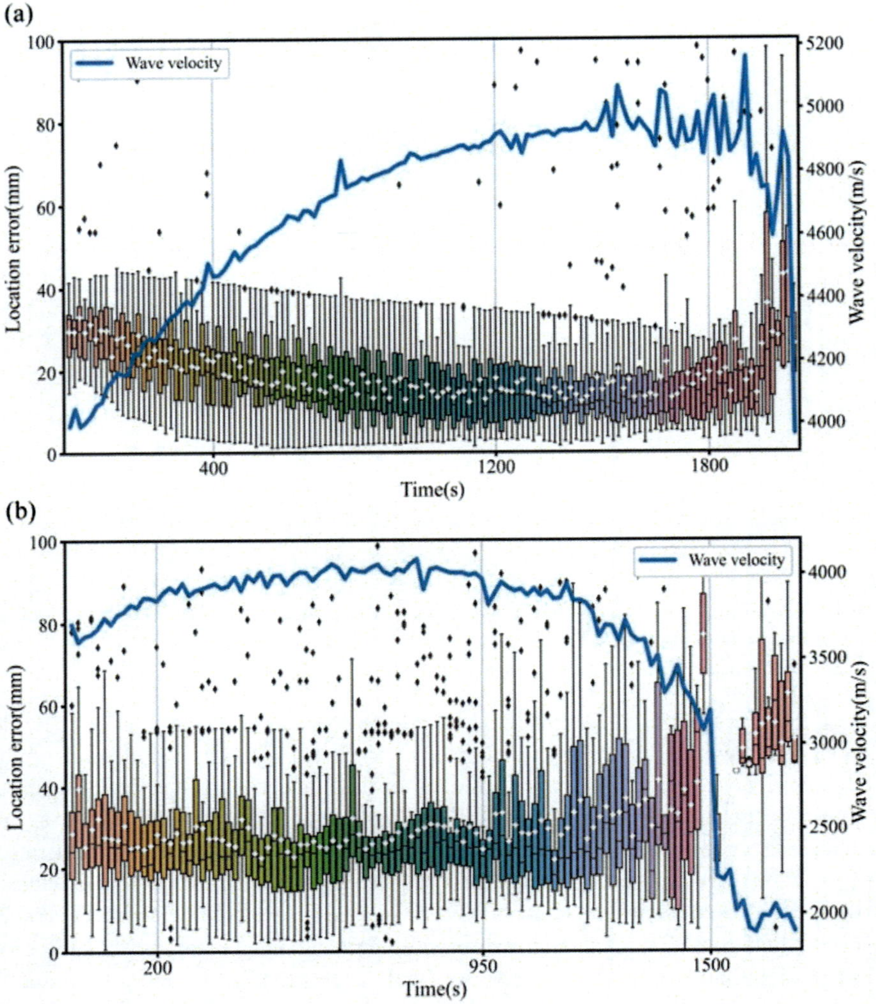

Fig. 3.19 Variations of AE event location error and wave velocity with time for granite sample **a** S-UCT and **b** S-BCT

III, while the location error of sources in the middle suddenly increases in stage IV and is much larger than that of sources at both ends. The location error of sources at the top of the sample ($z = 180$ mm) decreased continuously during the whole loading process, while the location error of sources at the bottom of the sample ($z = 20$ mm) increased slightly in stage IV, which show different variation trends.

Since the research shows that the location accuracy of the sources inside the sensor grid is higher than that outside the grid [11], it should be considered the effect of the sensor grid when analyzing the location error. In the early and medium loading stage,

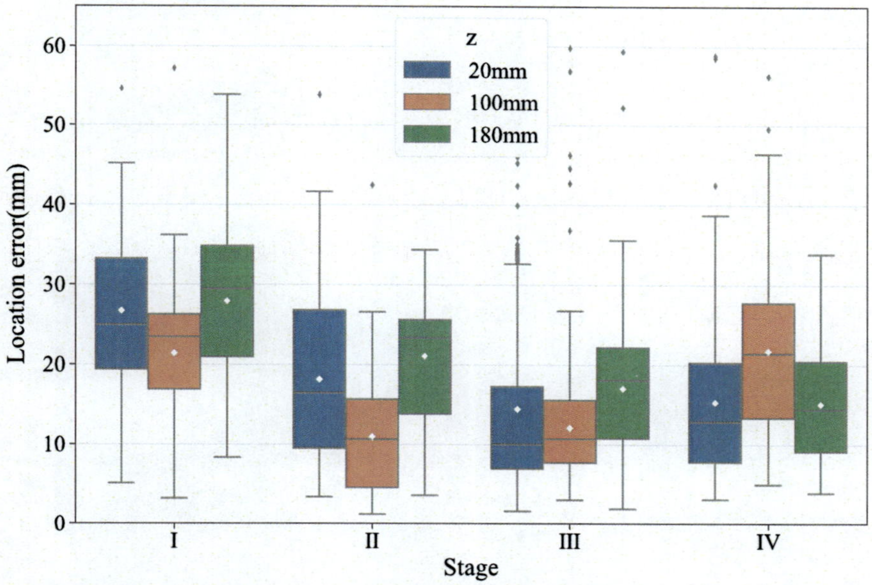

Fig. 3.20 Distribution of AE location error at different stages and heights of granite sample S-UCT

the internal structure of the rock is relatively homogeneous and the relative position of the sources in the sensor grid plays a dominant role in the location accuracy.

Therefore, the location error of the sources in the middle of the sample is smaller than that of the sources at both ends of the sample. In the later loading stage, as the internal damage of the rock intensifies, the local stress concentration of the rock makes the fracture damage in the middle of the sample more significant than at the two ends of the sample. In this case, the structural incompleteness plays a key role in the location accuracy. That is, the location error of the sources in the middle of the sample suddenly increases in the later loading stage and is much larger than the location error of the sources at both ends of the sample. The difference in location error between sources at the top and bottom of the sample is also related to the local fracture damage of the rock. The top and bottom fracture situation of the rock is shown in Figs. 3.21 and 3.22. The top structure of the sample is always relatively complete, while locally nucleated micro-cracks appeared at the bottom of the sample, which led to a small increase in the location error of the sources at the bottom of the sample.

3.3.3.2 Evaluation and Improvement of Localization Method

The location method without pre-velocity measurement converts the input of the wave velocity into an unknown value through mathematical derivation and then solves the source coordinates and the wave velocity through the subsequent iterative algorithm.

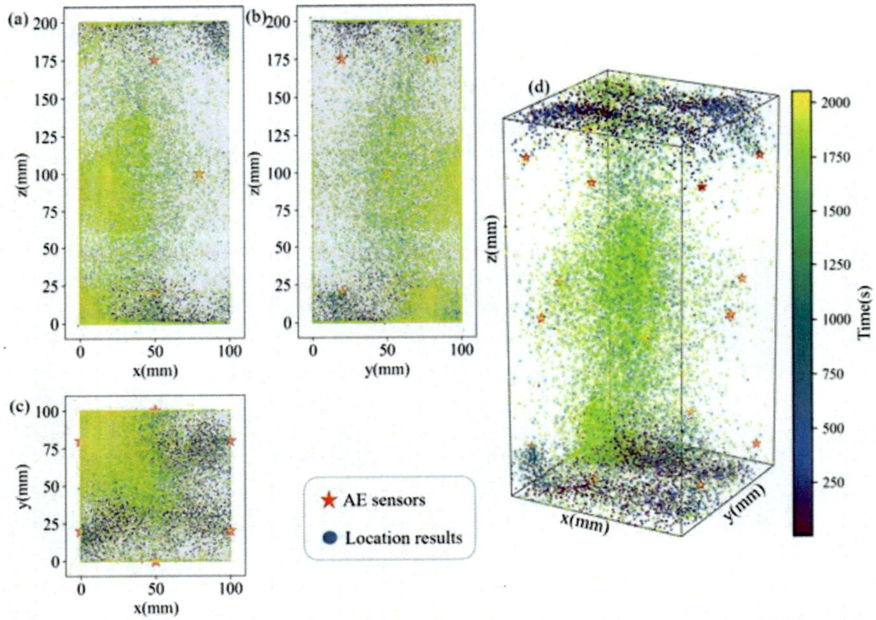

Fig. 3.21 **a** Front view, **b** Right view, **c** Top view and **d** 3-D view of the AE events whole-process location results of the granite sample S-UCT

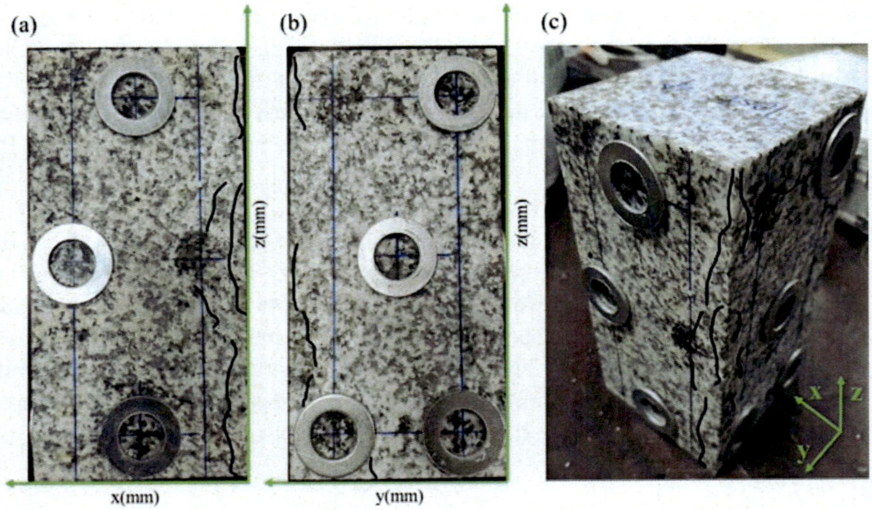

Fig. 3.22 **a** Rear view, **b** Left view and **c** Oblique rear view of macroscopic fracture in granite sample S-UCT

Analyzing the error of the wave velocity v_{error} is expected to optimize the location algorithm and improve the location accuracy. The calculation formula of v_{error} is as follows:

$$v_{error} = v_{cal} - v_{rta}, \tag{3.5}$$

where v_{cal} is the wave velocity value solved by the location method, and v_{rta} is the average of the measured wave velocity.

Taking the granite sample S-UCT as an example, the boxplot distribution of the location errors and the wave velocity error absolute values at different stages are shown in Figs. 3.23a, b, respectively. Figures 3.23a, b show similar trends, that is, there is a qualitative relationship between the location error and the wave velocity error. The larger the wave velocity error, the larger the location error. Figure 3.23c shows the distribution of wave velocity errors at different stages. It can be seen that the wave velocity used for location in stage I are generally small, and the wave velocity error in stage IV are widely distributed. Therefore, it is expected to improve the location accuracy by dynamically adjusting the value of wave velocity and constantly correcting the wave velocity used in the location.

The improvement of the source location method is a process of approaching the real source position, and the location error will decrease with the improvement of the technical level. The sample size of indoor experiment is often small and the variable is single compared with the complex environment of engineering, so it is not easy to give the reliability of the location method in engineering. However, we can give a qualitative evaluation of the location accuracy according to the stress state of the rock. Firstly, sensors are arranged in the monitoring area to monitor the acoustic parameters, such as MS/AE events. Secondly, the stress state and fracture state of the rock in the monitoring area are determined according to the variation characteristics of the monitoring parameters. Finally, the confidence level of the location results is obtained from the fracture stage of the rock. Taking the data of granite sample S-UCT as an example, if the rock is in stage I, the reliability of the source location results is about 82.5% and above; if it is in stage II, the reliability is about 87.5% and above; If it is in stage III, the confidence is about 90% and above, and if it is in stage IV, the confidence is about 85% and above. Besides, rocks in practical engineering are often affected by pre-compressive stress, so there is almost no micro-fracture compaction stage in the engineering rocks. The reliability of location result is maintained at more than 85%, which is relatively reliable.

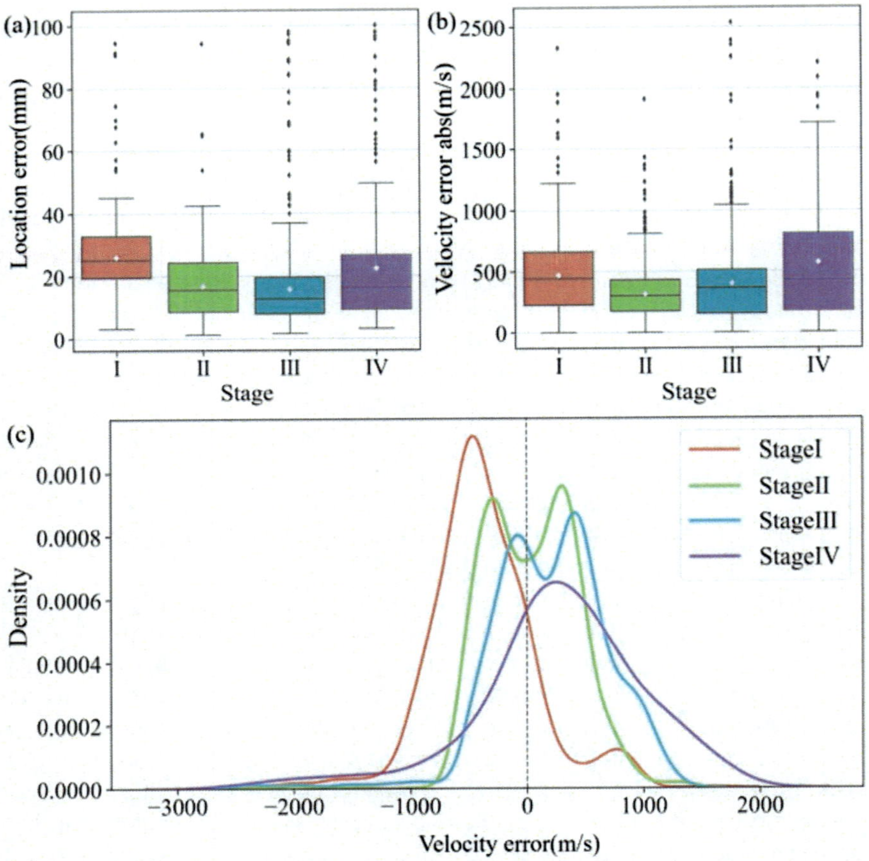

Fig. 3.23 Analysis of location error. **a** Location error at different stages. **b** Absolute value of wave velocity error at different stages; **c** The distribution of wave velocity errors at different stages

3.4 The Influence of Different Optimization Algorithms on Source Location Accuracy

Dong et al. [12] analyzed the influencing factor of different optimization algorithms (LM, SM, QN, the Max Inherit Optimization, SOMA as well as Global Optimization (GO) coupled with LM, SM, QN, the Max Inherit Optimization) on location accuracy using measured velocity. They found that it is easy to obtain a local optimum value using LM, SM, QN and SM alone and SOMA, SM coupled with GO, the MIO coupled with GO and the LM coupled with GO are more stable with a high location accuracy. In the section, the influencing factor on location accuracy of the velocity-free method was investigated. Base on the velocity-free theory, Eq. (2.21) (which is rewritten below) is the objective function of all optimization algorithm mentioned in the section.

$$(x, y, z, v) = \arg\min f(x, y, z, v) = \arg\min \sum_{i=2}^{m} \left(t_{i1} - \frac{R_i - R_1}{v} \right)^2$$

The nonlinear optimization algorithms (LM and QN) [13], global optimization algorithms (SM, SOMA and PSO) [14–16] as well as PSO algorithm coupled with LM, SM and QN (PSO-LM, PSO-SM and PSO-QN) are applied to solve Eq. (2.21). Note that the velocity is unknown in each algorithm.

3.4.1 Numerical Test

The numerical test is carried out in a cubic monitoring area with the size of 100 m (see Fig. 3.24) and eight sensors are employed in the vertex of cube area. Three sources are selected in P_1 (22, 34, 13), P_2 (51, 67, 48) and P_3 (159, 200, 188) (unit: m), where P_1 and P_2 are in the corner and center of sensor network respectively, and P_3 is outside the sensor network. In the simulation, the actual velocity is 4,500 m/s. The arrival times and coordinates of sensors are shown in Table 3.8. All the optimization algorithms used in the section are iterative methods so we need determine the initial value in advance. Theoretically, the initial source position can be determined by the point near where the sensor recorded the first arrival. To verify the velocity-free methods, initial velocity is taken as 2,000 m/s, which simulates the situation of a lack of wave velocity message. The location results of velocity-free methods using different optimization algorithm are shown in Table 3.9.

For the sources inside the sensor network (P_1 and P_2), velocity-free method can obtain accurate result. Comparing the calculation time of different algorithm, it can be noted that LM algorithm consumes the least time to converge to the actual solution (i.e. source position) and global optimization method PSO takes the longest time. The combination of PSO with LM, SM and QN can reduce the calculation time of PSO. For the source outside the sensor network (P_3), LM also consumes the least time but converges to the local optimum value. The joint methods PSO-SM, PSO-LM

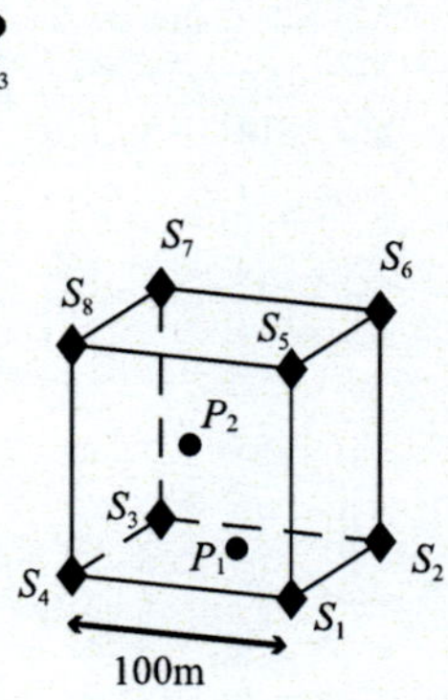

Fig. 3.24 Illustration of numerical test

Table 3.8 Arrival times and coordinates of sensors

Sensor no.	Coordinates of sensors (m)			Arrival times (s)		
	x	y	z	P_1	P_2	P_3
S_1	0	0	0	0.0294516	0.0415384	0.0904921
S_2	100	0	0	0.0391279	0.0413079	0.0823906
S_3	100	100	0	0.0428889	0.0369152	0.0691031
S_4	0	100	0	0.0357276	0.0372047	0.0790564
S_5	0	0	100	0.0413252	0.0419921	0.0800514
S_6	100	0	100	0.0470427	0.0417664	0.0702954
S_7	100	100	100	0.0498217	0.0374893	0.0523751
S_8	0	100	100	0.0447546	0.0377694	0.0660944

Table 3.9 Location result of three sources using different optimization algorithm

Algorithm	Number of iterations	Calculation time (s)	Absolute error (m)	Description of result	Location result (m)		
					x	y	z
Location result of P_1							
LM	41	0.024576	0	Convergence	22.00	34.00	13.00
SM	284	0.128736	0	Convergence	22.00	34.00	13.00
QN	37	0.665835	0	Convergence	22.00	34.00	13.00
SOMA	163	0.330101	0	Convergence	22.00	34.00	13.00
PSO	29,682	7.517369	0	Convergence	22.00	34.00	13.00
PSO-SM	4964	1.707394	0	Convergence	22.00	34.00	13.00
PSO-LM	1044	0.351326	0	Convergence	22.00	34.00	13.00
PSO-QN	149	0.727335	0	Convergence	22.00	34.00	13.00
Location result of P_2							
LM	50	0.008092	0	Convergence	51.00	67.00	48.00
SM	302	0.116727	0	Convergence	51.00	67.00	48.00
QN	49	0.676635	0	Convergence	51.00	67.00	48.00
SOMA	221	0.429135	0	Convergence	51.00	67.00	48.00
PSO	82,829	21.0165	0	Convergence	51.00	67.00	48.00
PSO-SM	10,179	2.617539	0	Convergence	51.00	67.00	48.00
PSO-LM	1054	0.396391	0	Convergence	51.00	67.00	48.00
PSO-QN	137	0.75502	0	Convergence	51.00	67.00	48.00

(continued)

Table 3.9 (continued)

Algorithm	Number of iterations	Calculation time (s)	Absolute error (m)	Description of result	Location result (m)		
					x	y	z
Location result of P_3							
LM*	36	0.003436	198.69	Convergence	65.30	71.06	69.37
SM	267	0.108431	0	Convergence	159.00	200.00	188.00
QN	44	0.707604	0	Convergence	159.00	200.00	188.00
SOMA	73	0.143393	0	Convergence	159.00	200.00	188.00
SOMA*	163	0.414202	198.69	Convergence	65.30	71.06	69.37
PSO	23,027	5.851894	0	Convergence	159.00	200.00	188.00
PSO-SM	3384	0.888631	0	Convergence	159.00	200.00	188.00
PSO-LM	580	0.251739	0	Convergence	159.00	200.00	188.00
PSO-QN	132	0.749757	0	Convergence	159.00	200.00	188.00

*The algorithm converges to a local optimum value

and PSO-QN consume less time than PSO and PSO-LM converge to the global optimum value. For all the sources, SM, QN and SOMA obtain the actual source position, and SM consumes the least time.

3.4.2 Blasting Test

The optimization algorithms are applied to source localization for blasting test in Sect. 2.4.2. The arrival times and coordinates of sensors are shown in Table 2.4. The information of blasting sources is shown in Table 2.5. Events 1–3 are indicated by E_1–E_3. The location results of three blasting sources using different algorithms are shown in Table 3.10. From Table 3.10, the location errors of different methods are similar, which means that the different algorithms have little influence on location accuracy. It can be seen that the location speed of different algorithms are consistent with that in numerical test.

After comparing the location result of numerical and blasting tests, LM is recommended to locate sources inside the sensor network because of fast location speed and accurate localization. The joint optimization algorithm is not necessary for velocity-free localization. However, when the monitoring area contains the outside of sensor network, the combination of global optimization method and LM is recommended.

Table 3.10 Location results of blasting sources using different algorithms

Algorithm	Number of iterations	Calculation time (s)	Absolute error (m)	Description of result	Location result (m)		
					x	y	z
Location result of E_1							
LM	34	0.014739	12.59	Convergence	84,529.79	22,548.07	−743.69
SM	210	0.113876	12.59	Convergence	84,529.80	22,548.08	−743.69
QN	37	0.951764	12.59	Convergence	84,529.80	22,548.08	−743.69
SOMA	45	0.525142	12.59	Convergence	84,529.80	22,548.08	−743.69
PSO	45,450	14.68433	12.59	Convergence	84,529.80	22,548.08	−743.69
PSO-SM	123	0.055937	12.59	Convergence	84,529.80	22,548.08	−743.69
PSO-LM	59	0.036547	12.59	Convergence	84,529.80	22,548.08	−743.69
PSO-QN	110	0.778199	12.59	Convergence	84,529.80	22,548.08	−743.69
Location result of E_2							
LM	33	0.0012	21.06	Convergence	84,494.98	22,572.58	−800.92
SM	225	0.112354	21.50	Convergence	84,495.39	22,572.80	−800.77
QN	43	1.054278	21.50	Convergence	84,495.39	22,572.80	−800.77
SOMA	39	0.436246	21.50	Convergence	84,495.39	22,572.80	−800.77
PSO	45,885	14.79083	21.50	Convergence	84,495.39	22,572.80	−800.77
PSO-SM	168	0.285231	21.50	Convergence	84,495.39	22,572.80	−800.77
PSO-LM	55	0.01237	21.50	Convergence	84,495.39	22,572.80	−800.77
PSO-QN	110	0.824067	21.50	Convergence	84,495.39	22,572.80	−800.77
Location result of E_3							
LM	33	0.001162	37.26	Convergence	84,352.22	22,678.57	−759.29
SM	228	0.093421	38.87	Convergence	84,352.23	22,679.06	−757.70
QN	41	1.041466	38.87	Convergence	84,352.23	22,679.06	−757.70
SOMA	41	0.46729	38.87	Convergence	84,352.23	22,679.06	−757.70
PSO	48,095	16.42709	38.87	Convergence	84,352.23	22,679.06	−757.70
PSO-SM	153	0.167927	38.87	Convergence	84,352.23	22,679.06	−757.70
PSO-LM	57	0.010996	38.87	Convergence	84,352.23	22,679.06	−757.70
PSO-QN	109	0.828115	38.87	Convergence	84,352.23	22,679.06	−757.70

3.5 Conclusions

Experiments and numerical test were carried out to investigate the influence of temperature, velocity, sensor position and stress stage on AE source location accuracy. Basing on the analysis of location results, the following conclusions can be drawn.

(1) For the research of how temperature affects location accuracy, the location results of each pulse point fluctuates in the middle period of heating because the thermal expansion inside the structure causes the change of the wave velocity in the structure. In the later period of heating, the internal expansion of the

structure leads to structural cracking, and the location error of the pulse point increases sharply. Since the acoustic wave cannot pass through the crack directly and needs to bypass the crack, the actual wave path will no longer be equal to the pre-calculated path.

To reduce the location error caused by the structural crack in the fire, it is of great importance to distinguish whether the calculated path between the sensor and the source is related to their actual travel time. Therefore, determining the time and position of structural cracking accurately and selecting sensors in more appropriate places will be meaningful to reduce the location error.

(2) For the research of how velocity and sensor positions affects location accuracy, it is found that the location error of traditional method with pre-measured velocity increases with the increase of the velocity error. In most cases, it is a liner relationship. However, the location error of MSLM-MV, a velocity-free localization method, is not affected by velocity error when initial velocity is low but fluctuates when initial velocity is close to the true value. The result of laboratory test shows that MSLM-MV has better location accuracy than traditional methods and the sources close to sensors have poor location accuracy.

(3) For the research of how the different stress stages affect location accuracy, the variation characteristics of the location accuracy in the whole process under complex stress conditions is analyzed. The results show that the location errors in stage II and III are less than that in stage I and IV, which is related to the uniformity of the wave velocity field and the uniformity of the wave velocity field is determined by the distribution of fractures. The location errors in different stress stages are then explained. It is expected to dynamically adjust the wave velocity used for location by analyzing the distribution of wave velocity error, thereby improving the location accuracy.

(4) Based on the numerical and blasting test, using LM, SM, QN, SOMA, PSO and PSO coupling with LM, SM and QN for optimization, it can be noted that LM algorithm produces a better location effect for sources inside sensor network and PSO-LM, SM and SOMA can be applied to localization for sources outside sensor network.

(5) In general, the location accuracy is significantly dominated by the velocity error and the uneven velocity field. The development of velocity-free methods is necessary. For localization method with measured velocity, using different optimization algorithm to solve the nonlinear objective function have different location accuracy but the different optimization algorithms only affect the localization for sources outside sensor network.

References

1. Xiao P, Hu Q-C, Tao Q, Dong L-J, Yang Z-F, Zhang W-B (2020) Acoustic emission location method for quasi-cylindrical structure with complex hole. IEEE Access 8:35263–35275. https://doi.org/10.1109/access.2020.2972411

2. Ji H-G (2004) Research and application of acoustic emission performance in concrete materials. Coal Industry Press, Beijing
3. Liu Y-J, Huang K, Zhang K-N, Chen K-P, Wen J-P, Chen Y-G, Jiang X-L, Huang B (2003) Experimentation and research on properties of simulated rock. Build Tech Dev 30(8):62–63
4. Zhou Z-L, Jing Z, Xin C, Rui Y-C, Chen L-J, Wang H-Q (2020) Acoustic emission source location considering refraction in layered media with cylindrical surface. Trans Nonferrous Metals Soc China 30(3):789–799
5. Dong L-J, Tao Q, Hu Q-C (2021) Influence of temperature on acoustic emission source location accuracy in underground structure. Trans Nonferrous Metals Soc China 31(8):2468–2478. https://doi.org/10.1016/S1003-6326(21)65667-4
6. Fatu D (2001) Kinetics of gypsum dehydration. J Therm Anal Calorim 65(1):213–220
7. Gollob S, Kocur GK, Schumacher T, Mhamdi L, Vogel T (2017) A novel multi-segment path analysis based on a heterogeneous velocity model for the localization of acoustic emission sources in complex propagation media. Ultrasonics 74:48–61
8. Dong L-J, Li X-B, Zhou Z-L, Chen G-H, Ma J (2015) Three-dimensional analytical solution of acoustic emission source location for cuboid monitoring network without pre-measured wave velocity. Trans Nonferrous Metals Soc China 25(1):293–302
9. Dong L-J, Tang Z, Li X-B, Chen Y-C, Xue J-C (2020) Discrimination of mining microseismic events and blasts using convolutional neural networks and original waveform. J Central South Univ 27(10):3078–3089
10. Dong L-J, Zhang Y-H, Sun D-Y, Chen Y-C, Tang Z (2022) Stage characteristics of acoustic emission and identification of unstable crack state for granite fractures. Chin J Rock Mechan Eng 41:120–131
11. Dong L-J, Li X-B, Tang L-Z, Gong F-Q (2011) Mathematical functions and parameters for microseismic source location without pre-measuring speed. Chin J Rock Mechan Eng 30(10):2057–2067
12. Dong L-J, Li X-B, Tang Z (2013) Main influencing factors for the accuracy of microseismic source location. Sci Technol Rev 31(24):26–32
13. Yuan Y-X, Sun W-Y (1997) Optimization theory and methods. Science Press, Beijing
14. Chong KPE, Żak HS (2013) An introduction to optimization. John Wiley & Sons Inc., Hoboken
15. Zelinka I (2004) SOMA—self-organizing migrating algorithm. in: new optimization techniques in engineering. Springer, Berlin, Heidelberg, Berlin, Heidelberg, pp 167–217. https://doi.org/10.1007/978-3-540-39930-8_7
16. Shi Y, Eberhart R (1998) A modified particle swarm optimizer. In: 1998 IEEE international conference on evolutionary computation proceedings. IEEE world congress on computational intelligence (Cat No 98TH8360), 1998. IEEE pp 69–73

Chapter 4
Three-Dimensional Analytical Solution Under the Cuboid, Rectangular Pyramid, and Random Sensor Networks

The basic thought of the analytical localization method is to solve the explicit formulas for source coordinates through the nonlinear governing equations [1–5]. Smith and Abel [6] presented three non-iterative methods for locating sources in the 3-D space, which were the spherical-interpolation method, spherical-intersection method, and plan-intersection method, respectively. By applying the model of arrival time difference, Mellen et al. [7] proposed the analytical solutions for the sensors network that contains greater than three sensors. Ge [8] summarized the main analytical localization methods including the Inglada method and USBM method. The P-wave velocity is usually taken as the known parameter in the localization process, which fails to characterize the temporal and spatial change of P-wave velocity. Therefore, it is difficult to locate sources with high accuracy in the media with unknown P-wave velocity or the heterogeneous and complex media.

In this chapter, three types of sensor networks are designed and the analytical solutions of nonlinear systems with unknown velocity are introduced.

Section 4.1 gives a general description about the problem in source localization. In Sect. 4.2, a cuboid sensor network used for localization is investigated. Different sensor arrangements on upper and lower surfaces were considered and used to establish nonlinear equations. Rectangular pyramid sensor network were designed in Sect. 4.3, and the AE/MS localization equations were established. The above two localization are based on six sensors and the specific sensor networks, which are only suitable for specific cases. Therefore, an analytic solution based on random sensor network which has a random number of sensors (at least six sensors) is descripted in Sect. 4.4. The proposed analytical solutions were validated using authentic data of numerical tests and experiments in Sect. 4.5.

L. Dong and X. Li, *Velocity-Free Localization Methodology for Acoustic and Microseismic Sources*, https://doi.org/10.1007/978-981-19-8610-9_4

4.1 Statement of the Problem

The AE/MS/seismic source location methods using P wave trigger time are widely used to calculate source coordinates for two reasons: the fastest propagation velocity of the P wave, and the easy identification of first trigger time. the AE/MS/seismic source location coordinate is (x, y, z); T_i (i = 1, 2,..., n) is the ith sensor, and its coordinate is (x_i, y_i, z_i) $(i = 1, 2, \ldots, n)$; l_i $(i = 1, 2, \ldots, n)$ is the distance from the AE/MS/seismic source to the station T_i; t_i $(i = 1, 2, \ldots, n)$ is trigger time recorded by sensor in the station T_i; t_0 is origin time of AE or seismic source. Variable v is the P-wave velocity. Then t_i can be expressed as

$$t_i = \frac{l_i}{v} + t_0. \tag{4.1}$$

By the spatial distance formula between two points (the source position and the sensor position), one can obtain

$$l_i = \sqrt{(x_i - x)^2 + (y_i - y)^2 + (z_i - z)^2}. \tag{4.2}$$

By taking Eq. (4.2) into Eq. (4.1), we have

$$(t_i - t_0)v = \sqrt{(x_i - x)^2 + (y_i - y)^2 + (z_i - z)^2}. \tag{4.3}$$

In Eq. (4.3), t_i $(i = 1, 2, \ldots, n)$ and (x_i, y_i, z_i) $(i = 1, 2, \ldots, n)$ are known; the seismic or AE source (x, y, z), v and origin time t_0 are unknown, which need to be solved. By taking each station data to Eq. (4.3), an equation can be obtained. Five stations correspond to five equations, and they can constitute a set of nonlinear equations. In order to find out the analytical solution of the AE/MS/seismic source location coordinates, the sensor location coordinates were optimized and simplified. Then, two networks are investigated. Generally, the greater the number of station is, the higher the positioning accuracy is. The location precision is greatly influenced by the error of the wave velocity and the intrinsic limitations of the iteration algorithm applied.

4.2 Analytical Method Under the Cuboid Sensor Network

A cuboid sensor network of sensor locations was selected, and the AE/MS/seismic source localization equations were established. The sensors are required to install at the vertices of the cuboid sensor network. There are two cases including four sensors installed on one surface and additional one sensor on another surface (Fig. 4.1) as well as three sensors installed on one surface and additional two sensors on another surface (Fig. 4.2).

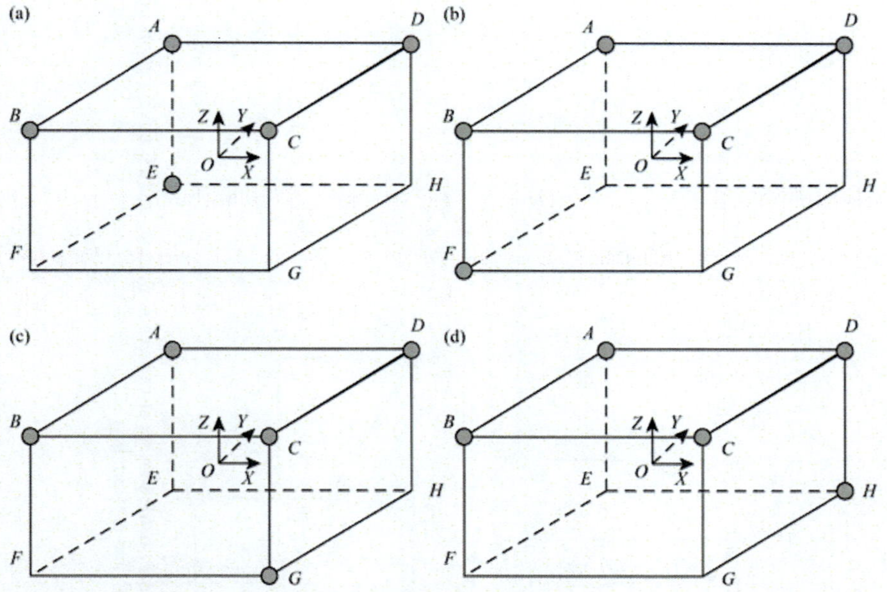

Fig. 4.1 3-D location schematic of cuboid network: **a** AE sensors at vertices *A*, *B*, *C*, *D*, and *E*; **b** AE sensors at vertices *A*, *B*, *C*, *D*, and *F*; **c** AE sensors at vertices *A*, *B*, *C*, *D*, and *G*; **d** AE sensors at vertices *A*, *B*, *C*, *D*, and *H* (from [9])

4.2.1 Analytic Solution I

For every surface of the first case (Fig. 4.1), there are four types of sensor networks including Figs. 4.1a, b, c, and d. The first type (Fig. 4.1a) is analyzed in this section, and the others are similar. Five sensors are installed at vertices *A*, *B*, *C*, *D*, and *E* of the cuboid network. The center of the cuboid is taken as the coordinate origin, and the coordinate direction is shown in Fig. 4.1

The lengths of three sides of the network cuboid are $2a$, $2b$, and $2c$, respectively. The first sensor *A* is taken as a reference. The travel time of the sensor *A* from an AE/MS/seismic event is expressed as t_{10}, and the arrival time of sensors *B*, *C*, *D*, and *E* is $t_{10}+\Delta t_2$, $t_{10}+\Delta t_3$, $t_{10}+\Delta t_4$, and $t_{10}+\Delta t_5$, respectively. According to Eq. (4.3), one can obtain Eqs. (4.4)–(4.8).

$$(a + x)^2 + (b - y)^2 + (c - z)^2 = v^2 t_{10}^2 \tag{4.4}$$

$$(a + x)^2 + (b + y)^2 + (c - z)^2 = v^2 (t_{10} + \Delta t_2)^2 \tag{4.5}$$

$$(a - x)^2 + (b + y)^2 + (c - z)^2 = v^2 (t_{10} + \Delta t_3)^2 \tag{4.6}$$

$$(a - x)^2 + (b - y)^2 + (c - z)^2 = v^2 (t_{10} + \Delta t_4)^2 \tag{4.7}$$

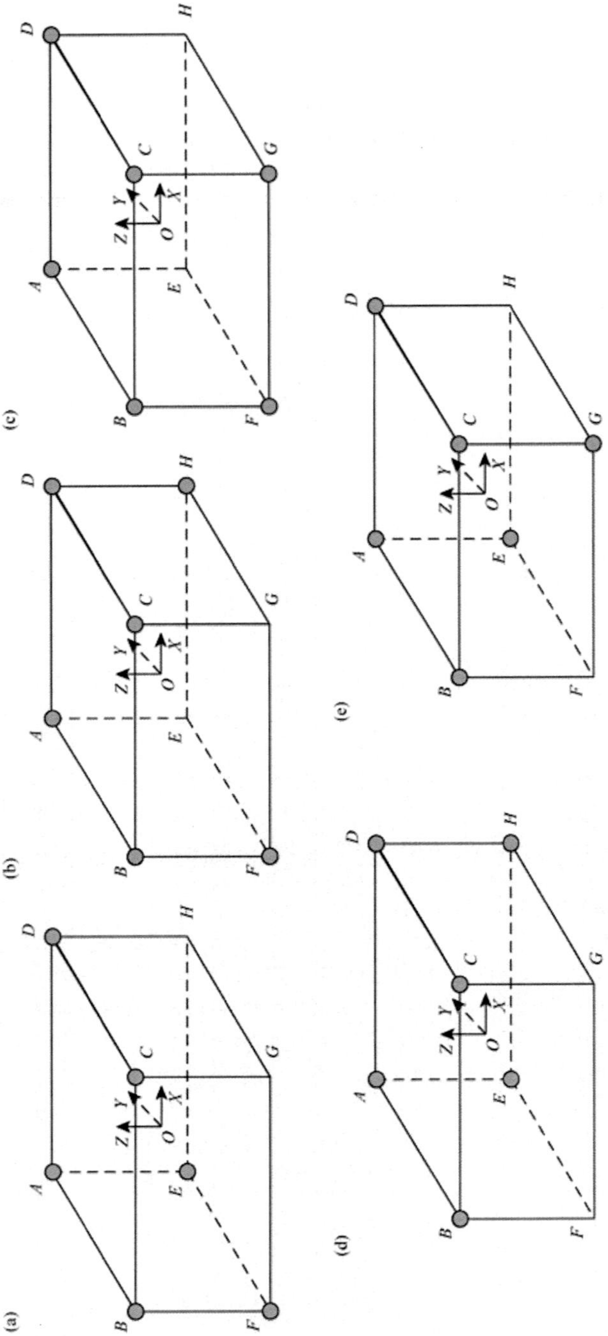

Fig. 4.2 3-D location schematic of cuboid network: **a** AE sensors at vertices *A*, *B*, *D*, *E*, and *F*; **b** AE sensors at vertices *A*, *B*, *D*, *F*, and *H*; **c** AE sensors at vertices *A*, *B*, *D*, *F*, and; **d** AE sensors at *A*, *B*, *D*, *E*, and *H*; **e** AE sensors at vertices *A*, *B*, *D*, *E*, and *G* (from [9])

$$(a+x)^2 + (b-y)^2 + (c+z)^2 = v^2(t_{10} + \Delta t_5)^2 \tag{4.8}$$

Taking subtraction of Eqs. (4.4) and (4.5), Eqs. (4.4) and (4.6), Eqs. (4.4) and (4.7), as well as Eqs. (4.4) and (4.8), we have Eqs. (4.9)–(4.12).

$$4by = v^2\left(2t_{10}\Delta t_2 + \Delta t_2^2\right) \tag{4.9}$$

$$-4ax + 4by = v^2\left(2t_{10}\Delta t_3 + \Delta t_3^2\right) \tag{4.10}$$

$$-4ax = v^2\left(2t_{10}\Delta t_4 + \Delta t_4^2\right) \tag{4.11}$$

$$-4cz = v^2\left(2t_{10}\Delta t_5 + \Delta t_5^2\right) \tag{4.12}$$

From Eqs. (4.9)–(4.11), one can easily obtain

$$\left(2t_{10}\Delta t_2 + \Delta t_2^2\right) + \left(2t_{10}\Delta t_4 + \Delta t_4^2\right) = \left(2t_{10}\Delta t_3 + \Delta t_3^2\right). \tag{4.13}$$

Resolving Eq. (4.13) yields

$$t_{10} = \frac{\Delta t_3^2 - \Delta t_2^2 - \Delta t_4^2}{2(\Delta t_2 + \Delta t_4 - \Delta t_3)}. \tag{4.14}$$

Taking the ratio of Eqs. (4.9) and (4.11), we have

$$y = \frac{-a\left(2t_{10}\Delta t_2 + \Delta t_2^2\right)}{b\left(2t_{10}\Delta t_4 + \Delta t_4^2\right)}x. \tag{4.15}$$

Supposing $l = \frac{2t_{10}\Delta t_2 + \Delta t_2^2}{2t_{10}\Delta t_4 + \Delta t_4^2}$, Eq. (4.15) can be rewritten as:

$$y = \frac{-al}{b}x. \tag{4.16}$$

From Eqs. (4.11) and (4.12) one can obtain

$$z = \frac{-a\left(2t_{10}\Delta t_5 + \Delta t_5^2\right)}{c\left(2t_{10}\Delta t_4 + \Delta t_4^2\right)}x. \tag{4.17}$$

Supposing $m = \frac{2t_{10}\Delta t_5 + \Delta t_5^2}{2t_{10}\Delta t_4 + \Delta t_4^2}$, Eq. (4.17) can be rewritten as:

$$z = \frac{-am}{c}x. \tag{4.18}$$

Equation (4.4) divided by Eq. (4.9), we have

$$\frac{(a+x)^2+(b-y)^2+(c+z)^2}{4by}=\frac{t_{10}^2}{2t_{10}\Delta t_2+\Delta t_2^2}. \tag{4.19}$$

Supposing $n=\frac{t_{10}^2}{2t_{10}\Delta t_2+\Delta t_2^2}$, and substituting Eqs. (4.16) and (4.18) into Eq. (4.19), we have

$$\left[\left(\frac{a}{b}l\right)^2+\left(\frac{a}{c}m\right)^2+1\right]x^2+(2a+2al+2am+4aln)x$$
$$+a^2+b^2+c^2=0. \tag{4.20}$$

Equation (4.20) can be rewritten as:

$$Ax^2+Bx+C=0, \tag{4.21}$$

where $A=\left(\frac{a}{b}l\right)^2+\left(\frac{a}{c}m\right)^2+1$, $B=2a+2al+2am+4aln$, and $C=a^2+b^2+c^2$.

Then, x, y, and z can be obtained by resolving Eqs. (4.20), (4.16), and (4.18). The solutions can be defined as analytical solution I (ASI).

4.2.2 Analytic Solution II

For every surface of the second case in Fig. 4.2, there are five types of sensor networks including Fig. 4.2a, b, c, d, and e. The first type (Fig. 4.2a) is analyzed below, and the others are similar. Five sensors are installed at vertices *A*, *B*, *D*, *E*, and *F* of the cuboid network. The center of the cuboid is taken as the coordinate origin, and the coordinate direction is shown in Fig. 4.2. The lengths of three sides of the network cuboid are $2a_s$, $2b_s$, and $2c_s$, respectively. The first sensor *A* is taken as a reference.

The travel time from an AE/MS/seismic source (x_s, y_s, z_s) to sensor *A* is expressed as t_{s10}, and the arrival times of sensors *B*, *D*, *E*, and *F* are $t_{s10}+\Delta t_{s2}$, $t_{s10}+\Delta t_{s3}$, $t_{s10}+\Delta t_{s4}$, and $t_{s10}+\Delta t_{s5}$, respectively. The P-wave velocity is expressed as v_s. According to Eq. (4.3), one can obtain Eqs. (4.22)–(4.26).

$$(a_s+x_s)^2+(b_s-y_s)^2+(c_s-z_s)^2=v_s^2t_{s10}^2 \tag{4.22}$$

$$(a_s+x_s)^2+(b_s+y_s)^2+(c_s-z_s)^2=v_s^2(t_{s10}+\Delta t_{s2})^2 \tag{4.23}$$

$$(a_s-x_s)^2+(b_s-y_s)^2+(c_s-z_s)^2=v_s^2(t_{s10}+\Delta t_{s3})^2 \tag{4.24}$$

$$(a_s+x_s)^2+(b_s-y_s)^2+(c_s+z_s)^2=v_s^2(t_{s10}+\Delta t_{s4})^2 \tag{4.25}$$

$$(a_s + x_s)^2 + (b_s + y_s)^2 + (c_s + z_s)^2 = v_s^2 (t_{s10} + \Delta t_{s5})^2 \tag{4.26}$$

Taking subtraction of Eqs. (4.22) and (4.23), Eqs. (4.22) and (4.24), Eqs. (4.22) and (4.25), as well as Eqs. (4.22) and (4.26), we have Eqs. (4.27)–(4.30).

$$4b_s y_s = v_s^2 \left(2t_{s10}\Delta t_{s2} + \Delta t_{s2}^2\right) \tag{4.27}$$

$$4a_s x_s = -v_s^2 \left(2t_{s10}\Delta t_{s3} + \Delta t_{s3}^2\right) \tag{4.28}$$

$$4c_s z_s = v_s^2 \left(2t_{s10}\Delta t_{s4} + \Delta t_{s4}^2\right) \tag{4.29}$$

$$4b_s y_s + 4c_s z_s = v_s^2 \left(2t_{s10}\Delta t_{s5} + \Delta t_{s5}^2\right) \tag{4.30}$$

From Eqs. (4.27), (4.29), and (4.30) one can easily obtain

$$\left(2t_{s10}\Delta t_{s2} + \Delta t_{s2}^2\right) + \left(2t_{s10}\Delta t_{s4} + \Delta t_{s4}^2\right) = 2t_{s10}\Delta t_{s5} + \Delta t_{s5}^2, \tag{4.31}$$

and then,

$$t_{s10} = \frac{\Delta t_{s5}^2 - \Delta t_{s2}^2 - t_{s4}^2}{2(t_{s2} + t_{s4} - t_{s5})}. \tag{4.32}$$

From Eqs. (4.27) and (4.28) one can obtain

$$y_s = -\frac{a_s}{b_s}\left(\frac{2t_{s10}\Delta t_{s2} + \Delta t_{s2}^2}{2t_{s10}\Delta t_{s3} + \Delta t_{s3}^2}\right) x_s. \tag{4.33}$$

Supposing $l_s = \frac{2t_{s10}\Delta t_{s2} + \Delta t_{s2}^2}{2t_{s10}\Delta t_{s3} + \Delta t_{s3}^2}$, Eq. (4.33) can be rewritten as:

$$y_s = -\frac{a_s l_s}{b_s} x_s. \tag{4.34}$$

From Eqs. (4.29) and (4.30) one can obtain

$$z_s = \frac{-a_s \left(2t_{s10}\Delta t_{s5} + \Delta t_{s5}^2\right)}{c_s \left(2t_{s10}\Delta t_{s4} + \Delta t_{s4}^2\right)} x. \tag{4.35}$$

Supposing $m_s = \frac{2t_{s10}\Delta t_{s5} + \Delta t_{s5}^2}{2t_{s10}\Delta t_{s4} + \Delta t_{s4}^2}$, Eq. (4.35) can be rewritten as

$$z_s = \frac{-a_s m_s}{c_s} x_s. \tag{4.36}$$

From Eqs. (4.22) and (4.27) one can obtain

$$\frac{(a_s + x_s)^2 + (b_s - y_s)^2 + (c_s - z_s)^2}{4b_s y_s} = \frac{t_{s10}^2}{2t_{s10}\Delta t_{s2} + \Delta t_{s2}^2}. \tag{4.37}$$

Supposing $n_s = \frac{t_{s10}^2}{2t_{s10}\Delta t_{s2}+\Delta t_{s2}^2}$, and substituting Eqs. (4.34) and (4.36) in Eq. (4.37), we have

$$\left[\left(\frac{a_s}{b_s}l_s\right)^2 + \left(\frac{a_s}{c_s}m_s\right)^2 + 1\right]x_s^2 + (2a_s + 2a_s l_s + 2a_s m_s + 4a_s l_s n_s)x_s + a_s^2 + b_s^2 + c_s^2 = 0. \tag{4.38}$$

Equation (4.38) can be rewritten as:

$$A_s x_s^2 + B_s x_s + C_s = 0, \tag{4.39}$$

where $A_s = \left(\frac{a_s}{b_s}l_s\right)^2 + \left(\frac{a_s}{c_s}m_s\right)^2 + 1$, $B_s = 2a_s + 2a_s l_s + 2a_s m_s + 4a_s l_s n_s$, and $C_s = a_s^2 + b_s^2 + c_s^2$.

Then, x_s, y_s, and z_s can be obtained by resolving Eqs. (4.38), (4.34), and (4.36), respectively. The solutions can be defined as analytical solution II (ASII).

4.2.3 Analytic Solution III

It is noted that the above two conditions are different networks which considered both upper and lower surfaces of the cuboid with five sensors. The first case is four sensors on upper surface and one sensor on the lower surface, while the second case is three sensors on upper surface and two sensors on the lower surface.

The first case has four types of networks and the second one has five types of networks. It is easy to find the different and significant characteristics between the two types of networks. If we consider six surfaces of the cuboid networks, we can see that the two selected conditions have the same characteristic, four and one sensors on two different surfaces. It is noted that the condition that two and three sensors on two different surfaces is not considered. To fix the problem systematically, the third condition, two sensors on one surface and three sensors on another surface (Fig. 4.2b), is analyzed and the analytical solution is also obtained.

The lengths of three sides of the network cuboid are $2a_p$, $2b_p$, and $2c_p$, respectively. The first sensor A is taken as a reference. The travel time from the AE/MS/seismic source (x_p, y_p, z_p) to the sensor A is expressed as t_{p10}, and the arrival time of sensors A, B, D, F, and H is $t_{p10} + \Delta t_{p2}$, $t_{p10} + \Delta t_{p3}$, $t_{p10} + \Delta t_{p4}$, and $t_{p10} + \Delta t_{p5}$, respectively. The P-wave velocity is expressed as v_p.

According to Eq. (4.3), one can obtain Eqs. (4.40)–(4.44).

$$(a_p + x_p)^2 + (b_p - y_p)^2 + (c_p - z_p)^2 = v_p^2 t_{p10}^2 \tag{4.40}$$

$$(a_p + x_p)^2 + (b_p + y_p)^2 + (c_p - z_p)^2 = v_p^2 (t_{p10} + \Delta t_{p2})^2 \tag{4.41}$$

$$(a_p - x_p)^2 + (b_p - y_p)^2 + (c_p - z_p)^2 = v_p^2 (t_{p10} + \Delta t_{p3})^2 \tag{4.42}$$

$$(a_p + x_p)^2 + (b_p - y_p)^2 + (c_p + z_p)^2 = v_p^2 (t_{p10} + \Delta t_{p4})^2 \tag{4.43}$$

$$(a_p + x_p)^2 + (b_p + y_p)^2 + (c_p + z_p)^2 = v_p^2 (t_{p10} + \Delta t_{p5})^2 \tag{4.44}$$

Taking subtraction of Eqs. (4.40) and (4.41), Eqs. (4.40) and (4.42), Eqs. (4.40) and (4.43), as well as Eqs. (4.40) and (4.44), we have Eqs. (4.45)–(4.48).

$$4b_p y_p = v_p^2 \left(2t_{p10}\Delta t_{p2} + \Delta t_{p2}^2\right) \tag{4.45}$$

$$4a_p x_p = -v_p^2 \left(2t_{p10}\Delta t_{p3} + \Delta t_{p3}^2\right) \tag{4.46}$$

$$4b_p y_p + 4c_p z_p = v_p^2 \left(2t_{p10}\Delta t_{p4} + \Delta t_{p4}^2\right) \tag{4.47}$$

$$4a_p x_p - 4c_p z_p = v_p^2 \left(2t_{p10}\Delta t_{p5} + \Delta t_{p5}^2\right) \tag{4.48}$$

From Eqs. (4.45)–(4.48) one can easily obtain

$$\begin{aligned} &\left(2t_{p10}\Delta t_{p2} + \Delta t_{p2}^2\right) - \left(2t_{p3}\Delta t_{p3} + \Delta t_{p3}^2\right) \\ &= \left(2t_{p4}\Delta t_{p4} + \Delta t_{p4}^2\right) - \left(2t_{p10}\Delta t_{p5} + \Delta t_{p5}^2\right), \end{aligned} \tag{4.49}$$

and then,

$$t_{p10} = \frac{\Delta t_{p4}^2 - \Delta t_{p5}^2 - \Delta t_{p2}^2 + \Delta t_{p3}^2}{2(-\Delta t_{p4} + \Delta t_{p5} + \Delta t_{p2} - \Delta t_{p3})}. \tag{4.50}$$

Taking the ratio of Eqs. (4.45) and (4.46), we have

$$y_p = -\frac{a_p}{b_p}\left(\frac{2t_{p10}\Delta t_{p2} + \Delta t_{p2}^2}{2t_{p10}\Delta t_{p3} + \Delta t_{p3}^2}\right)x_p. \tag{4.51}$$

Supposing $l_{\mathrm{p}} = \frac{2t_{\mathrm{p10}}\Delta t_{\mathrm{p2}} + \Delta t_{\mathrm{p2}}^2}{2t_{\mathrm{p10}}\Delta t_{\mathrm{p3}} + \Delta t_{\mathrm{p3}}^2}$, Eq. (4.51) can be rewritten

$$y_{\mathrm{p}} = -\frac{a_{\mathrm{p}} l_{\mathrm{p}}}{b_{\mathrm{p}}} x_{\mathrm{p}}. \tag{4.52}$$

Submitting Eq. (4.46) into Eq. (4.48) yields

$$-c_{\mathrm{p}} z_{\mathrm{p}} = -v_{\mathrm{p}}^2\left(2t_{\mathrm{p10}}\Delta t_{\mathrm{p5}} + \Delta t_{\mathrm{p5}}^2\right) + v_{\mathrm{p}}^2\left(2t_{\mathrm{p10}}\Delta t_{\mathrm{p3}} + \Delta t_{\mathrm{p3}}^2\right). \tag{4.53}$$

Taking the ratio of Eqs. (4.53) and (4.46), we have

$$z_{\mathrm{p}} = \frac{-a_{\mathrm{p}}\left[\left(2t_{\mathrm{p10}}\Delta t_{\mathrm{p5}} + \Delta t_{\mathrm{p5}}^2\right) - \left(2t_{\mathrm{p10}}\Delta t_{\mathrm{p3}} + \Delta t_{\mathrm{p3}}^2\right)\right]}{c_{\mathrm{p}}\left(2t_{\mathrm{p10}}\Delta t_{\mathrm{p3}} + \Delta t_{\mathrm{p3}}^2\right)} x_{\mathrm{p}}. \tag{4.54}$$

Supposing $m_{\mathrm{p}} = \frac{\left(2t_{\mathrm{p10}}\Delta t_{\mathrm{p5}} + \Delta t_{\mathrm{p5}}^2\right) - \left(2t_{\mathrm{p10}}\Delta t_{\mathrm{p3}} + \Delta t_{\mathrm{p3}}^2\right)}{2t_{\mathrm{p10}}\Delta t_{\mathrm{p3}} + \Delta t_{\mathrm{p3}}^2}$, Eq. (4.54) can be rewritten as:

$$z_{\mathrm{p}} = \frac{-a_{\mathrm{p}} m_{\mathrm{p}}}{c_{\mathrm{p}}} x_{\mathrm{p}}. \tag{4.55}$$

Taking the ratio of Eqs. (4.40) and (4.45) yields

$$\frac{\left(a_{\mathrm{p}} + x_{\mathrm{p}}\right)^2 + \left(b_{\mathrm{p}} - y_{\mathrm{p}}\right)^2 + \left(c_{\mathrm{p}} - z_{\mathrm{p}}\right)^2}{4b_{\mathrm{p}} y_{\mathrm{p}}} = \frac{t_{\mathrm{p10}}^2}{2t_{\mathrm{p10}}\Delta t_{\mathrm{p2}} + \Delta t_{\mathrm{p2}}^2}. \tag{4.56}$$

Supposing $n_{\mathrm{p}} = \frac{t_{\mathrm{p10}}^2}{2t_{\mathrm{p10}}\Delta t_{\mathrm{p2}} + \Delta t_{\mathrm{p2}}^2}$, and substituting Eqs. (4.52) and (4.54) in Eq. (4.56), we have

$$\left[\left(\frac{a_{\mathrm{p}}}{b_{\mathrm{p}}} l_{\mathrm{p}}\right)^2 + \left(\frac{a_{\mathrm{p}}}{c_{\mathrm{p}}} m_{\mathrm{p}}\right)^2 + 1\right] x_{\mathrm{p}}^2 + \left(2a_{\mathrm{p}} + 2a_{\mathrm{p}} l_{\mathrm{p}} + 2a_{\mathrm{p}} m_{\mathrm{p}} + 4a_{\mathrm{p}} l_{\mathrm{p}} n_{\mathrm{p}}\right) x_{\mathrm{p}} + a_{\mathrm{p}}^2 + b_{\mathrm{p}}^2 + c_{\mathrm{p}}^2 = 0. \tag{4.57}$$

Equation (4.57) can be rewritten as

$$A_{\mathrm{p}} x_{\mathrm{p}}^2 + B_{\mathrm{p}} x_{\mathrm{p}} + C_{\mathrm{p}} = 0 \tag{4.58}$$

where $A_{\mathrm{p}} = \left(\frac{a_{\mathrm{p}}}{b_{\mathrm{p}}} l_{\mathrm{p}}\right)^2 + \left(\frac{a_{\mathrm{p}}}{c_{\mathrm{p}}} m_{\mathrm{p}}\right)^2 + 1$, $B_{\mathrm{p}} = 2a_{\mathrm{p}} + 2a_{\mathrm{p}} l_{\mathrm{p}} + 2a_{\mathrm{p}} m_{\mathrm{p}} + 4a_{\mathrm{p}} l_{\mathrm{p}} n_{\mathrm{p}}$, and $C_{\mathrm{p}} = a_{\mathrm{p}}^2 + b_{\mathrm{p}}^2 + c_{\mathrm{p}}^2$.

Then, x_{p}, y_{p}, and z_{p} can be obtained by resolving Eqs. (4.58), (4.52), and (4.55), respectively. The solutions can be defined as analytical solution III (ASIII).

4.3 Analytical Method Under the Rectangular Pyramid Sensor Network

For rectangular pyramid network, the sensors are required to install at the vertices of the rectangular pyramid (Fig. 4.3). Five sensors are installed at the vertices *A*, *B*, *C*, *D*, and *I* of the rectangular pyramid network. The center of the rectangular pyramid is taken as the coordinate origin, and the coordinate direction is shown in Fig. 4.3. Define that the length of segment *FG* is $2a_r$, that of segment *GH* is $2b_r$ and that of segment *GC* is $2c_r$. The first sensor *A* is taken as a reference. The travel time of the sensor *A* from an AE/MS/seismic event (x_r, y_r, z_r) is expressed as t_{r10}, and the arrival time of sensors *B*, *C*, *D*, and *I* is $t_{r10} + \Delta t_{r2}$, $t_{r10} + \Delta t_{r3}$, $t_{r10} + \Delta t_{r4}$, and $t_{r10} + \Delta t_{r5}$, respectively. The P-wave velocity is expressed as v_r. The AE/MS/seismic source localization equations were given following.

$$(a_r + x_r)^2 + (b_r - y_r)^2 + (c_r - z_r)^2 = v^2 t_{r10}^2 \tag{4.59}$$

$$(a_r + x_r)^2 + (b_r + y_r)^2 + (c_r - z_r)^2 = v_r^2 (t_{r10} + \Delta t_{r2})^2 \tag{4.60}$$

$$(a_r - x_r)^2 + (b_r + y_r)^2 + (c_r - z_r)^2 = v_r^2 (t_{r10} + \Delta t_{r3})^2 \tag{4.61}$$

$$(a_r - x_r)^2 + (b_r - y_r)^2 + (c_r - z_r)^2 = v_r^2 (t_{r10} + \Delta t_{r4})^2 \tag{4.62}$$

$$x_r^2 + y_r^2 + (-c_r - z_r)^2 = v_r^2 (t_{r10} + \Delta t_{r5})^2 \tag{4.63}$$

Taking difference of Eqs. (4.59) and (4.60), Eqs. (4.59) and (4.61), Eqs. (4.59) and (4.61), as well as Eqs. (4.59) and (4.63), yield Eqs. (4.64)–(4.67), respectively.

$$-4b_r y_r = -v_r^2 (2t_{r10}\Delta t_{r2} + \Delta t_{r2}^2) \tag{4.64}$$

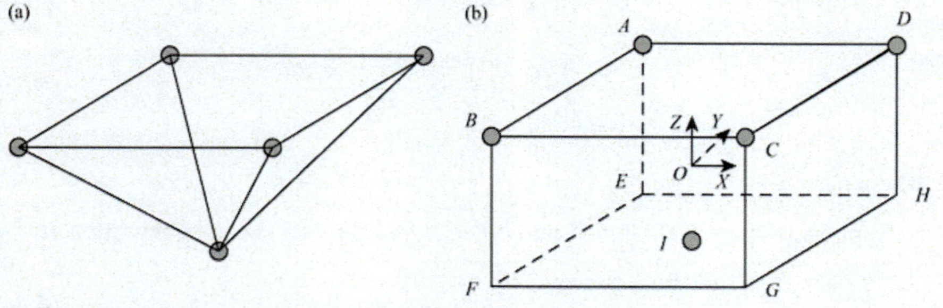

Fig. 4.3 Illustration of the rectangular pyramid network

$$4a_r x_r - 4b_r y_r = -v_r^2\,(2t_{r10}\Delta t_{r3} + \Delta t_{r3}^2) \tag{4.65}$$

$$4a_r x_r = -v_r^2(2t_{r10}\Delta t_{r4} + \Delta t_{r4}^2) \tag{4.66}$$

$$a_r^2 + 2a_r x_r + b_r^2 - 2b_r y_r - 4c_r z_r = -v_r^2(2t_{r10}\Delta t_{r5} + \Delta t_{r5}^2) \tag{4.67}$$

Equation (4.64) plus Eq. (4.66) equal Eq. (4.65), yields

$$(2t_{r10}\Delta t_{r2} + \Delta t_{r2}^2) + (2t_{r10}\Delta t_{r4} + \Delta t_{r4}^2) = (2t_{r10}\Delta t_{r3} + \Delta t_{r3}^2). \tag{4.68}$$

Resolving Eq. (4.68) yields

$$t_{r10} = \frac{\Delta t_{r3}^2 - \Delta t_{r2}^2 - \Delta t_{r4}^2}{2(\Delta t_{r2} + \Delta t_{r4} - \Delta t_{r3})}. \tag{4.69}$$

Taking the ratio of Eqs. (4.64) and (4.66), and it can be rewritten as:

$$y_r = -\frac{a_r}{b_r}\frac{2t_{r10}\Delta t_{r2} + \Delta t_{r2}^2}{2t_{r10}\Delta t_{r4} + \Delta t_{r4}^2}x_r. \tag{4.70}$$

Supposing $l_{r1} = \frac{2t_{r10}\Delta t_{r2} + \Delta t_{r2}^2}{2t_{r10}\Delta t_{r4} + \Delta t_{r4}^2}$, Eq. (4.70) can be rewritten as:

$$y_r = -\frac{a_r}{b_r} l_{r1} x_r. \tag{4.71}$$

Taking the ratio of Eqs. (4.67) and (4.66), and it can be rewritten as:

$$\frac{a_r^2 + 2a_r x_r + b_r^2 - 2b_r y_r - 4c_r z_r}{4a_r x_r} = \frac{2t_{r10}\Delta t_{r5} + \Delta t_{r5}^2}{2t_{r10}\Delta t_{r4} + \Delta t_{r4}^2}. \tag{4.72}$$

According to Eq. (4.72), z can be solved as:

$$z_r = \frac{a_r^2 + b_r^2}{4c_r} + \frac{a_r + a_r l_{r1} - 2l_{r2}a_r}{2c_r}x_r, \tag{4.73}$$

where $l_{r2} = \frac{2t_{r10}\Delta t_{r5} + \Delta t_{r5}^2}{2t_{r10}\Delta t_{r4} + \Delta t_{r4}^2}$.

Supposing $k_{r1} = \frac{a_r + a_r l_{r1} - 2l_{r2}a_r}{2c_r}$ and $k_{r2} = \frac{a_r^2 + b_r^2}{4c_r}$, Eq. (4.73) can be rewritten as:

$$z_r = k_{r1}x_r + k_{r2}. \tag{4.74}$$

Taking the ratio of Eqs. (4.60) and (4.65), and it can be rewritten as

$$\frac{(a_r + x_r)^2 + (b_r - y_r)^2 + (c_r - z_r)^2}{4b_r y_r} = n_r \tag{4.75}$$

where $n_r = \frac{t_{r10}^2}{2t_{r10}\Delta t_{r2} + \Delta t_{r2}^2}$, and Eq. (4.75) can be converted to

$$(a_r + x_r)^2 + (b_r - y_r)^2 + (c_r - z_r)^2 = 4n_r b_r y_r. \tag{4.76}$$

Substituting Eqs. (4.71) and (4.74) into Eq. (4.76), it can obtained

$$A_r x_r^2 + B_r x_r + C_r = 0 \tag{4.77}$$

where $A_r = 1 + \left(\frac{a_r}{b_r} l_{r1}\right)^2 + k_{r1}^2$, $B_r = 2a_r + 2a_r l_{r1} - 2(c_r - k_{r2})k_{r1} + 4a_r n_r l_{r1}$, and $C_r = a_r^2 + b_r^2 + (c_r - k_{r2})^2$.

Then, x_r, y_r, and z_r can be obtained by resolving Eqs. (4.77), (4.71), and (4.74).The above method was called AS-UWRPS (analytical solution for unknown wave velocity rectangular pyramid system).

4.4 Analytical Method Under the Random Sensors Network

4.4.1 Analytical Method for Six Sensors

The coordinates of the MS/AE source and the six sensors are assumed as P (x, y, z) and S_i (x_i, y_i, z_i) (i = 1, 2, 3, 4, 5, 6) respectively. The governing equation for the coordinates of the MS/AE source is shown below:

$$(x_i - x)^2 + (y_i - y)^2 + (z_i - z)^2 = v^2(t_i - t_0)^2 \tag{4.78}$$

where t_0 is the trigger time of the MS/AE source. t_i is the arrival time corresponding to the sensor S_i. The average velocity of P-wave is represented as v.

From the point of space geometry, every equation from Eq. (4.78) represents a sphere with the center locating at the coordinates of own sensor. Any two spheres intersect to form a circle. The coordinates of source are located in these intersecting circles. By taking difference between the equation with i = 1 and the others (i = 2, 3, 4, 5, 6), we can obtain Eq. (4.79) as below:

$$\begin{gathered} 2x(x_m - x_1) + 2y(y_m - y_1) + 2z(z_m - z_1) \\ +2v^2(t_m - t_1)t_0 + v^2(t_1^2 - t_m^2) = l_{m-1} \end{gathered} \tag{4.79}$$

where $l_{m-1} = \left(x_m^2 - x_1^2\right) + \left(y_m^2 - y_1^2\right) + \left(z_m^2 - z_1^2\right)$, m = (2, 3, 4, 5, 6).

By substituting V, S for v^2, Vt_0 respectively, Eq. (4.79) can be transformed into Eq. (4.80).

$$\begin{aligned} 2x(x_m - x_1) + 2y(y_m - y_1) + 2z(z_m - z_1) \\ +2(t_m - t_1)S + V(t_1^2 - t_m^2) = l_{m-1} \end{aligned} \tag{4.80}$$

Equation (4.80) can also rewrite as:

$$\mathbf{AS} = \mathbf{B} \tag{4.81}$$

where $\mathbf{A} = \begin{bmatrix} 2(x_2 - x_1) & 2(y_2 - y_1) & 2(z_2 - z_1) & 2(t_2 - t_1) & t_1^2 - t_2^2 \\ 2(x_3 - x_1) & 2(y_3 - y_1) & 2(z_3 - z_1) & 2(t_3 - t_1) & t_1^2 - t_3^2 \\ 2(x_4 - x_1) & 2(y_4 - y_1) & 2(z_4 - z_1) & 2(t_4 - t_1) & t_1^2 - t_4^2 \\ 2(x_5 - x_1) & 2(y_5 - y_1) & 2(z_5 - z_1) & 2(t_5 - t_1) & t_1^2 - t_5^2 \\ 2(x_6 - x_1) & 2(y_6 - y_1) & 2(z_6 - z_1) & 2(t_6 - t_1) & t_1^2 - t_6^2 \end{bmatrix}$, $\mathbf{S} = \begin{bmatrix} x \\ y \\ z \\ S \\ V \end{bmatrix}$,

$\mathbf{B} = \begin{bmatrix} l_2 \\ l_3 \\ l_4 \\ l_5 \\ l_6 \end{bmatrix}$

The important parameters including the coordinates of source (x, y, z), the average velocity v of P-wave, and the trigger time t_0 can be obtained easily through the proposed analytical method. Furthermore, there is no need to measure the velocity before monitoring. The form of calculating formulas is explicit. A set of unique coordinates for MS/AE sources can be determined by every six sensors.

4.4.2 Analytical Method for Greater Than Six Sensors

A unique solution can be obtained through the proposed analytical localization method with the coordinates and the arrival times of six sensors under the unknown velocity system. The sensor network is not only a critical factor to the locating accuracy [10–12], but also the basis for accurately locating and real-time identifying sources in the practical applications of engineering projects [13–17]. In fact, it is common that greater than six sensors are used to improve the locating accuracy in the practical engineering. Therefore, t is a vital problem to take full advantage of the remaining sensors.

As described in the previous section, every six sensors can make up a set of sensor network to calculate the source coordinates. It is feasible to obtain C_m^6 groups of analytical solutions in the locating system with m triggered sensors. Thus, the

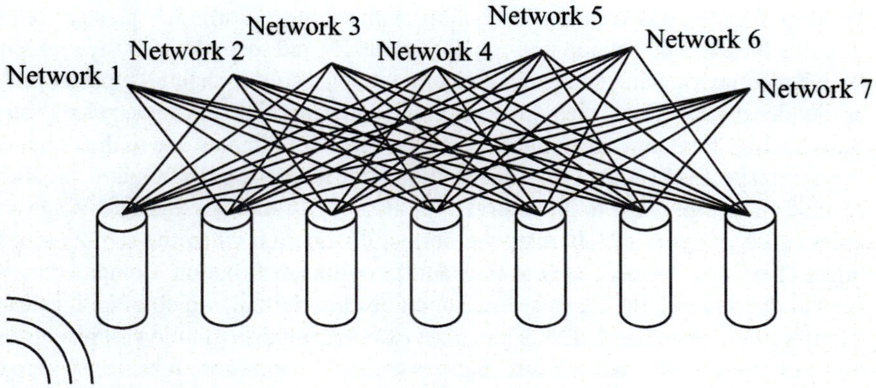

Fig. 4.4 Illustration of an example for the formation of sensor networks, where m is assumed as seven. Every six triggered sensors can constitute a set of sensor network and seven kinds of networks are represented

number of analytical solutions is exactly equal to the number of sensor networks, which is equal to C_m^6. Figure 4.4 shows the example for the formation of sensor networks, where the parameter m is assumed as seven. All the C_m^6 groups of analytical solutions obtained from their own sensor networks can be regard as the same when the propagation medium is homogeneous and the arrival time error is inexistent. However, the propagation medium such as rock is hardly ever homogeneous, a comprehensive analytical localization method can be applied to determine the reasonable and reliable source coordinates. The detailed steps are stated below:

Firstly, the invalid sensors should be excluded. There are three main characteristics of the invalid sensors. One is that the recorded data are noise or the SNR is very low, another is the recorded data have no relation with the event to be located, the third one is locating result using the sensor S_x is a lot different from the others.

Secondly, all the analytical solutions will be analyzed statistically. Six sensors are selected randomly from m triggered sensors to combine $n = C_m^6$ groups of analytical solutions. Numerous kinds of probability density functions are applied to fit the whole solutions. The coordinates of MS/AE source are exactly the abscissa corresponding to the maximum value of the probability density function which fits the data best.

After comparing and analyzing the commonly used more than 60 types of probability density functions, the logistic, normal, and generalized extreme value probability density functions are applied to fit the source coordinates due to their characteristics respectively. The normal distribution is the theoretic basis of many statistical methods. It is useful due to the central limit theorem. In addition, the analyzing variables usually approximately obey the normal distribution when the sample size is huge. Therefore, it is necessary to consider the normal distribution. Compared to the normal distribution, the logistic distribution has heavier tail, which often increases the robustness of analysis. The generalized extreme value distribution represents the

Gumbel, Fréchet, and Weibull distributions with a uniform form. It is mostly used to describe the extreme variability of random variables and solve the stability problem.

After comparing and analyzing the commonly used more than 60 types of probability density functions, the logistic, normal, and generalized extreme value probability density functions are applied to fit the source coordinates due to their characteristics respectively. The normal distribution is the theoretic basis of many statistical methods. It is useful due to the central limit theorem. In addition, the analyzing variables usually approximately obey the normal distribution when the sample size is huge. Therefore, it is necessary to consider the normal distribution. Compared to the normal distribution, the logistic distribution has heavier tail, which often increases the robustness of analysis. The generalized extreme value distribution represents the Gumbel, Fréchet, and Weibull distributions with a uniform form. It is mostly used to describe the extreme variability of random variables and solve the stability problem.

The three functions are shown below, in which the function f represents the probability density function and the function F represents the cumulative distribution function.

$$f_L(x;\mu,s)=\frac{e^{-\frac{x-\mu}{s}}}{s\left(1+e^{-\frac{x-\mu}{s}}\right)^2} \tag{4.82}$$

$$F_L(x;\mu,s)=\frac{1}{1+e^{-\frac{x-\mu}{s}}} \tag{4.83}$$

In Eqs. (4.82) and (4.83), x is the random variable, μ is the mean value, and s is a scale parameter proportional to the standard deviation. The fitting error decreases as the decrease of scale parameter s.

$$f_N(x;\mu,\sigma)=\frac{1}{\sqrt{2\pi}\sigma}e^{-\frac{(x-\mu)^2}{2\sigma^2}} \tag{4.84}$$

$$F_N(x;\mu,\sigma)=\frac{1}{2}\left[1+erf\left(\frac{x-\mu}{\sigma\sqrt{2}}\right)\right] \tag{4.85}$$

In Eqs. (4.84) and (4.85), μ is the mean value of the normal distribution, and σ is the standard deviation. The fitting degree is better when the parameter σ is smaller.

$$f_G(x;\mu,\sigma,\xi)=\frac{1}{\sigma}t(x)^{\xi+1}e^{-t(x)} \tag{4.86}$$

$$F_G(x;\mu,\sigma,\xi)=e^{-t(x)} \tag{4.87}$$

where $t(x)=\begin{cases}\left[1+\xi\left(\frac{x-\mu}{\sigma}\right)\right]^{-1/\xi} & \xi\neq 0\\ e^{-(x-\mu)/\sigma} & \xi=0\end{cases}$

In Eqs. (4.86) and (4.87), μ is the location parameter, σ is the scale parameter and ξ is the shape parameter. The possible source coordinates can be represented with μ.

4.5 Validated Examples and Discussion

4.5.1 Numerical Examples and Experimental Validation Under the Cuboid Sensor Network

4.5.1.1 Numerical Examples

In the first example, a positioning system includes five sensors at the five cuboid vertices, and the coordinates are *A* (−130, 165, 220), *B* (−130, −165, 220), *C* (130, −165, 220), *D* (130, 165, 220), *E* (−130, 165, −220). The average equivalent P-wave velocity in the medium is expressed as *v*, and $v = 5{,}000$ m/s. The AE/MS sources are *O* (110, 200, 180), *P* (210, 97, −89), *Q* (−77, −89, 190), *R* (−98, 22, 168), and *S* (99, −289, 190) (all coordinates have the length unit of m). The arrival time recorded by sensors is listed in Table 4.1, and the accuracy of time is 10^{-6} s. By using the proposed analytical solution to calculate the AE/MS source coordinates, coordinate values of the five sensors and arrival time of five sensors for five events are taken into Eqs. (4.21), (4.16), and (4.18), and the coordinate values (*x*, *y*, *z*) of five AE events can be resolved. The actual position and calculated results are listed in Table 4.2. It can be seen from Table 4.2, one set of the location results of the proposed analytical solutions are fully consistent with the actual coordinates.

In the second example, a positioning system includes five sensors at the five cuboid vertices, and the coordinates are *A* (−130, 165, 220), *B* (−130, −165, 220), *D* (130, 165, 220), *E* (−130, 165, −220), and *F* (−130, −165, −220), and average equivalent P-wave velocity in the medium is expressed as *v*, and $v = 5000$ m/s. Assume that AE/MS sources *O*, *P*, *Q*, *R*, and *S* are as the same as the first example (all coordinates have the length unit of m). The arrival time recorded by sensors is listed in Table 4.3. By using the proposed analytical solution to calculate the AE/MS source coordinates, the coordinate values of the five sensors and arrival time of the five sensors for the

Table 4.1 Arrival time recorded by sensors *O*, *P*, *Q*, *R*, and *S* in the first example

Sensor	Arrival time recorded by sensor/s				
	O	*P*	*Q*	*R*	*S*
A	0.049163	0.092888	0.05224	0.031098	0.101874
B	0.087733	0.105778	0.019478	0.039343	0.052428
C	0.073546	0.082589	0.044508	0.059886	0.026258
D	0.011358	0.06527	0.065807	0.054822	0.091209
E	0.093557	0.074131	0.097041	0.08295	0.130638

Table 4.2 Comparison between actual and calculated coordinates and errors of absolute distance in the first example

AE event	Actual coordinate/m			Calculated coordinate/m						Error of absolute distance/m	
				Solution 1			Solution 2				
	x	y	z	x_1	y_1	z_1	x_2	y_2	z_2	Solution 1	Solution 2
O	110	200	180	120.46	219.01	197.10	110.00	199.99	179.99	27.63	0.02
P	210	97	−89	316.37	146.13	−134.08	209.95	96.98	−88.98	125.55	0.06
Q	−77	−89	190	77.00	−89.00	190.01	−142.62	164.86	351.94	0.01	308.18
R	−98	22	168	−97.97	22.00	167.94	−236.76	−53.15	405.85	0.06	285.44
S	99	−289	190	99.00	−289.01	190.00	70.78	206.61	135.83	0.01	499.36

Table 4.3 Arrival time recorded by sensors *A*, *B*, *D*, *E*, and *F* in the second example

Sensor	Arrival time recorded by sensor/s				
	O	*P*	*Q*	*R*	*S*
A	0.049163	0.092888	0.05224	0.031098	0.101874
B	0.087733	0.105778	0.019478	0.039343	0.052428
D	0.011358	0.06527	0.065807	0.054822	0.091209
E	0.093557	0.074131	0.097041	0.08295	0.130638
F	0.118461	0.089756	0.084068	0.08638	0.097143

five events are taken into Eqs. (4.39), (4.34), and (4.36), and the coordinate values (x_s, y_s, z_s) of five AE events can be resolved. The actual and calculated results are listed in Table 4.4. It can be seen from Table 4.4 that one set of the location results of the proposed analytical solutions are fully consistent with the authentic coordinates.

In the third example, a positioning system includes five sensors at the five cuboid vertices, and the coordinates are *A* (−130, 165, 220), *B* (−130, −165, 220), *D* (130, 165, 220), *F* (−130, −165, −220), and *H* (130, 165, −220). The average equivalent P-wave velocity in the medium is expressed as *v*, and $v = 5{,}000$ m/s. Assume that AE/MS sources *O*, *P*, *Q*, *R*, and *S* are as the same as the first example (all coordinates have the length unit of m).The arrival time recorded by sensors is listed in Table 4.5. By using the proposed analytical solution to calculate the AE/MS source coordinates, the coordinate values (x_p, y_p, z_p) of the five sensor and arrival time of five sensors for the five events are taken into Eqs. 4.57, 4.51, and 4.54, and the coordinate values of the five AE events can be resolved. The actual and calculated results are listed in Table 4.6. It can be seen from Table 4.6 that one set of the location results of the proposed analytical solutions are fully consistent with the actual coordinates.

4.5.1.2 Experimental Validation

The AE tests were carried out in a cuboid of granite rock using five AE sensors. The sensor coordinates are *A* (−60, 80, 90), *B* (−60, −80, 90), *C* (60, −80, 90), *D* (60, 80, 90), *E* (−60, 80, −90). AE/MS sources are *O* (60, 30, 40), *P* (28, −80, 38), and *Q* (−29, 80, 58) (all coordinates have the length unit of cm). The arrival time recorded by sensors is listed in Table 4.7, and the accuracy of time is 10^{-6} s. By using the proposed analytical solution to calculate the AE/MS source coordinates, coordinate values of the five sensor and arrival time of the five sensors for five events are taken into equations of ASI, ASII, and ASIII. The coordinate values of AE events can be resolved. The actual and calculated results are listed in Table 4.8. It can be seen from Table 4.8 that one set of the location results of the proposed analytical solutions are fully consistent with the actual coordinates.

Table 4.4 Comparison between actual and calculated coordinates and errors of absolute distance in the second example

AE event	Actual coordinate/m			Calculated coordinate/m						Error of absolute distance/m	
				Solution 1			Solution 2				
	x	y	z	x_{s1}	y_{s1}	z_{s1}	x_{s2}	y_{s2}	z_{s2}	Solution 1	Solution 2
O	110	200	180	120.45	219.00	197.09	110.00	200.00	180.00	27.61	0.00
P	210	97	−89	316.55	146.21	−134.15	209.84	96.92	−88.93	125.75	0.20
Q	−77	−89	190	−77.00	−89.00	190.01	−142.62	−164.85	351.93	0.01	190.47
R	−98	22	168	−98.00	22.00	167.99	−236.68	53.13	405.74	0.01	276.99
S	99	−289	190	98.99	−288.96	189.97	70.79	−206.64	135.85	0.05	102.52

Table 4.5 Arrival time recorded by sensors in the third example

Sensor	Arrival time recorded by sensor/s				
	O	*P*	*Q*	*R*	*S*
A	0.049163	0.092888	0.05224	0.031098	0.101874
B	0.087733	0.105778	0.019478	0.039343	0.052428
D	0.011358	0.06527	0.065807	0.054822	0.091209
F	0.118461	0.089756	0.084068	0.08638	0.097143
H	0.080405	0.033577	0.10497	0.094441	0.122503
G	0.108374	0.060731	0.093107	0.097467	0.085892

It can be seen from the above validated examples that there are two groups of solutions using the proposed analytical solutions. One is the real and correct solution, the other one is meaningless solution. The problem is how to select the real solution and cancel the meaningless solution. The checking calculation with arrival time of the sixth sensor is an efficient approach to select a reasonable solution. For example, the arrival time of six sensors (i.e. sensor *G*) is listed in Table 4.5. The solutions in Table 4.6 are obtained only using the arrival time of the first five sensors. We can use arrival time t_{p6} and coordinates of the sixth sensor *G* to select the reasonable solution. The distance *D* between the solved source and sensor *G* can be calculated according to the distance formula between two points in space. v_p can be solved using Eq. (4.40). According to t_{p10}, the original time $t_{original}$ of the event can be obtained by taking the subtraction of t_{p1} and t_{p10}, then the reasonable solution should meet the following criterion:

$$D = v_p\left(t_{p6} - t_{original}\right) \tag{4.88}$$

Taking the values of sensor *G* into Eq. (4.88), we can easily get the reasonable solutions which are listed in Table 4.6.

4.5.2 Numerical Examples Under the Rectangular Pyramid Sensor Network

In the example, a localization system includes five sensors at the five vertices of rectangular pyramid network, and the coordinates are *A* (−130, 165, 220), *B* (−130, −165, 220), *C* (130, −165, 220), *D* (130, 165, 220), *I* (0, 0, −220), and average equivalent P-wave velocity propagation in the medium is expressed as v, and v = 5,000 m/s. Assumed that AE/MS sources are *O* (110, 200, 180), *P* (210, 97, −89), *Q* (−77, −89, 190), *R* (−98, 22, 168), *S* (99, −289, 190), *T* (−120, 170, −130),

Table 4.6 Comparison between actual and calculated coordinates and errors of absolute distance in the third example

AE event	Actual coordinate/m			Calculated coordinate/m						Error of absolute distance/m	
				Solution 1			Solution 2				
	x	y	z	x_{p1}	y_{p1}	z_{p1}	x_{p2}	y_{p2}	z_{p2}	Solution 1	Solution 2
O	110	200	180	120.44	218.98	197.08	110.01*	200.01*	180.01*	27.59	0.02
P	210	97	−89	316.34	146.12	−134.07	209.97*	96.99*	−88.99*	125.51	0.03
Q	−77	−89	190	−76.99*	−89.00*	189.99*	−142.63	164.87	351.96	0.01	308.20
R	−98	22	168	−98.00*	22.00*	168.00*	−236.68	−53.13	405.74	0.01	285.30
S	99	−289	190	98.97*	−288.91*	189.94*	70.80	206.67	135.88	0.11	499.42

*: Reasonable solutions

Table 4.7 Arrival time recorded by sensors for three groups of AE tests

Sensor	Arrival time for ASI/s			Sensor	Arrival time for ASII/s		
	O	*P*	*Q*		*O*	*P*	*Q*
A	0.060279	0.09038	0.16009	*A*	0.080279	0.100381	0.19009
B	0.060342	0.090205	0.160333	*B*	0.080341	0.100205	0.190333
C	0.060242	0.090122	0.160372	*D*	0.080142	0.100343	0.190189
D	0.060142	0.090343	0.16019	*E*	0.080368	0.100447	0.190303
E	0.060369	0.090447	0.160303	*F*	0.080418	0.100311	0.190441
Sensor	Arrival time for ASIII/s						
	O	*P*	*Q*				
A	0.010279	0.11038	0.280089				
B	0.010341	0.110205	0.280332				
D	0.010142	0.110343	0.28019				
F	0.010417	0.110312	0.280441				
H	0.010279	0.110415	0.280347				

U (−200, −300, −230), and V (190, −169, 111) (all coordinates are the length unit: m). The arrival times recorded by sensors were listed in Table 4.9, and the time is accurate to 10^{-6} s. Figure 4.5 shows the position of sensors and sources. By proposed AS-UWRPS to calculate the AE/MS source coordinates, coordinates of the five sensor values and triggered time of five sensors for five events are taken into Eqs. (4.77), (4.71), and (4.74), and the coordinate values of eight AE events can be resolved. The calculated results are listed in Table 4.10. As can be seen from Table 4.10 and Fig. 4.6, one set of the location results of proposed AS-UWRPS are fully consistent with the actual coordinates. The sixth sensor is used to determine a reasonable solution from solutions 1 and 2, and the equation is similar to Eq. (4.88).

4.5.3 Blasting Tests Under the Random Sensor Network

As the probability density of source coordinate is considered in this method, the fitting results will not be affected seriously by a small amount of special values. A total of six blasting tests in two mines proved the accuracy and effectiveness of the proposed three dimensional comprehensive analytical and probability density function localization method.

Table 4.8 Results and comparison of AE experiments

Solution	Event	Actual coordinate/cm			Calculated coordinate/cm					
					Solution 1			Solution 2		
		x	*y*	*z*	*x*	*y*	*z*	*x*	*y*	*z*
ASI	*O*	60	30	40	185.39	92.97	123.03	57.71*	28.94*	38.29*
	P	28	80	38	57.41	−164.31	79.39	28.39*	81.27*	39.27*
	Q	−29	80	58	29.32*	81.54*	58.94*	48.31	134.39	97.14
ASII	*O*	60	30	40	205.63	97.98	129.91	54.13*	25.79*	34.19*
	P	28	−80	38	58.6	−164.69	77.62	28.99*	81.48*	38.40*
	Q	−29	80	58	28.69*	80.18*	58.03*	48.89	136.66	98.9
ASIII	*O*	60	30	40	178.14	89.12	118.76	59.95*	29.99*	39.97*
	P	28	−80	38	54.51	−155.35	74.25	30.25*	86.20*	41.20*
	Q	−29	80	58	29.19*	79.58*	58.33*	49.90	136.06	99.72

* Reasonable solutions

Table 4.9 Trigger times recorded by sensors

Sensor	Arrival times recorded by sensor/s							
	O	*P*	*Q*	*R*	*S*	*T*	*U*	*V*
A	0.050436	0.084444	0.096313	0.048430	0.085554	0.041983	0.101467	0.082881
B	0.068743	0.096162	0.104880	0.053155	0.031845	0.090065	0.067347	0.056408
C	0.061901	0.075081	0.111011	0.084398	0.034438	0.099063	0.089295	0.019046
D	0.040619	0.059337	0.102955	0.081504	0.086553	0.058859	0.117190	0.063640
I	0.048374	0.048334	0.024777	0.064407	0.091221	0.071924	0.062403	0.075348

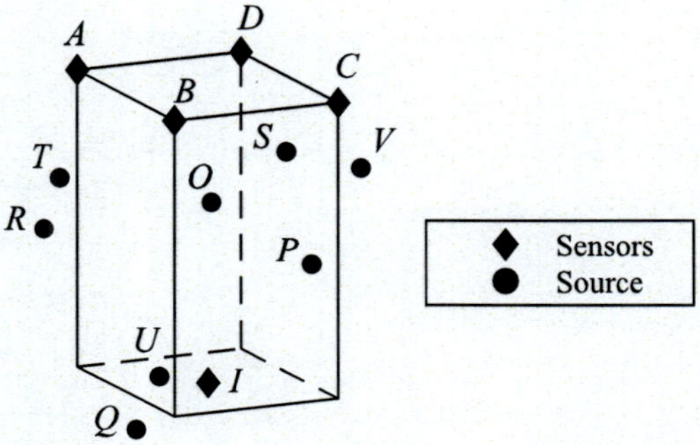

Fig. 4.5 Sensors and sources position under the rectangular pyramid network

4.5.3.1 Yongshaba Mine

The ore body of Yongshaba mine is controlled by ten obvious faults and the stability of stope roof is poor. Some underground goafs are reserved due to the open stope mining methods in the early stage of mining engineering. As a result, the rockburst or rock instability for large areas may be induced by the local stress concentration. To avoid these disasters, a 32-channel digital microseismic monitoring system was established. There are totally 26 single-component sensors and two three-component sensors distributed on the transport tunnels in 930, 1,080, and 1,120 levels to detect signals every day. Figure 4.7 shows the layout of microseismic monitoring system and the relationship between geological structures. Figure 4.8 shows the positions of sensors under the three dimensional structure of Yongshaba mine.

The data of three blasting tests in the Yongshaba mine were calculated to verify the proposed TDCAS-PDF. By substituting the coordinates and the arrival times of triggered sensors into Eqs. (4.78)–(4.81), the source coordinates corresponding to different kinds of combinations can be obtained. Then, the logistic, normal, and generalized extreme value probability density functions are applied to fit the source coordinates. In the case of the event No.1, the fitting results of three probability density functions are compared in Table 4.11. The logistic probability density function shows the best fitting degree because the standard deviations for *X*, *Y*, and *Z* of logistic are the minimum among three functions. Therefore, the source coordinates will be selected according to the logistic probability density function. Figures 4.9,

Table 4.10 Result comparisons (unit: m)

AE event	Actual coordinate			Calculated coordinate						Error of absolute distance	
				Solution 1			Solution 2				
	x	y	z	x_1	y_1	z_1	x_2	y_2	z_2	Solution 1	Solution 2
O	52	100	21	279.99	538.44	−106.77	52.00	100.00	21.00	510.43	0.00
P	210	97	−89	210.31	97.14	−89.20	210.00	97.00	−89.00	0.40	0.00
Q	−77	79	−300	−41.69	42.78	−139.46	−77.00	79.00	−300.00	168.32	0.00
R	−250	22	30	−250.00	22.00	30.00	−287.81	25.33	26.95	0.00	38.08
S	−10	−289	190	−7.07	−204.40	149.06	−10.00	−289.00	190.00	94.03	0.00
T	−99	291	29	−76.11	223.71	33.89	−99.00	291.00	29.00	71.25	0.00
U	−200	−264	−130	−102.68	−135.53	−42.34	−200.00	−264.00	−130.00	183.46	0.00
V	164	−169	121	197.89	−203.93	135.64	164.00	−169.00	121.00	50.82	0.00

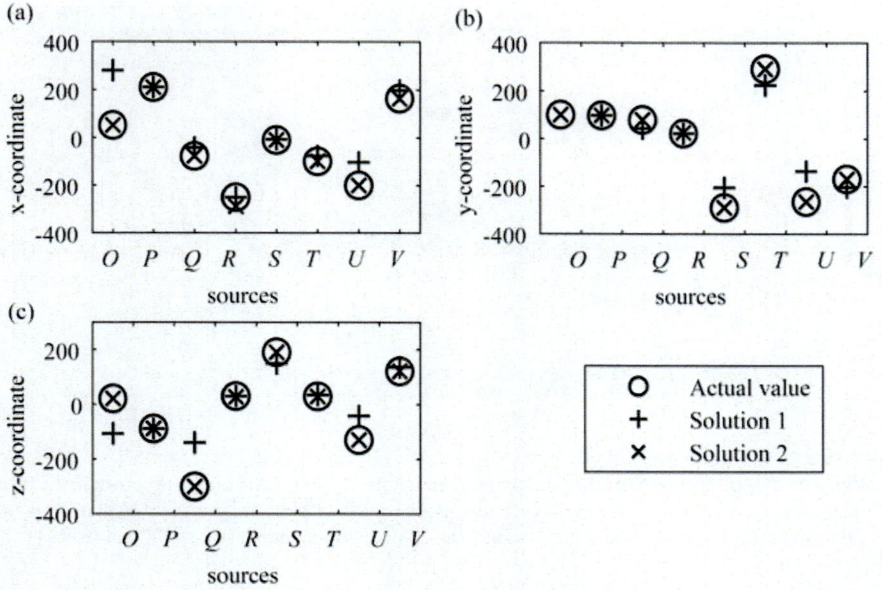

Fig. 4.6 Location results and actual value of **a** x-coordinate, **b** y-coordinate, and **c** z-coordinate of sources under rectangular pyramid network

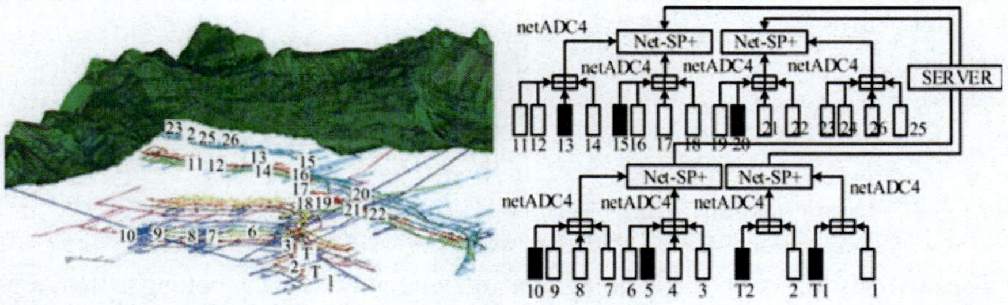

Fig. 4.7 The established microseismic monitoring system in Yongshaba mine, where T1 as well as T2 are the three-component sensors and the remaining sensors are the single-component sensors. This system consists of sensors, data collectors, signal processors, underground data centers, communication cables and surface monitoring center (from [18])

4.10, and 4.11 show probability density functions and cumulative distribution functions of the logistic, normal, and generalized extreme value for the events No.1–3 in the Yongshaba mine respectively. The abscissa corresponding to the maximum value of the logistic probability density function is the locating coordinate. The location results are listed in the Table 4.12. It is clear to see the ADEs of the events No.1–3 are 28.7 m, 52.4 m, and 61.1 m respectively, which can commonly meet the requirements of engineering application in mines.

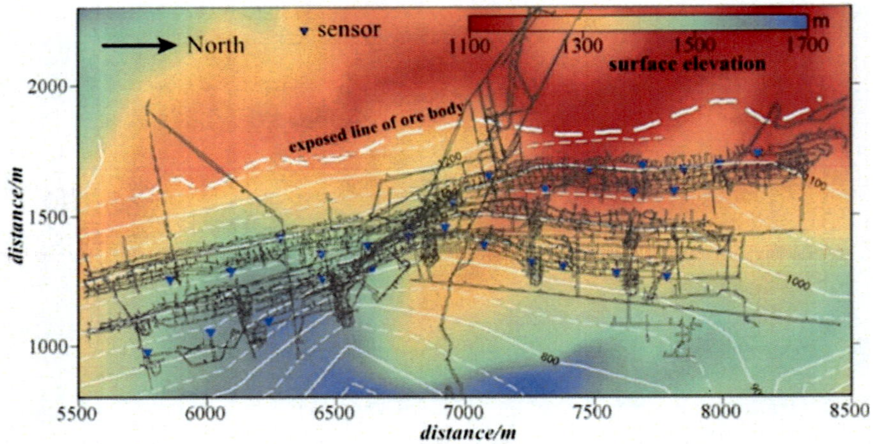

Fig. 4.8 The positions of sensors and the main structure of Yongshaba mine are shown in the three dimensional cloud map. It is clear to see that the elevation of mine surface is more than 1,500 m. The current mining level is about 800 m and the relative mining depth reaches 700 m (from [18])

Table 4.11 Comparison results of the standard deviation error in the fitting for event No.1

Probability density function	*X*		*Y*		*Z*	
	Std. Err	Rank	Std. Err	Rank	Std. Err	Rank
Logistic	1.56	1	2.73	1	2.99	1
Normal	5.44	2	5.56	2	6.24	3
Generalized Extreme Value	8.63	3	5.94	3	5.88	2

4.5.3.2 Dongguashan Mine

The ore body and main surrounding rock of Dongguashan mine belong to hard rock, where exists the possibility of rockburst in the deep mining process [19]. The micro-seismic monitoring system was established to ensure the safety mining in the condition of multiple underground goafs. There are 12 single-component sensors working in the 514 and 558 levels, and other six single-component sensors are distributed in the 630 level.

The data of three blasting tests were calculated according to the proposed TDCAS-PDF. Similarly, it is feasible to obtain the source coordinates by applying the three probability density functions. Figures 4.12, 4.13, and 4.14 show probability density functions and cumulative distribution functions of the logistic, normal, and generalized extreme value for the events No.4–6 in the Dongguashan mine respectively. Also, the locating results and the ADEs are listed in the Table 4.13.

After comparing and analyzing the results comprehensively, it can be concluded that the proposed TDCAS-PDF has a high locating accuracy. Furthermore, there is no need to solve the iterative solutions when applying the proposed TDCAS-PDF.

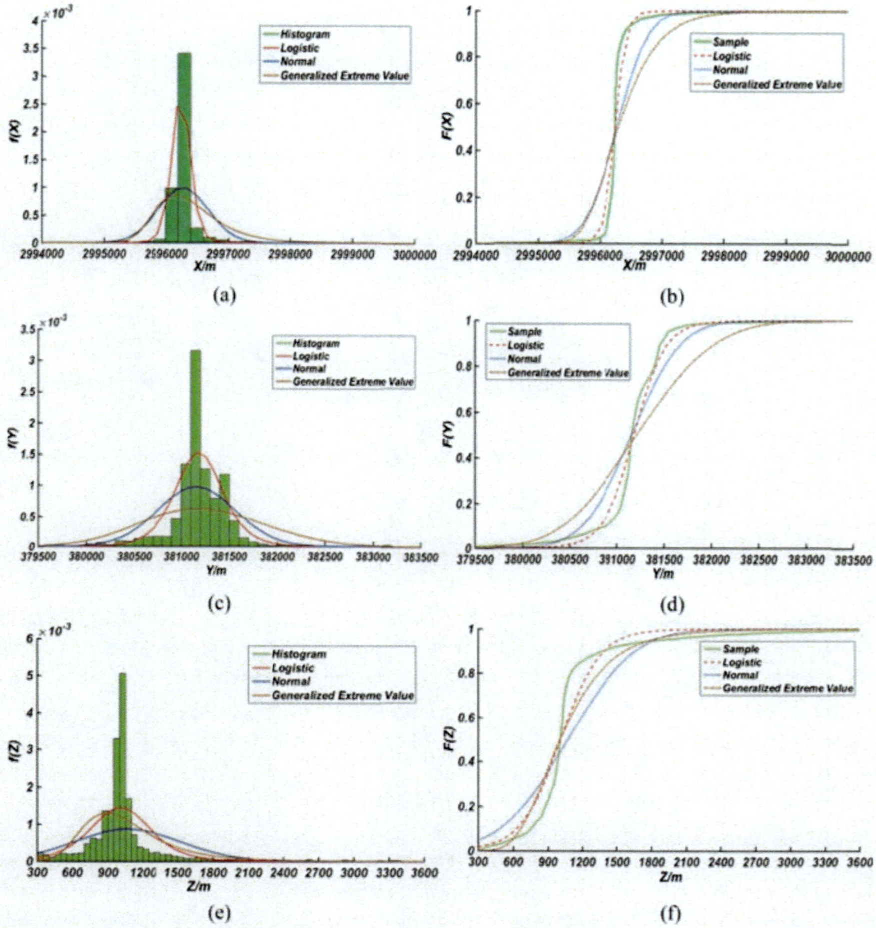

Fig. 4.9 The fitting results for the event No.1, where graphs **a** and **b** show the fitting results of the logistic, normal, and generalized extreme value distributions for the coordinate X. It is similar in graphs **c** and **d**, as well as graphs **e** and **f** for the coordinate Y and Z respectively. The best fitting coordinates for the event No.1 are $X = 2996{,}250$ m, $Y = 381{,}183$ m, and $Z = 1{,}009$ m (from [18])

The calculating formulas are explicit and the physical significance is clear, where the velocity of P-wave and the iterative algorithm make no influence to the source coordinates. All of these advantages make the TDCAS-PDF achieve the real-time locating and become more practical than traditional localization methods in the aspect of engineering applications.

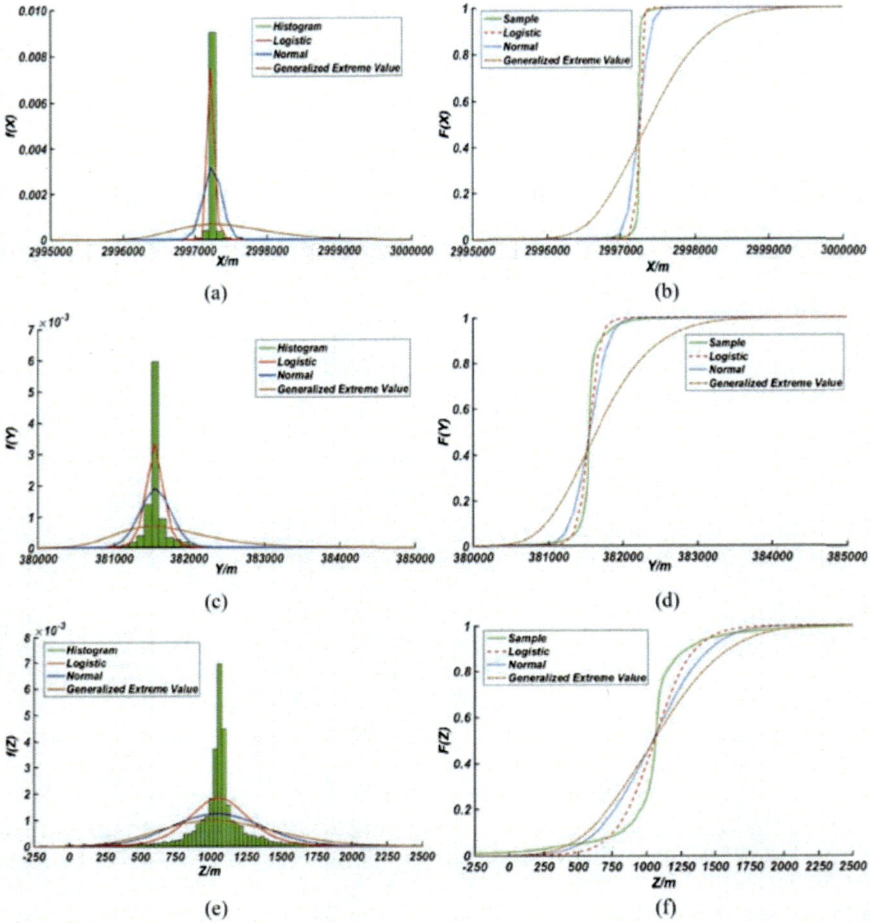

Fig. 4.10 The fitting results for the event No.2, where graphs **a** and **b** show the fitting results of the logistic, normal, and generalized extreme value distributions for the coordinate X. It is similar in graphs **c** and **d**, as well as graphs **e** and **f** for the coordinate Y and Z respectively. The best fitting coordinates for the event No.2 are $X = 2997{,}250$ m, $Y = 381{,}546$ m, and $Z = 1{,}058$ m (from [18])

4.6 Conclusions

(1) Based on the proposed functions of time difference of arrivals, the analytical solutions were obtained using six sensors under two networks, cuboid network and rectangular pyramid network. The proposed analytical solutions were validated using authentic data. The results show that the proposed analytical solution is reasonable and a set of the resolved solutions are consistent with the authentic results.

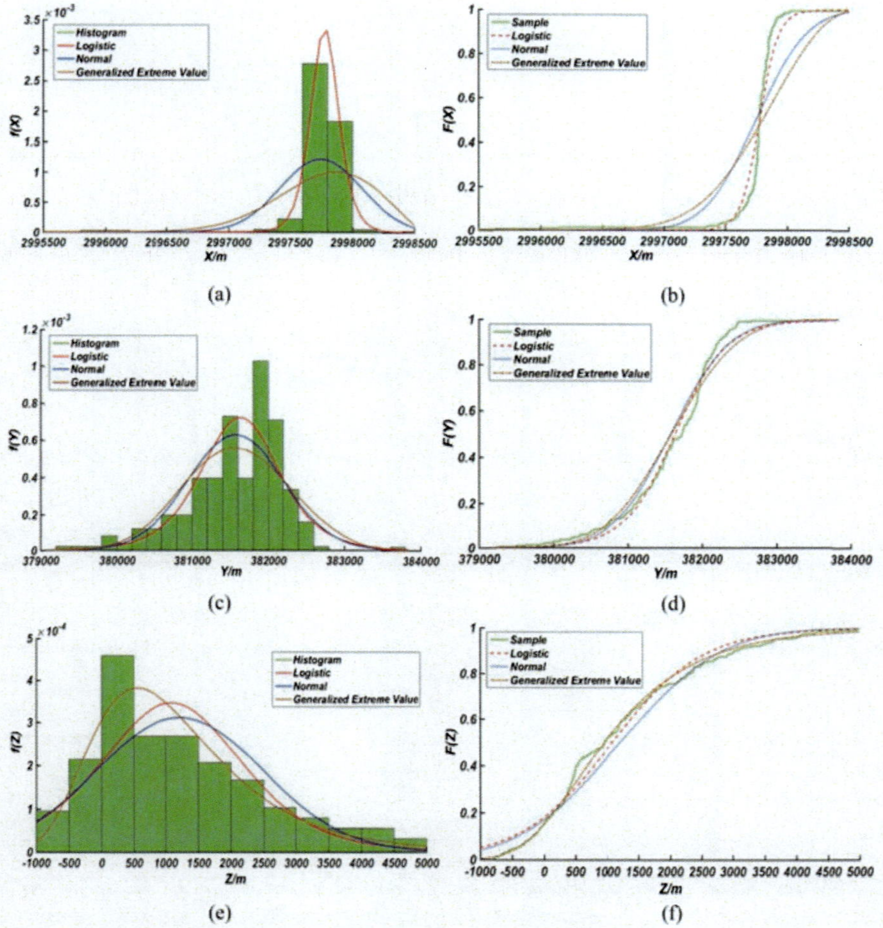

Fig. 4.11 The fitting results for the event No.3, where graphs **a** and **b** show the fitting results of the logistic, normal, and generalized extreme value distributions for the coordinate X. It is similar in graphs **c** and **d**, as well as graphs **e** and **f** for the coordinate Y and Z respectively. The best fitting coordinates for the event No.3 are $X = 2997{,}780$ m, $Y = 381{,}631$ m, and $Z = 1{,}082$ m (from [18])

Table 4.12 Location results using TDCAS-PDF in the Yongshaba mine

Event No	TDCAS-PDF			Blasting coordinates			Error
	X/m	Y/m	Z/m	X/m	Y/m	Z/m	D^1/m
1	2996,250	381,183	1,009	2996,224	381,194	1,014	28.7
2	2997,250	381,546	1,058	2997,278	381,590	1,053	52.4
3	2997,780	381,631	1,082	2997,760	381,683	1,107	61.1
Average value							47.4

D^1 is the ADE (the distances between blasting coordinates and locating coordinates).

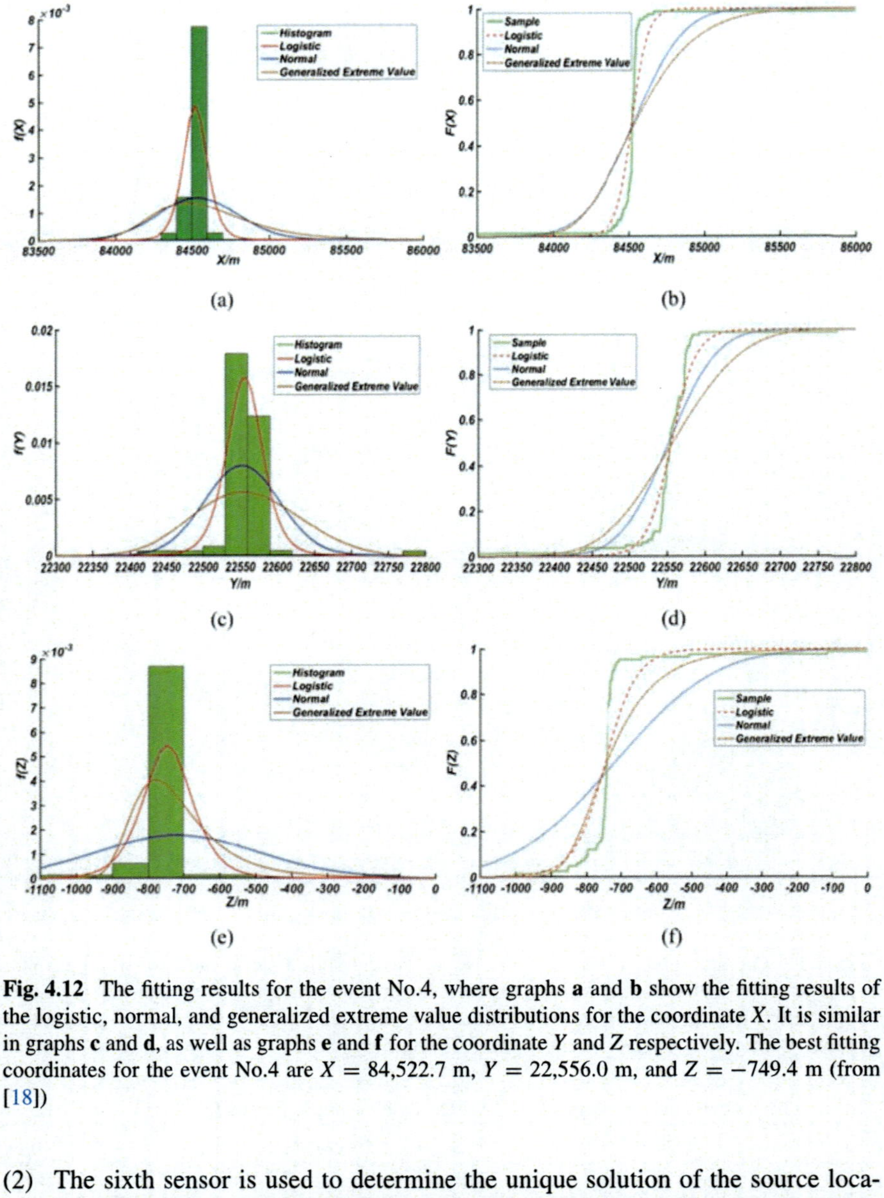

Fig. 4.12 The fitting results for the event No.4, where graphs **a** and **b** show the fitting results of the logistic, normal, and generalized extreme value distributions for the coordinate X. It is similar in graphs **c** and **d**, as well as graphs **e** and **f** for the coordinate Y and Z respectively. The best fitting coordinates for the event No.4 are $X = 84{,}522.7$ m, $Y = 22{,}556.0$ m, and $Z = -749.4$ m (from [18])

(2) The sixth sensor is used to determine the unique solution of the source location. Based on a cuboid network of sensor location, the method can locate the coordinates of AE/MS source only using simple four arithmetic operations. The method highlights three outstanding advantages of without using iterative solution, without initial evaluated hypocenter coordinates and without pre-measured velocity or pre-given velocity boundary conditions.

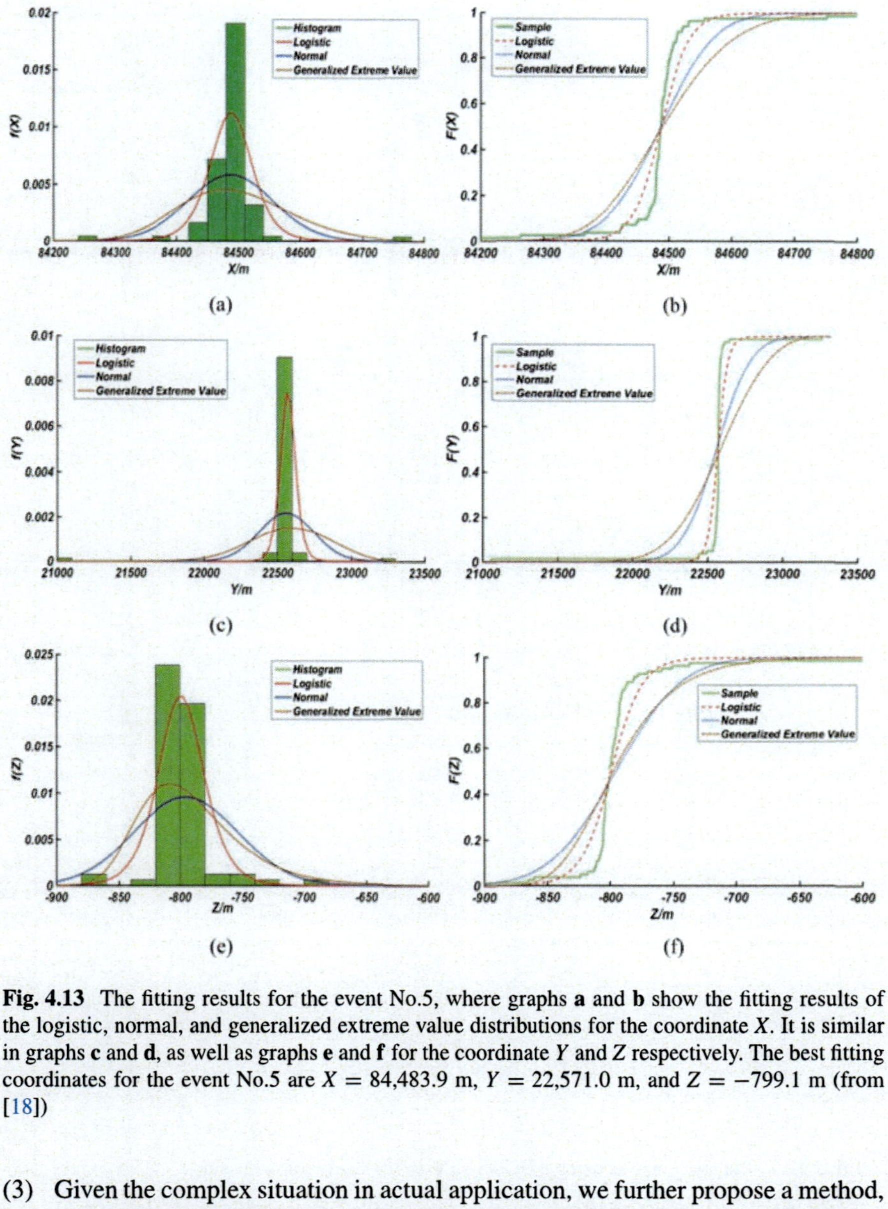

Fig. 4.13 The fitting results for the event No.5, where graphs **a** and **b** show the fitting results of the logistic, normal, and generalized extreme value distributions for the coordinate X. It is similar in graphs **c** and **d**, as well as graphs **e** and **f** for the coordinate Y and Z respectively. The best fitting coordinates for the event No.5 are $X = 84{,}483.9$ m, $Y = 22{,}571.0$ m, and $Z = -799.1$ m (from [18])

(3) Given the complex situation in actual application, we further propose a method, called EDCAS-PDF, which can be applied to random network and random number of sensors (at least six sensors). The method has a high locating accuracy. The calculating formulas are explicit and the physical significance is clear, where the velocity of P-wave and the iterative algorithm make no influence to the source coordinates. All of these advantages make the TDCAS-PDF achieve the real-time locating and become practical in the aspect of engineering applications.

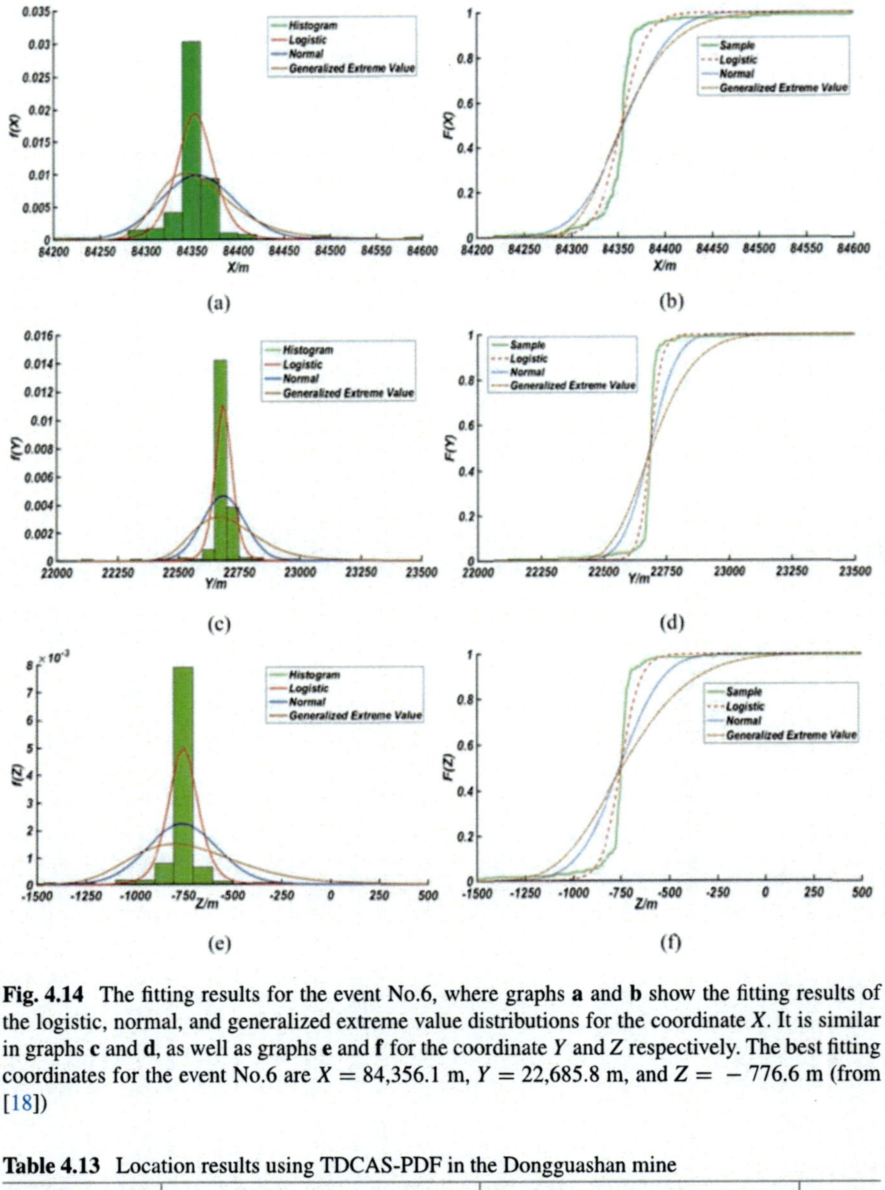

Fig. 4.14 The fitting results for the event No.6, where graphs **a** and **b** show the fitting results of the logistic, normal, and generalized extreme value distributions for the coordinate X. It is similar in graphs **c** and **d**, as well as graphs **e** and **f** for the coordinate Y and Z respectively. The best fitting coordinates for the event No.6 are $X = 84{,}356.1$ m, $Y = 22{,}685.8$ m, and $Z = -776.6$ m (from [18])

Table 4.13 Location results using TDCAS-PDF in the Dongguashan mine

Event No	TDCAS-PDF			Blasting coordinates			Error
	X/m	Y/m	Z/m	X/m	Y/m	Z/m	D^1/m
1	84,522.7	22,556.0	−749.4	84,528.4	22,556.2	−753.2	6.85
2	84,483.9	22,571.0	−799.1	84,479.0	22,570.0	−814.4	16.09
3	84,356.1	22,685.8	−776.6	84,359.0	22,673.0	−795.5	23.01
Average value							15.32

References

1. Chan Y-T, Ho K (1994) An efficient closed-form localization solution from time difference of arrival measurements. In: Proceedings of ICASSP'94. IEEE international conference on acoustics, speech and signal processing, 1994. IEEE, vol 392, pp II/393–II/396
2. Liu H, Milios E (2005) Acoustic positioning using multiple microphone arrays. J Acoust Soc Am 117(5):2772–2782
3. Brandstein MS, Adcock JE, Silverman HF (1997) A closed-form location estimator for use with room environment microphone arrays. IEEE Trans Speech Audio Process 5(1):45–50
4. Dong L, Zou W, Li X, Shu W, Wang Z (2019) Collaborative localization method using analytical and iterative solutions for microseismic/acoustic emission sources in the rockmass structure for underground mining. Eng Fract Mech 210(2019):95-112
5. Duraiswami R, Zotkin D, Davis L (1999) Exact solutions for the problem of source location from measured time differences of arrival. J Acoust Soc Am 106(4):2277–2277
6. Smith J, Abel J (1987) Closed-form least-squares source location estimation from range-difference measurements. IEEE Trans Acoust Speech Signal Process 35(12):1661–1669
7. Mellen G, Pachter M, Raquet J (2003) Closed-form solution for determining emitter location using time difference of arrival measurements. IEEE Trans Aerosp Electron Syst 39(3):1056–1058
8. Ge M (2003) Analysis of source location algorithms—part I: overview and non-iterative methods. J Acoust Emiss 21(1):14–28
9. Dong L-J, Li X-B, Zhou Z-L, Chen G-H, Ma J (2015) Three-dimensional analytical solution of acoustic emission source location for cuboid monitoring network without pre-measured wave velocity. Trans Nonferrous Metals Soc China 25(1):293–302. https://doi.org/10.1016/S1003-6326(15)63604-4
10. Dong L-J, Li X-B, Tang L-Z, Gong F-Q (2011) Mathematical functions and parameters for microseismic source location without pre-measuring speed. Chin J Rock Mech Eng 30(10):2057–2067
11. Han G, Yang X, Liu L, Zhang W, Guizani M (2017) A disaster management-oriented path planning for mobile anchor node-based localization in wireless sensor networks. IEEE Trans Emerg Top Comput 8(1):115–125
12. Xie S, Wang Y (2014) Construction of tree network with limited delivery latency in homogeneous wireless sensor networks. Wireless Pers Commun 78(1):231–246
13. Shen J, Tan HW, Wang J, Wang JW, Lee S (2015) A novel routing protocol providing good transmission reliability in underwater sensor networks. J Internet Technol 16(1):171–178
14. Zhang Y, Sun X, Wang B (2016) Efficient algorithm for k-barrier coverage based on integer linear programming. China Commun 13(7):16–23
15. Han G, Liu L, Chan S, Yu R, Yang Y (2017) HySense: A hybrid mobile crowdsensing framework for sensing opportunities compensation under dynamic coverage constraint. IEEE Commun Mag 55(3):93–99
16. Han G, Liu L, Jiang J, Shu L, Hancke G (2015) Analysis of energy-efficient connected target coverage algorithms for industrial wireless sensor networks. IEEE Trans Industr Inf 13(1):135–143
17. Dong L-J, Wesseloo J, Potvin Y, Li X-B (2016) Discriminant models of blasts and seismic events in mine seismology. Int J Rock Mech Min Sci 86:282–291
18. Dong LJ, Shu WW, Li XB, Han GJ, Zou W (2017) Three dimensional comprehensive analytical solutions for locating sources of sensor networks in unknown velocity mining system. IEEE Access 5:11337–11351. https://doi.org/10.1109/Access.2017.2710142
19. Dong L, Wesseloo J, Potvin Y, Li X (2016) Discrimination of mine seismic events and blasts using the fisher classifier, naive bayesian classifier and logistic regression. Rock Mech Rock Eng 49(1):183–211

Chapter 5
Iterative Method for Velocity-Free Model

Compared to the analytical localization method, the iterative method is more accurate but time-consuming, since it seeks the optimal results in the whole range with the advantage of multiple sensors. Generally, iterative method is more suitable and accurate for locating numerous sources with multiple sensors. Most of iterative methods are mostly developed on account of the thoughts of Geiger. Based on the Geiger algorithm, many researchers presented numerous optimization methods including the parameter separation, the joint inversion of 3-D velocity structure and seismic source, as well as the separate calculation for coupled velocity and seismic source [1–5]. However, the application scope of the assumption that the P-wave velocity is fixed is relatively limited.

In this chapter, three iterative methods for source localization with unknown velocity are introduced. Based on TD method which has mentioned in Chap. 2, a multi-step localization method (MLM) without premeasured velocity is demonstrated in Sect. 5.1. The method optimizes and narrows the velocity interval in the localization process for heterogeneous and complex media. The MSLM-MV is based on TD method and improves the location accuracy combining Levenberg–Marquardt algorithm, which is introduced in Sect. 5.2. Both methods only use the information of P-wave, so the method using P-wave and S-wave arrivals for localization is explored in Sect. 5.3.

5.1 Multi-step Localization Method

The coordinates of a microseismic source and the triggered sensors are assumed as P (x, y, z) and S_j (x_j, y_j, z_j) $(j = 1, 2, 3, \ldots, n)$, respectively. The average propagation velocity of P-wave in the media is represented with parameter v. Then, the governing equation for the source coordinates can be established in the Cartesian coordinate system:

L. Dong and X. Li, *Velocity-Free Localization Methodology for Acoustic and Microseismic Sources*, https://doi.org/10.1007/978-981-19-8610-9_5

$$t_{ij}^{\text{mea}} - t_{ij}^{\text{cal}} = \left(\frac{\partial t}{\partial x}\right)_{ij} \Delta x + \left(\frac{\partial t}{\partial y}\right)_{ij} \Delta y + \left(\frac{\partial t}{\partial z}\right)_{ij} \Delta z + \left(\frac{\partial t}{\partial v}\right)_{ij} \Delta v + \Delta t_{0i} + e_{ij}, \tag{5.1}$$

where t_{ij}^{mea} and t_{ij}^{cal} are the measured and calculated travel times from the *i*th event to the *j*th sensor. t_{0i} is the origin time of the *i*th event. Δ denotes the perturbation of a parameter. e_{ij} represents higher-order terms of perturbations and data error.

The difference between the measured travel time t_{ij}^{mea} and the calculated travel time t_{ij}^{cal} can be used to describe the deviation degree. The fitting degree and locating accuracy will be better when the deviation degree is smaller. Based on the quadratic sum of differences between all the regression values and all the measured values, the objective function with the model of arrival time difference can be established as Eq. (5.2). Obviously, the unknowns *x*, *y*, *z*, and *v* should minimize the function value, to obtain the accurate and stable localization results.

$$f(x, y, z, v) = \sum_{i,j=1}^{n} \left(t_{ij}^{\text{mea}} - t_{ij}^{\text{cal}}\right)^2 = \min \tag{5.2}$$

Since the above equation is a non-negative quadratic function with the independent variables *x*, *y*, *z*, and *v*, the minimum value is always there for sure. It is feasible to obtain the source coordinates (*x*, *y*, *z*) and average velocity value *v* for an arbitrary source locating problem, as long as the number of triggered sensors is greater than four.

However, it is common that plenty of microseismic sources in the monitoring area need to be localized in the meanwhile for many practical applications. As the propagation media is usually heterogeneous and the P-waves triggered by different microseismic sources have various travel paths, then there must be differences for the average propagation velocity of different microseismic sources. The locating accuracy will be affected seriously by performing the TD locating algorithm only once due to the inaccuracy of P-wave velocity interval. Thus, the P-wave velocity interval should be optimized by performing the TD algorithm for many times, to improve the locating accuracy in the heterogeneous propagation media.

Figure 5.1 shows the flowchart of the whole localization process for the microseismic sources in the heterogeneous media using the MLM. According to the characteristic of P-wave velocity and propagation media, the velocity interval is set as [$v_{\min}^0$, $v_{\max}^0$] in the first localization process, where $v_{\min}^0$ and $v_{\max}^0$ are the lower limit and the upper limit, respectively. For example, the velocity interval can be set as [1, 5,000] in a masonry structure building. As mentioned before, the source coordinates (x_0, y_0, z_0) and corresponding average velocity value *v* of different microseismic sources can be obtained easily. Then, we can find the maximum velocity value $v_{\max}^1$ and the minimum velocity value $v_{\min}^1$ among all the velocity values in the first localization process, which is shown as:

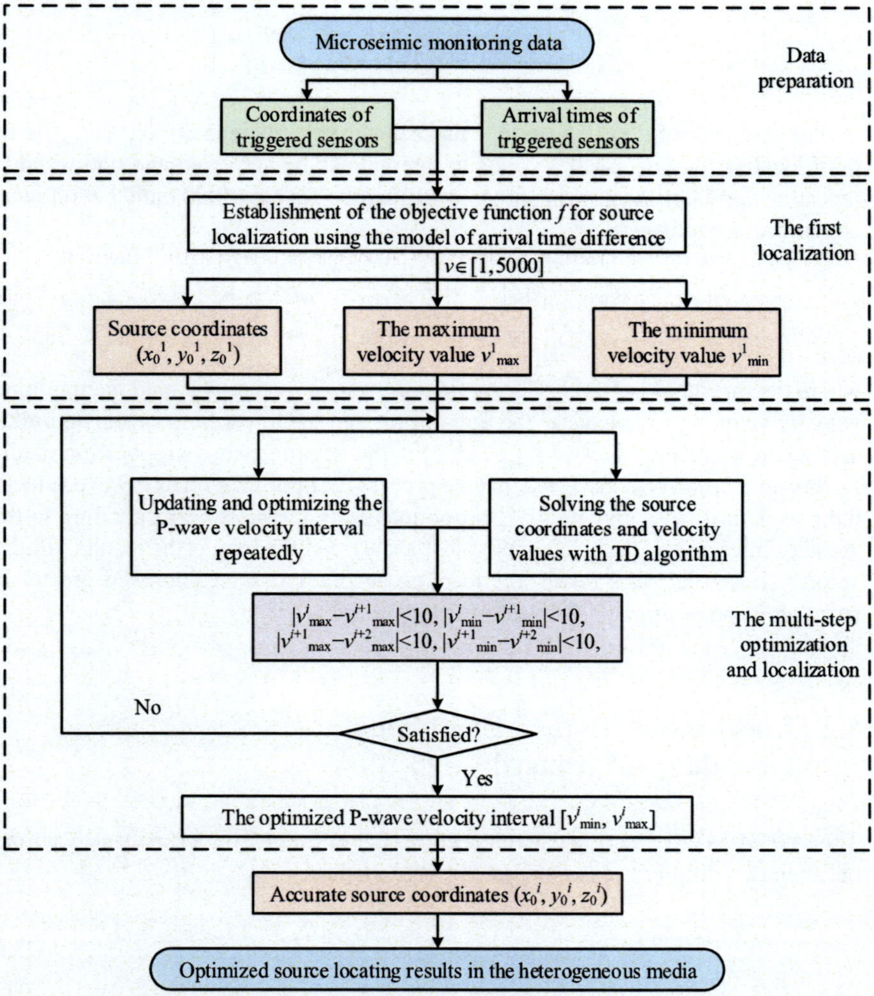

Fig. 5.1 The flowchart for locating microseismic sources in the heterogeneous media using the MLM (from [6])

$$
\begin{cases} v_{\min}^{1} = I[f(x, y, z, v)] & v \in \left[v_{\min}^{0}, v_{\min}^{0}\right] \\ v_{\max}^{1} = I[f(x, y, z, v)] & v \in \left[v_{\min}^{0}, v_{\min}^{0}\right] \end{cases}, \tag{5.3}
$$

where the function I is the inversion function used to solve the velocity value and source coordinates. Therefore, the maximum velocity value $v_{\max}^{1}$ and the minimum velocity value $v_{\min}^{1}$ can obtained and selected as the upper limit and the lower limit for the velocity interval $[v_{\min}^{1}, v_{\max}^{1}]$ in the second localization process.

Similarly, the maximum velocity value $v_{\max}^{2}$ and the minimum velocity value $v_{\min}^{2}$ can also be obtained from the second localization:

$$\begin{cases} v_{\min}^{2} = I[f(x,y,z,v)] \;\; v \in \left[v_{\min}^{1}, v_{\min}^{1}\right] \\ v_{\max}^{2} = I[f(x,y,z,v)] \;\; v \in \left[v_{\min}^{1}, v_{\min}^{1}\right] \end{cases}. \quad (5.4)$$

The obtained results can be used to make up the velocity interval $[v_{\min}^{2}, v_{\max}^{2}]$ in the third localization process. The velocity interval can be narrowed and optimized by repeating the above localization processes using the TD algorithm until the following conditions are satisfied:

$$\begin{cases} \left|v_{\max}^{i} - v_{\max}^{i+1}\right| < 10, \left|v_{\min}^{i} - v_{\min}^{i+1}\right| < 10 \\ \left|v_{\max}^{i} - v_{\max}^{i+2}\right| < 10, \left|v_{\min}^{i} - v_{\min}^{i+2}\right| < 10 \end{cases}, \quad (5.5)$$

where the threshold is 10 m/s. The maximum velocity value $v_{\max}^{i}$ and the minimum velocity value $v_{\min}^{i}$ are exactly the upper limit and the lower limit of the optimized and superior velocity interval $[v_{\min}^{i}, v_{\max}^{i}]$ in the heterogeneous propagation media. At this time, it can be considered that the process for optimization of P-wave velocity interval is finished completely. Also, the localization results corresponding to this velocity interval $[v_{\min}^{i}, v_{\max}^{i}]$ are the most accurate compared to the results of other velocity intervals. As a result, the locating accuracy can be improved greatly by narrowing and optimizing the velocity interval.

5.2 Localization Method Combining Levenberg–Marquardt Algorithm

The MSLM-MV is descripted in this section. The difference of observed arrival times between two different sensors i and j is expressed as:

$$\Delta t_{ij} = t_i - t_j = \frac{R_i - R_j}{v}. \quad (5.6)$$

where the average equivalent wave velocity v is an unknown parameter.

Equation (5.6) can be rewritten as:

$$\Delta t_{ij} = f(x_i, y_i, z_i, x_j, y_j, z_j, x_0, y_0, z_0, v). \quad (5.7)$$

In Eq. (5.7), the independent variables (x_i, y_i, z_i), (x_j, y_j, z_j), and the dependent variable Δt_{ij} are known parameters. The variables x_0, y_0, z_0, and v are unknown parameters. Since there are four unknown parameters, more than four sensors are needed to obtain the solutions by solving the nonlinear equations.

The solving methods of MSLM-MV is a nonlinear fitting problem with a single dependent variable. According to all the observed data $(x_i, y_i, z_i; x_j, y_j, z_j)$, the Eq. (5.7) can determine a regression value:

$$\Delta\hat{t}_{ij} = f(x_i, y_i, z_i, x_j, y_j, z_j, x_0, y_0, z_0, v). \tag{5.8}$$

The difference between $\Delta\hat{t}_{ij}$ and Δt_{ij} can describe the degree of deviation between the regression value and observed values.

Due to (x_i, y_i, z_i; x_j, y_j, z_j), if the degree of deviation between Δt_{ij} and $\Delta\hat{t}_{ij}$ is smaller, it shows that the fitted line and experimental points could fit better. The sum of squared deviations of all observations and fitted values is

$$Q(x_0, y_0, z_0, v) = \sum_{i,j=1}^{n} \left[\Delta\hat{t}_{ij} - \Delta t_{ij}\right]^2. \tag{5.9}$$

Equation (5.9) describes the deviations between all observed and experimental values. Therefore, the source parameters (x_0, y_0, z_0, v) can be solved out if the Q (x_0, y_0, z_0, v) reaches a minimum:

Since Eq. (5.9) is the second non-negative function of x_0, y_0, z_0, v, it always has the minimum. The coordinates of the MS/AE source location and the P wave velocity (i.e. x_0, y_0, z_0, v) can be obtained by minimizing Eq. (5.9). The common nonlinear fitting methods include LM, SM, the gradient descent method, as well as the Gauss–Newton method.

Simplex method is a direct search method [7]. Its computational process is simple and it does not require calculation of derivatives [8]. However, SM does not rely on the gradient so it may converge slowly or may not converge at all. This scenario is not uncommon and remarkably reduces the efficiency of SM in solving complicated problems such as optimizing multi-dimensional source locations or training neural networks. In the gradient descent method, the sum of the squared errors is reduced by updating the parameters in the direction of the greatest reduction of the least squares objective. In the Gauss–Newton method, the sum of the squared errors is reduced by assuming that the least squares function is locally quadratic, and finding the minimum of the quadratic.

The LM algorithm is an iterative technique that locates the minimum of a multivariate function that is expressed as the sum of squares of non-linear real-valued functions [9, 10]. It has become a standard technique for non-linear least-squares problems, widely adopted in a broad spectrum of disciplines. The LM algorithm acts more like a gradient-descent method when the parameters are far from their optimal value, and acts more like the Gauss–Newton method when the parameters are close to their optimal value. Therefore, the LM algorithm [11] was applied to solve the optimum MS/AE source positions in the method. The flowchart of LM algorithm is shown in Fig. 5.2. Some symbols in Fig. 5.2 and their calculation formulas are described in detail below. The objective function can be rewritten as

$$f(\mathbf{x}) = \frac{1}{2} r(\mathbf{x})^{\mathrm{T}} r(\mathbf{x}). \tag{5.10}$$

By comparing Eq. (5.10) with Eqs. (5.6) and (5.9), $r(\mathbf{x})$ can be calculated by

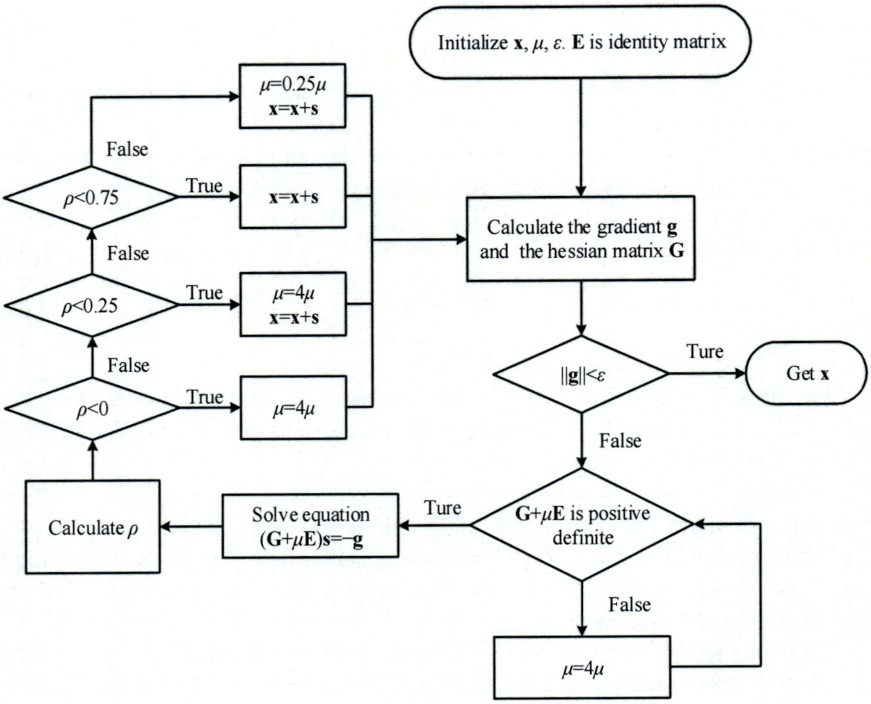

Fig. 5.2 Flowchart of LM algorithm

$$r(\mathbf{x}) = \begin{bmatrix} t_{21} - \dfrac{R_2 - R_1}{v} \\ t_{31} - \dfrac{R_3 - R_1}{v} \\ \vdots \\ t_{m1} - \dfrac{R_m - R_1}{v} \end{bmatrix}. \tag{5.11}$$

Note that the added coefficient in Eq. (5.10) does not affect the solution and the coefficient is defined as 0.5 for the convenience of calculation. Thus, gradient and Hess matrix of the objective function are derived by Eqs. (5.12) and (5.13), respectively.

$$\mathbf{g} = \nabla f(\mathbf{x}) = J(\mathbf{x})^{\mathrm{T}} r(\mathbf{x}) \tag{5.12}$$

$$\mathbf{G} = J(\mathbf{x})^{\mathrm{T}} J(\mathbf{x}) + S(\mathbf{x}) \approx J(\mathbf{x})^{\mathrm{T}} J(\mathbf{x}) \tag{5.13}$$

where $J(\mathbf{x}) = \begin{bmatrix} \frac{\partial r_1(\mathbf{x})}{\partial x_1} & \cdots & \frac{\partial r_1(\mathbf{x})}{\partial x_q} \\ \vdots & \ddots & \vdots \\ \frac{\partial r_p(\mathbf{x})}{\partial x_1} & \cdots & \frac{\partial r_p(\mathbf{x})}{\partial x_q} \end{bmatrix}$, $r(\mathbf{x}) = \begin{bmatrix} r_1(\mathbf{x}) \\ \vdots \\ r_p(\mathbf{x}) \end{bmatrix}$, $\mathbf{x} = \left(x_1, x_2, \cdots, x_q\right)$.

LM method approximates the objective function with quadratic function at the iteration point. Variable ρ measures the degree of approximation between the quadratic model and the objective function. It can be derived by

$$\rho = \frac{\|r(\mathbf{x})\|_2^2 - \|r(\mathbf{x} + \mathbf{s})\|_2^2}{\|r(\mathbf{x})\|_2^2 - \|r(\mathbf{x}) + J(\mathbf{x})\mathbf{s}\|_2^2} \tag{5.14}$$

where $\|\cdot\|_2$ means 2-norm.

5.3 Localization Method Using P-wave and S-wave Arrivals

Assume that n sensors are placed in a location area, and they are not in the same plane. 3-D coordinates of the microseismic sensors are known, which are (x_1, y_1, z_1), (x_2, y_2, z_2), …, (x_i, y_i, z_i). When the microseismic event occurs, the sensor i receive the MS/AE source signals and arrival times of P-wave and S-wave (PS waves) are recorded as $t_{\alpha i}$ and $t_{\beta i}$, respectively. The onset time is represented by t_0. The distance between MS/AE source and received station (sensor) i is R_i. The average velocities of PS waves are expressed as α and β, respectively.

The distance from MS/AE source to sensor i can be expressed as

$$R_i = \alpha(t_{\alpha i} - t_0) = \beta\left(t_{\beta i} - t_0\right). \tag{5.15}$$

The time difference of arrival times between PS waves is given by

$$t_{\beta i} - t_{\alpha i} = \frac{R_i}{\beta} - \frac{R_i}{\alpha} = R_i\left(\frac{1}{\beta} - \frac{1}{\alpha}\right). \tag{5.16}$$

Supposing an equivalent parameter consider

$$\tilde{v} = \left(\frac{1}{\beta} - \frac{1}{\alpha}\right)^{-1} = \frac{\alpha\beta}{\alpha - \beta}. \tag{5.17}$$

Equation (5.16) can be rewritten as

$$t_{\beta i} - t_{\alpha i} = \frac{R_i}{\beta} - \frac{R_i}{\alpha} = \frac{R_i}{\tilde{v}}. \tag{5.18}$$

Then, the distance between MS/AE source and received stations i, as well as the distance between MS/AE source and received stations j are obtained as follows:

$$R_i = \tilde{v}\left(t_{\beta i} - t_{\alpha i}\right), \tag{5.19}$$

$$R_j = \tilde{v}\left(t_{\beta j} - t_{\alpha j}\right). \tag{5.20}$$

According to the distance formula of two point in the space, for sensors i and j, Eqs. (5.19) and (5.20) are given as follows:

$$(x_0 - x_i)^2 + (y_0 - y_i)^2 + (z_0 - z_i)^2 = R_i^2, \tag{5.21}$$

$$\left(x_0 - x_j\right)^2 + \left(y_0 - y_j\right)^2 + \left(z_0 - z_j\right)^2 = R_j^2. \tag{5.22}$$

The left side and the right side of Eq. (5.21) added the left side and the right side of Eq. (5.22), it can be expressed as

$$\begin{aligned}&(x_0 - x_i)^2 + (y_0 - y_i)^2 + (z_0 - z_i)^2\\&+ \left(x_0 - x_j\right)^2 + \left(y_0 - y_j\right)^2 + \left(z_0 - z_j\right)^2 = R_i^2 + R_j^2.\end{aligned} \tag{5.23}$$

Submitting Eqs. (5.19) and Eq. (5.20) into Eq. (5.23), it can be rewritten as

$$\begin{aligned}&\left[(x_0 - x_i)^2 + (y_0 - y_i)^2 + (z_0 - z_i)^2 + \left(x_0 - x_j\right)^2 + \left(y_0 - y_j\right)^2\right.\\&\left.+\left(z_0 - z_j\right)^2\right] \times \left(\tilde{v}^2\right)^{-1} = \left(t_{\beta i} - t_{\alpha i}\right)^2 + \left(t_{\beta j} - t_{\alpha j}\right)^2.\end{aligned} \tag{5.24}$$

Equation (5.24) can be expressed as

$$\Delta t_{ij} = \left(t_{\beta i} - t_{\alpha i}\right)^2 + \left(t_{\beta j} - t_{\alpha j}\right)^2 = f\left(x_i, y_i, z_i, x_j, y_j, z_j, x_0, y_0, z_0, \tilde{v}\right). \tag{5.25}$$

In Eq. (5.25), sensor coordinates (x_i, y_i, z_i), (x_j, y_j, z_j), as well as dependent variable $t_{\alpha i}$, $t_{\beta i}$, $t_{\alpha j}$, and $t_{\beta j}$ are known parameters. The variables x_0, y_0, z_0, and $\tilde{v}$ are unknown parameters. There are four unknown parameters; therefore, it needs more than four sensors to get the solutions by solving the nonlinear equations. This method was called PSAFUVS (MS/AE source location method using of PS waves for arrivals unknown velocity system).

Equation (5.25) is a nonlinear fitting problem with a single dependent variable. According to all the observed data (x_i, y_i, z_i; x_j, y_j, z_j), Eq. (5.25) can determine a regression value:

$$\Delta \hat{t}_{ij} = f\left(x_i, y_i, z_i, x_j, y_j, z_j, x_0, y_0, z_0, \tilde{v}\right). \tag{5.26}$$

The difference between $\Delta \hat{t}_{ij}$ and Δt_{ij} can describe the degree of deviation between the regression value and observed values.

Due to (x_i, y_i, z_i; x_j, y_j, z_j), if the degree of deviation between Δt_{ij} and $\Delta \hat{t}_{ij}$ is smaller, it shows that the fitted line and experimental points could fit better. The sum of squared deviations of all observations and fitted values is

$$Q(x_0, y_0, z_0, \tilde{v}) = \sum_{i,j=1}^{n} \left[\Delta \hat{t}_{ij} - \Delta t_{ij}\right]^2. \tag{5.27}$$

Equation (5.27) describes the deviations between all observed and experimental values. Therefore, the MS/AE source parameters (x_0, y_0, z_0,$\tilde{v}$) can have values such that the $Q(x_0, y_0, z_0,\tilde{v})$ will reach a minimum as follows:

$$Q(x_0, y_0, z_0, \tilde{v}) = \sum_{i,j=1}^{n} \left[\Delta \hat{t}_{ij} - \Delta t_{ij}\right]^2 = \min. \tag{5.28}$$

$P_{ij}(x_0, y_0, z_0,\tilde{v})$ is a parameter to express the fitting error for each set of observations and fitted values as follows:

$$P_{ij}(x_0, y_0, z_0, \tilde{v}) = \left(\Delta \hat{t}_{ij} - \Delta t_{ij}\right)^2 = \min_{ij}. \tag{5.29}$$

Equations (5.28) and (5.29) are the nonnegative function of (x_0, y_0, z_0,$\tilde{v}$), so it always has the minimum. The coordinates of the MS/AE source location and the equivalent velocity (x_0, y_0, z_0,$\tilde{v}$) can be obtained by solving Eqs. (5.28) and (5.29). According to $\alpha = \sqrt{3}\beta$, Eq. (5.17) can be rewritten as

$$\tilde{v} = \frac{\sqrt{3}\beta^2}{\sqrt{3}\beta - \beta} = \frac{\sqrt{3}\beta}{\sqrt{3} - 1}, \tag{5.30}$$

α and β can be obtained by resolving Eq. (5.30).

Equation (5.28) is a nonlinear fitting problem. The commonly used nonlinear fitting methods include LM, SM, Quasi-Newton Method (QN), Max Inherit Optimization (MIO), Self-Organizing Migrating Algorithms (SOMA), and Global Optimization (GO) [12–14]. Dong et al. discussed the location error of the different methods [14], and analyzed results show that it is easy to trap a local optimum value using MIO, QN, and LM simply. It is suggested that the selection of MIO, QN, and LM simply should be careful. The joint use of GO as well as MIO, QN, and LM can significantly improve positioning accuracy. The SOMA, SM coupled with GO, MIO coupled with GO, and LM coupled with GO, which are more stable with the high positioning accuracy, should be recommended preferred methods to locate MS/AE source coordinates. Therefore, the SOMA coupled with GO was used to fit the MS/AE source locations using Eqs. (5.28) and (5.29) in this method.

5.4 Verification of Three Methods

5.4.1 Verification of MLM

The microseismic sources locating test was performed at a building in Kaiyang County, Guizhou Province, China, which was built in the mountainside. Several tunnels have been excavated in this mountain, which was surrounded by multiple main roads and railways. The test building is a masonry structure with tiles covered on the surface, which mainly consists of bricks and concrete. Hence, the velocity value of P-wave will not be the same due to the difference of propagation media, the thickness of concrete, and the masonry structure itself. Besides, the building has been constructed and used for decades. The possible invisible cracks and slight damage will also affect the velocity structure, which causes the greater difficulty for the accurate localization of microseismic sources in the heterogeneous and complex media.

A portable microseismic monitoring system with eight channels was laid in the building, which is composed of six sensors, a data acquisition system, and an analytical computer. There were five single-component sensors and one three-component sensor. Figure 5.3 shows the pictures of this portable microseismic monitoring system, where the graphs d and e explain the structure and hierarchies, as well as the working principle of this system, respectively. The simplified model of test building is shown in Fig. 5.4, where the blue triangles and green spheres indicate the sensors and microseismic sources, respectively. The sensors S_1 and S_3 were laid out at the staircase between the third floor and the fourth floor, as well as the staircase between the fourth floor and the fifth floor, respectively. The sensors S_2 and S_6 were distributed on the fourth floor, while the sensors S_4 and S_5 were distributed on the third floor. A total of 25 points were randomly selected as the microseismic sources for localization.

Table 5.1 lists the coordinates of sensors S_1 to S_6. Considering the components of the masonry structure building, the initial velocity interval is set as [1, 5,000], which can surely include the maximum velocity value. As long as the coordinates and arrival times of triggered sensors are obtained, the localization results can be solved using the TD algorithm. Table 5.2 lists the authentic coordinates, the localization results and errors, as well as the velocity values of selected microseismic sources for the first localization process.

Since the localization results and errors, as well as the velocity values have been solved, it is feasible to find the minimum velocity value v_{min}^1 and the maximum velocity value v_{max}^1 in the first localization process, which are 183.2 m/s and 1,590.8 m/s, respectively. As clarified in the theory of MLM, v_{min}^1 and v_{max}^1 should be taken as the lower limit and upper limit of the velocity interval for the second localization process. Similarly, the localization results, localization errors, the minimum velocity values, and the maximum velocity values can be obtained using the TD algorithm for the subsequent localization processes, until the velocity differences are less than the threshold 10 m/s. At this time, it can be regarded as the wave

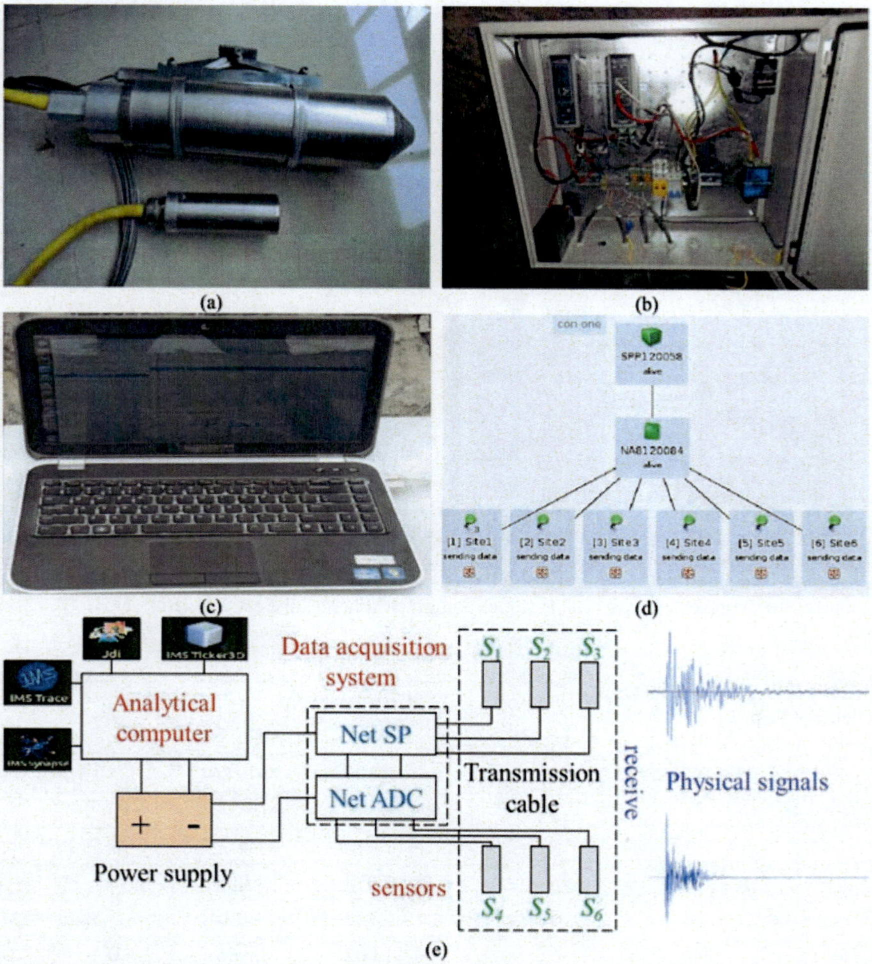

Fig. 5.3 The portable microseismic monitoring system used at the test building. The graph **a** shows the sensors, where the bigger one is the three-component sensor and the smaller one is the single-component sensor. The graphs **b** and **c** show the data acquisition system and the analytical computer, respectively. The graph **d** exhibits the structure and hierarchies of this system. The graph **e** explains the working principle of this system, where the physical signals are received by sensors and transmitted to the analytical computer. Then, the analytical results are used to guide the actual work, which achieves the effect of a cyber-physical system (from [15])

velocity and the localization algorithm are both stable, which is a significant issue for microseismic sources localization. Table 5.3 lists the velocity values and average localization errors for the localization processes from 1 to 11st. Figure 5.5 shows the change and comparison for the minimum velocity values, the maximum velocity values, the average velocity values, and the average localization errors Err_{xyz} of all the 11 localization processes. Generally, the minimum velocity value increases

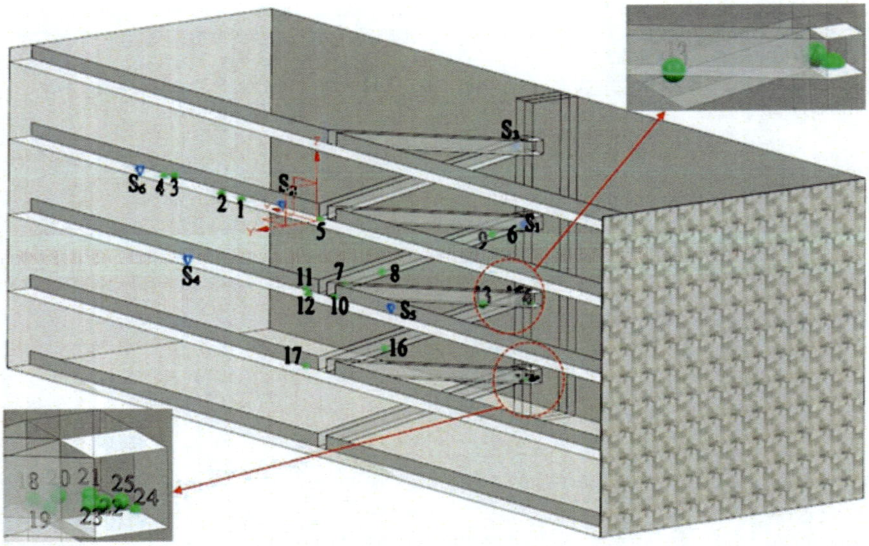

Fig. 5.4 The simplified model of test building. The locations of six sensors and 25 microseismic sources are represented with blue triangles and green spheres, respectively (from [15])

Table 5.1 The coordinates of sensors S_1 to S_6

Sensors	Coordinates/m		
	X	Y	Z
S_1	−0.500	−5.861	−1.706
S_2	2.490	0	0
S_3	1.310	−4.750	1.530
S_4	7.978	1.009	−3.278
S_5	−5.162	−0.141	−3.278
S_6	11.002	0.966	0

continuously and the maximum velocity value keeps decreasing, while the increasing and decreasing rates change from fast to slow. Besides, the minimum velocity value and the maximum velocity value tend to be stable after the 9th localization process. Obviously, it can be considered that the velocity interval composed of the minimum velocity value $v_{\min}^9$ and the maximum velocity value $v_{\max}^9$ is exactly the superior range for the actual P-wave velocity in the test building, which contains complex and heterogeneous propagation media. Therefore, the localization results corresponding to the velocity interval $[v_{\min}^9, v_{\max}^9]$ are more accurate than other localization results.

As shown in Table 5.3, the average localization error of the 9th process is the minimum among the 11 localization processes. Thus, the optimal localization results are exactly solved from the 9th process, which are listed in Table 5.4. Figure 5.6 shows the 3-D localization results for the 9th process, as well as the locations of 25 authentic coordinates. The sizes of two symbols indicate the serial numbers to distinguish

Table 5.2 The authentic coordinates, as well as the first localization results and errors of each sources

No	Authentic coordinates/m			Localization results/m			Velocity/m•s^{-1}	Localization errors/m		
	X	*Y*	*Z*	*X*	*Y*	*Z*	*v*	*X*	*Y*	*Z*
1	4.500	0	0	4.646	1.956	1.500	1,325.2	0.146	1.956	1.500
2	5.500	0	0	5.284	0.712	1.500	1,048.6	0.216	0.712	1.500
3	7.475	1.103	1.100	6.727	−2.134	1.479	353.9	0.748	3.237	0.379
4	10.003	−0.500	0	7.491	−2.883	1.384	183.2	2.512	2.383	1.384
5	0	−1.500	0	2.398	−2.657	−0.250	1,590.8	2.398	1.157	0.250
6	1.000	−5.811	−1.706	−0.602	−12.714	−5.947	1,138.5	1.602	6.903	4.241
7	2.458	−1.771	−2.943	0.071	−3.766	−7.674	1,169.1	2.387	1.995	4.731
8	2.458	−2.841	−2.476	1.185	−3.264	−6.944	637.4	1.273	0.423	4.468
9	2.458	−4.071	−1.879	0.701	−5.706	−4.767	1,187.4	1.757	1.635	2.888
10	−0.514	−0.481	−3.278	0.121	0.473	−9.464	795.7	0.635	0.954	6.186
11	1.278	−1.141	−3.278	0.439	−1.082	−8.939	618.3	0.839	0.059	5.661
12	−0.014	0.519	−3.278	0.849	0.940	−9.850	665.0	0.863	0.421	6.572
13	1.288	−3.371	−4.364	0.253	−2.525	−10.000	883.0	1.035	0.846	5.636
14	1.608	−4.851	−4.836	−1.904	−8.946	−9.319	1,369.4	3.512	4.095	4.483
15	2.508	−4.551	−4.836	2.205	−4.296	−4.615	1,021.0	0.303	0.255	0.221
16	2.508	−1.971	−5.900	1.091	−2.223	−6.364	808.6	1.417	0.252	0.464
17	1.278	−0.544	−6.375	1.149	−1.859	−6.336	579.6	0.129	1.315	0.039
18	1.938	−6.584	−8.065	−1.984	3.000	−2.820	498.3	3.922	9.584	5.245
19	1.338	−6.566	−8.065	−1.453	−10.622	−9.706	798.0	2.791	4.056	1.641
20	1.938	−7.124	−8.065	2.319	−5.596	−3.759	750.1	0.381	1.528	4.306

(continued)

Table 5.2 (continued)

No	Authentic coordinates/m			Localization results/m			Velocity/m•s^{-1}	Localization errors/m		
	X	*Y*	*Z*	*X*	*Y*	*Z*	*v*	*X*	*Y*	*Z*
21	1.338	−8.254	−8.065	3.558	−7.472	−4.775	490.6	2.220	0.782	3.29
22	1.278	−8.254	−8.065	1.768	−6.343	−0.01	764.3	0.490	1.911	1.935
23	1.278	−8.104	−8.065	−0.467	−5.237	−5.174	778.3	1.745	2.867	2.891
24	0.618	−10.234	−8.065	3.547	−9.047	−8.120	627.8	2.929	1.187	0.055
25	1.278	−10.234	−8.065	0.425	−8.983	−9.839	768.5	0.853	1.251	1.774
Average values							834.0	1.484	2.071	2.870

Note The bold values indicate the maximum velocity value v^1_{max} and the minimum velocity value v^1_{min} in the first localization process

Table 5.3 The velocity values and average localization errors for each localization process

Localization process	Velocity value (m/s)			Average location errors (m)			Err_{xyz} (m)
	v_{min}	v_{max}	v_{ave}	X	Y	Z	
1st	183.2	1,590.8	834.0	1.484	2.071	2.870	2.142
2nd	220.2	1,577.2	783.5	1.567	2.122	3.023	2.237
3rd	301.0	1,533.9	800.5	1.493	2.109	3.026	2.209
4th	420.5	1,483.2	854.9	1.511	2.026	2.986	2.174
5th	448.5	1,475.7	863.6	1.427	2.037	2.885	2.116
6th	477.2	1,474.3	854.3	1.428	2.071	2.827	2.109
7th	520.8	1,468.5	846.5	1.423	1.978	2.970	2.124
8th	535.8	1,459.3	859.6	1.576	1.800	2.908	2.095
9th	572.1	1,433.2	869.4	1.465	1.750	2.902	2.039
10th	577.3	1,424.9	863.5	1.534	1.802	2.837	2.058
11st	580.0	1,420.2	882.9	1.566	2.068	2.780	2.138

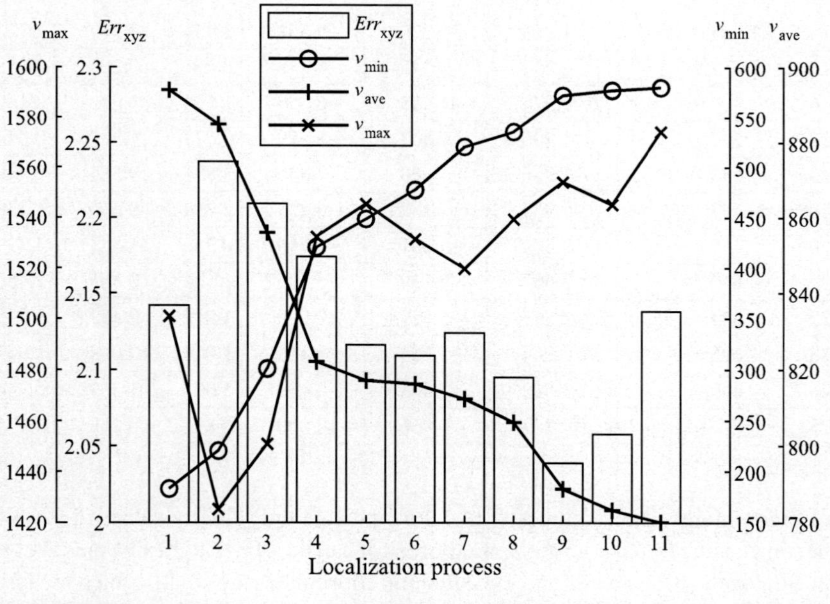

Fig. 5.5 The change and comparison of velocity values for the 11 localization processes. The broken lines with markers "o", "x", and "+" indicate the minimum velocity, the maximum velocity, and the average velocity, respectively. The histograms indicate the Err_{xyz} of each localization process

Table 5.4 The velocity values, localization results and errors for the 9th localization process

No	Velocity (m/s)	Localization results (m)			Localization errors (m)		
	V	*X*	*Y*	*Z*	*X*	*Y*	*Z*
1	1,261.1	4.892	1.168	1.498	0.392	1.168	1.498
2	916.0	5.563	0.054	1.492	0.063	0.054	1.492
3	849.7	6.084	−0.249	1.500	1.391	1.352	0.400
4	723.1	8.062	−0.850	1.500	1.941	0.350	1.500
5	1,433.2	2.590	−3.005	−0.797	2.590	1.505	0.797
6	1,117.8	−0.502	−13.287	−6.204	1.502	7.476	4.498
7	1,169.5	0.067	−3.765	−7.676	2.391	1.994	4.733
8	679.4	0.964	−3.180	−6.333	1.494	0.339	3.857
9	1,186.4	0.701	−5.677	−4.711	1.757	1.606	2.832
10	766.7	0.103	0.642	−9.982	0.617	1.123	6.704
11	596.2	0.562	−1.048	−9.258	0.716	0.093	5.980
12	655.2	0.894	0.943	−9.956	0.908	0.424	6.678
13	1,025.6	−0.475	−2.799	−10.000	1.763	0.572	5.636
14	1,345.0	−1.846	−9.270	−9.558	3.454	4.419	4.722
15	1,017.7	2.257	−4.345	−4.478	0.251	0.206	0.358
16	806.9	0.974	−1.983	−6.059	1.534	0.012	0.159
17	572.1	1.206	−1.863	−6.419	0.072	1.319	0.044
18	643.7	−1.936	3.000	−2.586	3.874	9.584	5.479
19	847.1	−0.952	−7.182	−6.829	2.290	0.616	1.236
20	752.8	2.304	−5.601	−3.746	0.366	1.523	4.319
21	643.2	2.662	−9.139	−5.605	1.324	0.885	2.460
22	644.0	2.297	−5.821	−10.000	1.019	2.433	1.935
23	695.6	0.264	−5.055	−5.118	1.014	3.049	2.947
24	609.4	3.622	−8.865	−8.409	3.004	1.369	0.344
25	777.9	0.382	−9.966	−10.000	0.896	0.268	1.935

different microseismic sources. The colors indicate the velocity values of P-wave in the travel paths for 25 microseismic sources. It can be seen that the distances between the 9th localization results and the authentic coordinates are small, which means that the localization method is optimized and the locating accuracy is improved effectively. Compared to the 1st localization process without optimization (TD method), the locating accuracy is improved by 5% using the MLM with narrowing velocity interval, which is a significant improvement for sources localization of small scale in the heterogeneous and complex media. Through the comparison between the MLM and TD methods, the novelty of MLM lies in that it can solve the global optimum rather than a local optimum, in the optimized velocity range, instead of the whole velocity range. The computation time is reduced greatly, as well as the computation

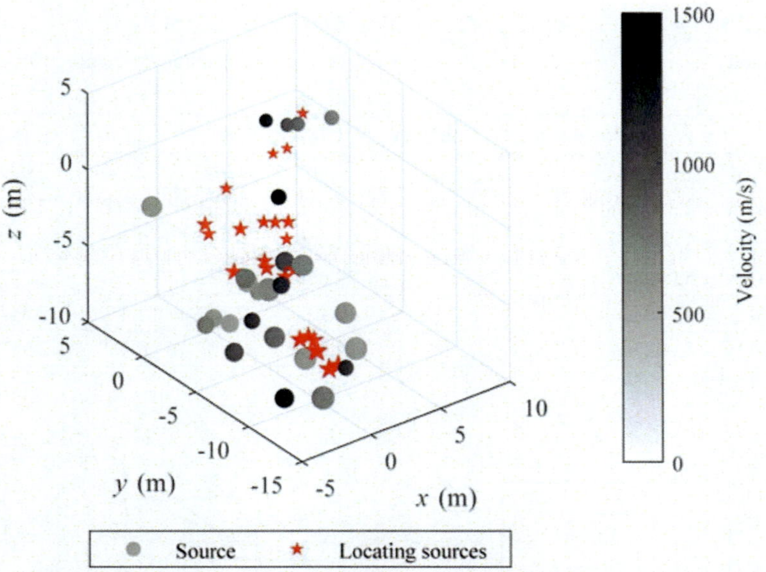

Fig. 5.6 The 3-D distributions for the 9th localization results and the authentic coordinates. The star symbols and circulars with different grayscale indicate the authentic coordinates and the 9th localization coordinates, respectively. The sizes of both two symbols indicate the serial numbers of the microseismic sources. The grayscale indicate the velocity values of P-wave in the travel paths for 25 microseismic sources

efficiency and locating accuracy is improved only with a part increase of localization program.

All the localization processes were performed with narrowing velocity intervals rather than premeasured velocity. However, the traditional method (TM) with premeasured velocity is still applied widely in many fields. It is worth comparing that the locating accuracy of the one-step localization method (the 1st localization process), the MLM (the 9th localization process), and the TM, to select the localization method with higher accuracy. As known to all, the P-wave velocity in the common media such as granite is around 3,000–4,000 m/s. Considering the heterogeneity and complexity of the propagation media in buildings, as well as the measuring accuracy for velocity, the premeasured velocity values for the TM are set as 2,000, 3,000, 4,000, and 5,000 m/s. In addition, we can add another premeasured velocity value for the TM, which is 1,350 m/s due to the optimal velocity interval is [572.1, 1,433.2]. Therefore, the localization results and errors can be solved using the TD algorithm. The comparison results of the one-step localization method without premeasured velocity, the MLM without premeasured velocity, and the TM with different premeasured velocity values are listed in Table 5.5. Figure 5.7 shows the comparison results for the locating accuracy of the above localization methods.

As clarified before, the MLM is superior to the one-step localization method without premeasured velocity. Obviously, it can be seen that the locating accuracy of

Table 5.5 The comparison results of localization errors for the three methods

Method	Average localization errors/m			Err_{xyz}/m
	x	y	z	
1.One-step localization method (TD method)	1.484	2.071	2.870	2.142
2.MLM	1.465	1.750	2.902	2.039
3.TM (v = 1,350 m/s)	1.931	2.061	3.205	2.399
4.TM (v = 2,000 m/s)	2.409	2.154	3.342	2.635
5.TM (v = 3,000 m/s)	2.662	2.332	3.217	2.737
6.TM (v = 4,000 m/s)	2.722	2.363	3.447	2.844
7.TM (v = 5,000 m/s)	2.850	2.461	3.439	2.917

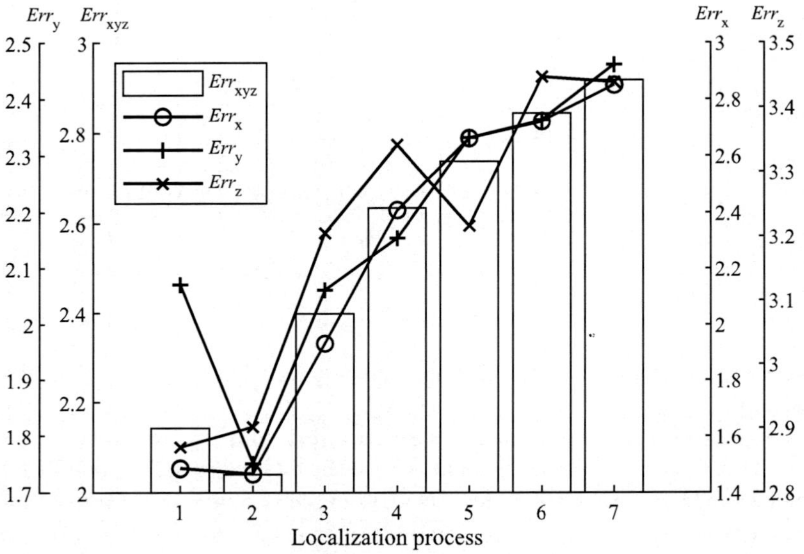

Fig. 5.7 The comparison results of locating accuracy for the three methods. The broken lines with markers "o", " + ", and "x" indicate the average localization errors for x, y, and z, respectively. The histograms indicate the Err_{xyz} of each localization method

the MLM is much higher than the TM with different values of premeasured velocity. Through the comparison between the MLM and the TM, the locating accuracy is improved by 15% for the TM with the premeasured velocity of 1,350 m/s, which is the most accurate results among all the results of TM. In addition, the one-step localization method without premeasured velocity is also more accurate than the TM. Therefore, the order for locating accuracy of three methods can be sorted from better to poorer, which is the MLM, the one-step localization method without premeasured velocity, and the TM. As the premeasured velocity is corresponding to the specific travel path of the manual source, it will not equal to the velocity values of other sources in the heterogeneous and complex media, which caused great temporal and spatial errors. It is summarized that the MLM can not only eliminate the locating errors causes by the premeasured velocity, but also can improve the locating accuracy and computation efficiency by narrowing the velocity interval, which is an efficient and effective method for locating invisible cracks and monitoring potential damage in buildings.

However, there are also several limitations of the MLM. Firstly, as for the MLM, one more sensor must be triggered compared to the TM. It can be satisfied for many common situations, but the localization difficulty will be caused for some small-scale microseismic monitoring system, where the number of sensors is limited. Secondly, the initial velocity interval should be determined appropriately when using the MLM. The determination of velocity interval depends on the characteristics of P-wave in the propagation media and the experience of calculation person. The accurate localization results will be obtained only with the reasonable and correct initial velocity interval, which include both the minimum velocity value and the maximum velocity value. Otherwise, the locating accuracy may not be improved significantly.

5.4.2 *Verification of MSLM-MV*

5.4.2.1 Acoustic Emission Experiments in Rocks

A cuboid size granite sample was used to conduct the experiments, with the dimension of 63 × 43 × 10 cm. Before starting the experiment, the upper surface was meshed into 35 grids and the grid lines were extended to the side faces F_1 to F_4. Other four lines were marked out 4 cm under the upper surface forming 48 grids, of which six points were used to arrange the AE sensors. Another six randomly selected points of the grids, 3, 9, 10, 14, 15, and 17, were used to carry out pencil lead break tests. The end vertex in lower left corner of the upper surface was set as the origin of the coordinate, the actual arrangements are shown in Fig. 5.8.

After connecting and opening all the devices, a fire of liquefied gas was lit to heat the lower surface of the sample. Five minutes later, a crack about 20 cm from F_1 to the center of the sample occurs with a loud noise. With the heating process continues, the temperature of the sample increases gradually, but the crack extends rapidly. Another ten minutes later, the second crack occurs. After heating about fifty minutes, the second crack forks into two directions. In the next two hours, the cracks extend slowly. The heating progress was terminated until the extending cracks were invisible. The actual occurrence and the extended faults are shown in Fig. 5.9.

In the experiment, pencil lead break tests were used to validate the MSLM-WV method and analyze location errors. The six pencil lead break tests source location were calculated by the MSLM-WV method using the arrivals and coordinates of sensors, and the located results and errors are listed in Table 5.6.

From the Table 5.6, it shows that the calculated coordinates using the MSLM-WV method are more consistent with practical coordinates. The minimum and maximum

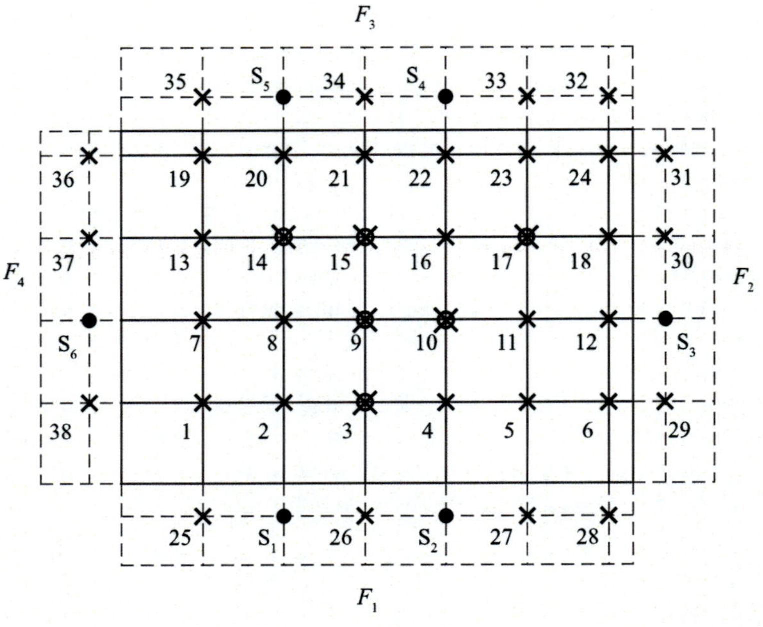

Fig. 5.8 Location of the AE sensors and pencil lead break points (the solid line shows the layout of the upper surface and the dotted line shows the layout of the side faces F_1 to F_4, S_1 to S_6 are the location of the AE sensors, points 3, 9, 10, 14, 15, and 17 are the location of pencil lead break tests)

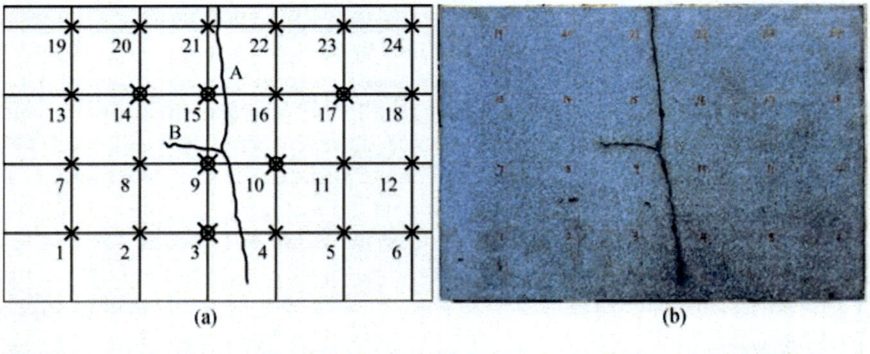

Fig. 5.9 Actual occurrence and extension of cracks (A indicates the first crack and B indicates the second one). **a** shows the fault location in grids, and **b** shows practical faults in the granite sample (from [16])

absolute errors of X coordinates are 0.56 and 4.82 cm, respectively, and the minimum and maximum absolute errors of Y coordinates are 0.66 and 4.69 cm, respectively. The average errors of X and Y using MSLM-WV are 2.29 and 2.95 cm, respectively, which are smaller than the errors of 7.04 and 4.82 cm using the TM.

Table 5.6 Real coordinates and located results using methods MSLM-WV and TM

Event No	Real coordinates (cm)		MSLM-WV (cm)				TM (cm)			
	X	*Y*	*X*	*Y*	E_X	E_Y	*X*	*Y*	E_X	E_Y
1	30	10	32.04	14.69	2.04	4.69	30.35	6.34	0.35	3.66
2	30	20	30.85	20.66	0.85	990	37.60	21.78	7.60	1.78
3	40	20	41.79	23.64	1.79	3.64	46.49	15.45	6.49	4.55
4	20	30	20.56	27.09	0.56	2.91	28.50	32.98	8.50	2.98
5	30	30	34.82	26.34	4.82	3.66	34.68	27.63	4.68	2.37
6	50	30	53.69	32.14	3.69	2.14	35.37	34.21	14.63	4.21
Average					2.29	2.95			7.04	3.26

The located results for fault location of the thermal fracture experiment are shown in Fig. 5.10. The points indicate AE events of thermal fracture. Figure 5.10 clearly indicates that most AE events of the thermal fracture occur in the center of the granite sample. It has the tendency to extend to the direction of F1 and F4, and the tendency is consistent with the actual situation shown in Fig. 5.8. From Fig. 5.10, graphs (a) and (b) show the coordinates *X* and *Y* with the colorbar, which indicates the amplitude of the events using the TM and MSLM-WV, respectively; Graphs (c) and (d) show the coordinates *X* and *Y* with the colorbar, which indicates coordinates *Z* using the TM and MSLM-WV, respectively; graphs (e) and (f) show the coordinates *X* and *Y* with the colorbar, which indicates the amplitude of events using the TM and MSLM-WV, respectively. They show that the located events of the MSLM-WV are more accurate than the results of the TM, which is much closer to the practical faults in the granite sample shown in Fig. 5.9b.

The visible and simple fact is the fire of liquefied gas heat the lower surface of the granite sample (the plane with $Z = 0$), which can induce numerous AE events in the lower surface. It is consistent with the results of the graph (f) using the MSLM-WV, while the AE events of graph (e) are very scare in the lower surface. Both pencil lead break tests and the thermal fracture experiment have proved that the MSLM-MV can improve the location accuracy significantly, which is superior to results of the TM using pre-measured wave velocity.

5.4.2.2 Blasting Experiments and Results

To validate the MSLM-WV, blasting tests were carried out at Dongguashan copper mine. No permits were required to access the mine. The field studies did not involve endangered or protected species. The copper deposit is about 1,000 m beneath the surface and it is controlled by an anticline. Its strike is NE35°–40° and the dip along the strike is about 10°. The maximum in-situ stress is 30–35 MPa, approximately

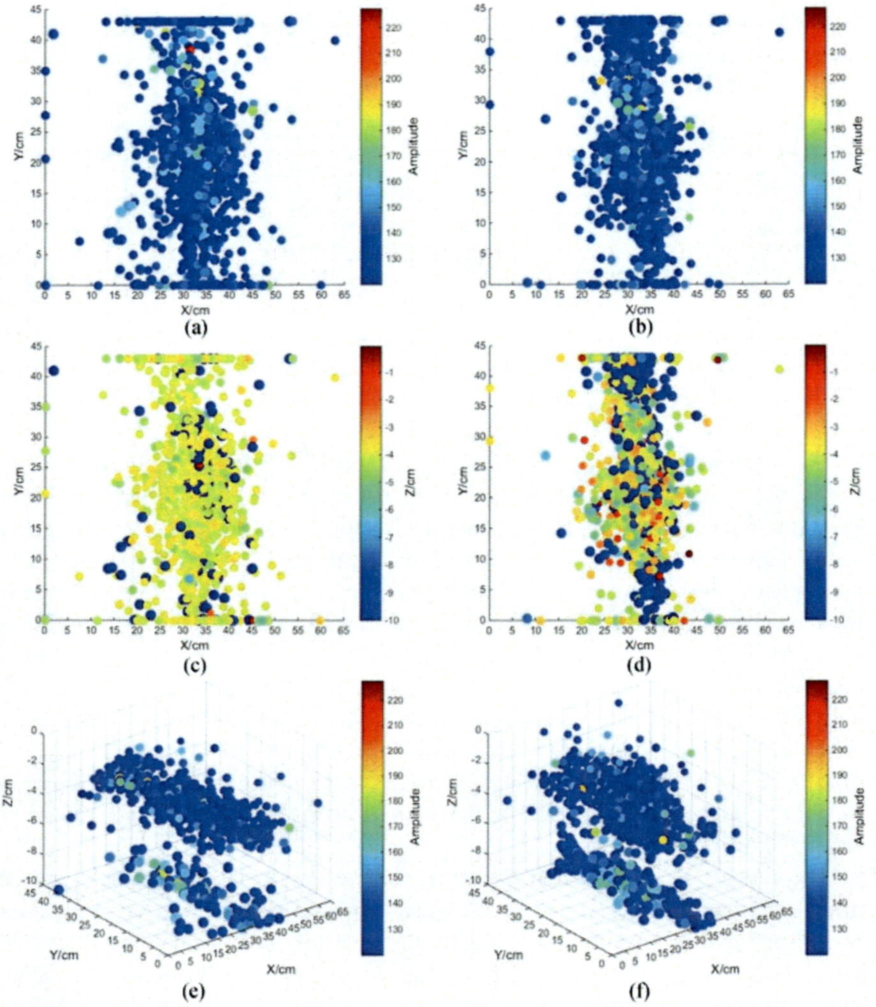

Fig. 5.10 The located results for fault location of the thermal fracture experiment, where graphs **a** and **b** show the coordinates *X* and *Y* with the colorbar indicating the amplitude of events using the methods TM and MSLM-WV, respectively; graphs **c** and **d** show the coordinates *X* and *Y* with the colorbar indicating coordinates *Z* using the methods TM and MSLM-WV, respectively; and graphs **e** and **f** show the coordinates *X*, *Y*, and *Y* with the colorbar indicating the amplitude of events using the TM and MSLM-WV, respectively (from [16])

parallel to the strike direction of the orebody, and the minimum in-situ stress is 9–16 MPa, approximately vertical. The ore and its surrounding rocks are very hard and prone to rockburst during mining operation [1].

To verify the location accuracy and practicality of the MSLM-WV, blasting tests that have mentioned in Chap. 2 were used to simulate the microseismic sources. The sensor coordinates and trigger times are shown in Table 2.4. According to records of trigger times and the sensor location coordinates, the source locations of the blasts were calculated by the MSLM-WV. To compare with TM, the calculated results of MSLM-WV and TM are listed in Table 5.7, and the actual coordinates are also listed in Table 5.7. The P-wave velocity is 5,732 m/s in calculation using the TM, which is the average value of three blasting tests. The comparisons of location errors using MSLM-WV and TM are shown in Table 5.8.

Table 5.8 shows that the three direction coordinate errors and ADEs from MSLM-WV are less than that from the TM. The average errors for the three axes are about 5 m, and the maximum is 11.14 m. The average errors of the three axes by the TM are about 8 m, the maximum is 24.90 m. The average error of absolute distances of MSLM-WV is 10.16 m, whereas the error by the TM is 17.55 m. It is concluded that the MSLM-WV is superior to the TM, and the prediction accuracy is higher than that of the TM, even when the pre-measured velocity is used in the TM.

Furthermore, to compare the fitting processes between observations and fitting values, the fitting curves from three blast events are shown in Fig. 5.11. Figure 5.11 with lines can be used to indicate the fitted differences of the two methods. Figure 5.12 gives measured value, fitted values of MSLM-WV and TM for blasting events No.1, No.2, and No.3. It can be seen that the properties of approximation for MSLM-WV is better than that of the TM. The reasons may be that the MSLM-WV can more accurately fit the relationships between the sensor coordinates and the difference of arrival times through the nonlinear algorithm. Although the basic theory of the MSLM-WV is also based on average wave velocity, it can dynamically adjust in real time, and it searches for the real time best value to meet the nonlinear relationship between the sensor coordinates and the difference of arrival times. In the calculating process of TM, the constant value of the wave velocity may induce large errors between the measured wave velocity in field and the real value, which would strongly affect the fitting accuracy.

5.4.2.3 Published Example and Comparisons

The published data in Refs. [17, 18] were also used to verify the MSLM-WV. The results were compared with the results of TM and reported results in Refs. [17, 18]. The sensor coordinates and arrival times for four events are listed in Table 5.9. The four major microseismic events, which caused by a nearby blast (5,725 m, 5,660 m, 2,088 m), were recorded during a 6 s period at a mine site. Source location results using the MSLM-WV, TM, and the methods in Refs. [17, 18] are shown in Table 5.10. Location errors of X, Y, Z as well as ADE are listed in Table 5.11.

In Table 5.11, the source location results in Refs. [17, 18] were obtained on the condition that the blast site (5,725 m, 5,660 m, 2,088 m) was used as an approximation of the origin of these events. However, in practical applications, the real microseismic source location is unknown. Source location results are also calculated by MSLM-WV and TM using (0.01 m, 0.01 m, 0.01 m) as an approximation of the origin of these events.

Table 5.7 Real coordinates and location results of methods MSLM-WV and TM

Blasting event No	Real coordinates/m			MSLM-WV/m			TM/m		
	X	*Y*	*Z*	*X*	*Y*	*Z*	*X*	*Y*	*Z*
1	84,528.4	22,556.2	−753.2	84,524.8881	22,548.5462	−746.2285	84,519.8508	22,547.4364	−749.0477
2	84,479.0	22,570.0	−814.4	84,477.8613	22,566.9001	−806.6764	84,488.3244	22,570.7159	−802.8833
3	84,359.0	22,673.0	−795.5	84,350.9794	22,667.3735	−800.7970	84,351.9548	22,676.1976	−771.8323

Table 5.8 The comparisons of location errors using MSLM-WV and TM

	MSLM-WV/m				TM/m			
Event No	X_{erro}	Y_{erro}	Z_{erro}	D_{erro}	X_{erro}	Y_{erro}	Z_{erro}	D_{erro}
1	3.5119	7.6538	6.9715	10.93233	8.5492	8.7636	4.1523	12.9279
2	1.1387	3.0999	7.7236	8.400001	9.3244	0.7159	11.5167	14.8355
3	8.0206	5.6265	5.297	11.13758	7.0452	3.1976	23.6677	24.9002

Table 5.11 shows that the calculated locations of MSLM-WV are nearest to the blast site with very similar locations. ADEs for four events 39, 40, 41, and 43 using MSLM-WV are smaller than the result of using TM, and the maximum deviations are 13.83 m and 23.80 m, respectively. ADEs of MSLM-WV for events 39, 41, and 43 are also smaller than the results reported in Refs [17, 18], and the maximum deviations are 13.83 m and 18.60 m, respectively.

It can be clearly seen in Table 5.11 that source location errors are within 12 m for three events 39, 41, and 43 using MSLM-WV, while source location errors are within 12 m for only one event using TM or the method in Refs [17, 18]. Therefore, the MSLM-WV gives a better accuracy of source location compared to any of the existing methods.

To clarify the differences between the methods TM and MSLM-WV, Table 5.12 lists the characteristics, advantages and disadvantages of two methods. It also shows the MSLM-WV is a beneficial complement to the method TM in mines.

5.4.3 Verification of PSAFUVS

In numerical experimentations, the location network includes six sensors S_1, S_2, S_3, S_4, S_5, and S_6. Their coordinates are (600, 100, 500), (300,−500, 200), (400, 300,−250), (−300,−200, 400), (200, 600,−300), and (−350,−450,−550), respectively. The average P-wave velocity and S-wave velocity in the medium are 5,000 and 2,887 m/s, respectively. The original times of MS/AE sources are supposed as 0.01 s. According to the relationship between distance formulas, the coordinates of MS/AE sources *A*, *B*, *C*, *D*, and *E* as well as their arrivals are obtained and listed in Table 5.13. The spatial distribution of MS/AE source and sensor locations are shown in Fig. 5.13 (all coordinates are in the length unit: meter). The MS/AE source locations were calculated by the PSAFUVS using arrivals in Table 5.13 and coordinates of sensors. The arrivals are accurate result with an accuracy 10^{-6} s. The location results calculated by accurate arrivals were expressed as E0 (the error is approximately equal to 0) and listed in Table 5.14. Compared results between authentic values and fitting results of dependent variables were listed in Table 5.15. It shows that the calculated coordinates (C) of *A*, *B*, *C*, *D*, and *E* using PSAFUVS are consistent with authentic results (T), and the fitting values are approximately equal to authentic values. To

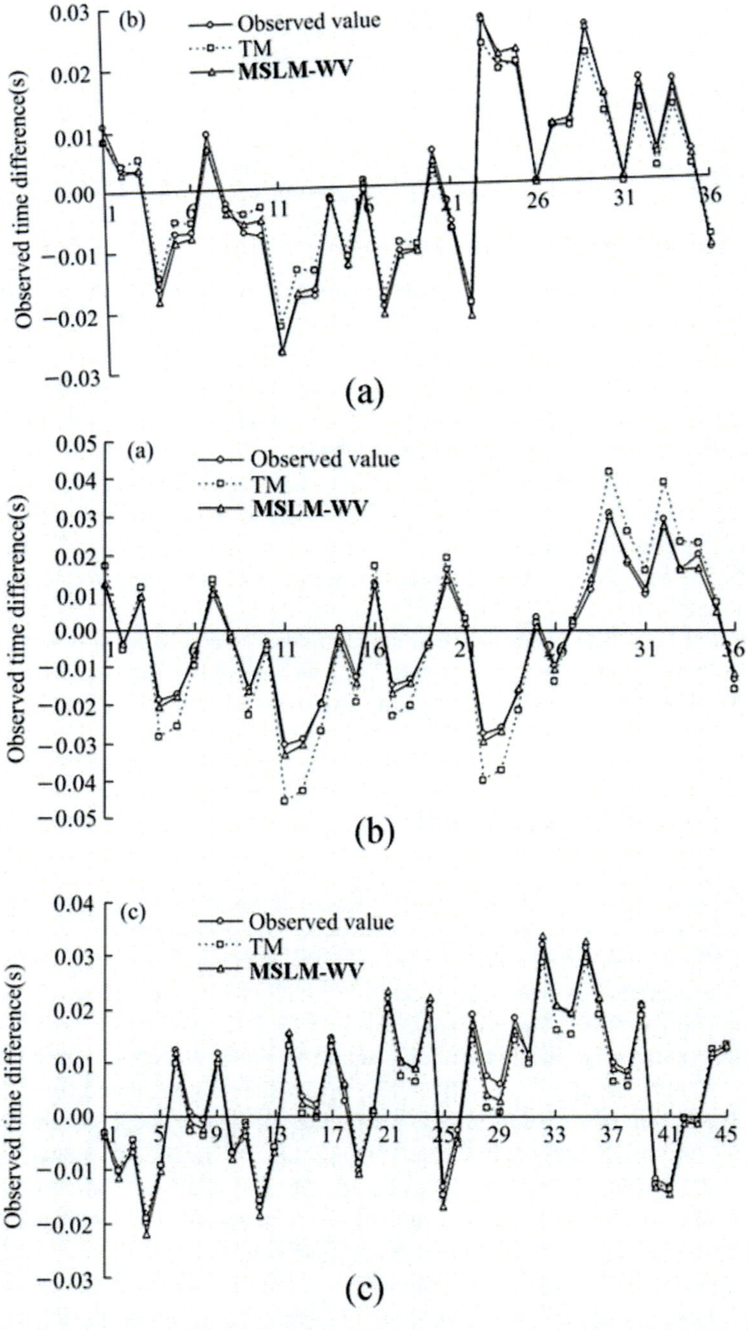

Fig. 5.11 Approximations of fitting curves for blasting events **a** No.1, **b** No.2, and **c** No.3, respectively. Note that the horizontal coordinates are the serial number for differences between arrival times (from [16])

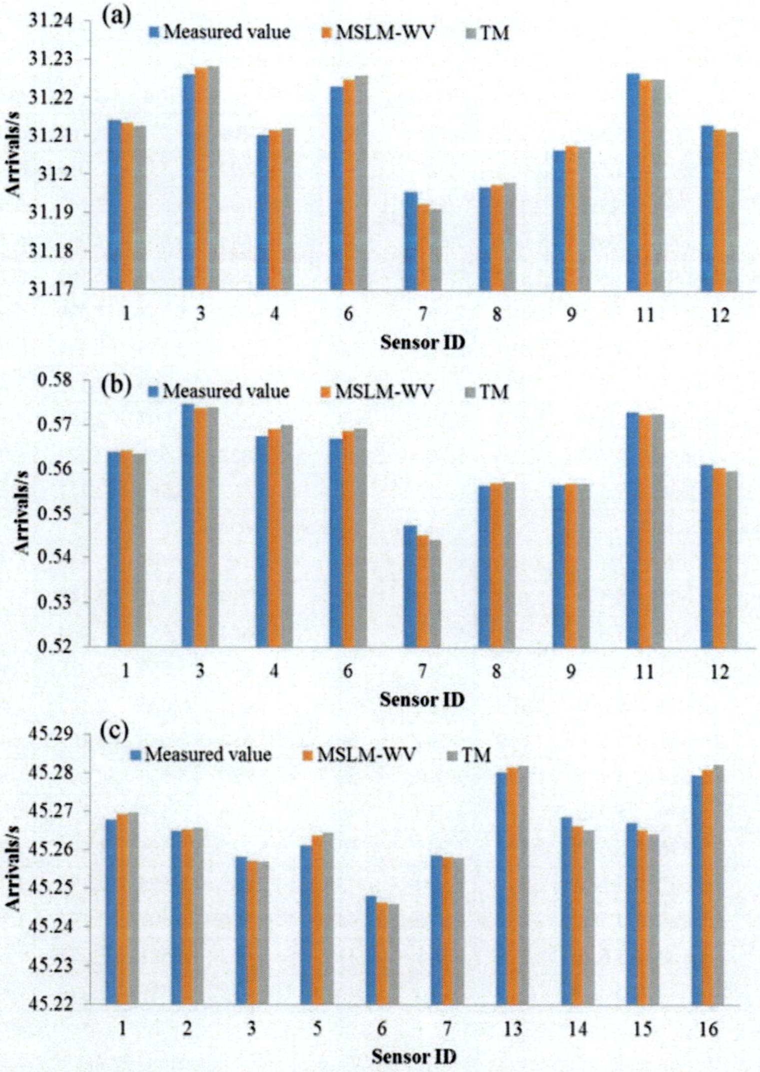

Fig. 5.12 The measured value, fitted values of MSLM-WV and TM for blasting events **a** No.1, **b** No.2, and **c** No.3, respectively (from [16])

further verify the method, α and β were recalculated using Eq. (5.30), and the calculated results were compared with the authentic α and β. $\tilde{v}$ is 6,830.127 m/s. α and β are 5,000.00 and 2,887.00 m/s which are equal to authentic values, respectively. Results prove that the PSAFUVS is reasonable and reliable with a high precision for the unknown velocity system.

Table 5.9 Sensor coordinates and arrival times for events

Sensor	Event 39				Sensor	Event 40			
	X (m)	*Y* (m)	*Z* (m)	*t* (10 μs)		*X* (m)	*Y* (m)	*Z* (m)	*t* (10 μs)
52	5,709	5,572	2,102	6,820	52	5,709	5,572	2,102	6,055
57	5,692	5,559	2,038	7,325	57	5,692	5,559	2,038	6,130
49	5,714	5,555	2,140	7,555	47	5,604	5,635	2,142	6,308
59	5,684	5,654	1,964	7,620	59	5,684	5,654	1,964	6,445
47	5,604	5,635	2,142	7,940	50	5,584	5,629	2,102	6,790
50	5,584	5,629	2,102	7,960	53	5,639	5,533	2,099	6,885
45	5,706	5,559	2,197	8,125	60	5,724	5,555	1,964	6,920
53	5,639	5,533	2,099	8,130	45	5,706	5,559	2,197	7,245
60	5,724	5,555	1,964	8,380	44	5,596	5,607	2,197	7,390
43	5,639	5,555	2,196	8,715	40	5,628	5,559	2,276	8,030
44	5,596	5,607	2,197	8,810	41	5,662	5,581	2,271	8,185
Sensor	Event 41				Sensor	Event 43			
	X (m)	*Y* (m)	*Z* (m)	*t* (10 μs)		*X* (m)	*Y* (m)	*Z* (m)	*t* (10 μs)
57	5,692	5,559	2,038	0	49	5,714	5,555	2,140	0
59	5,684	5,654	1,964	185	59	5,684	5,654	1,964	15
47	5,604	5,635	2,142	245	57	5,692	5,559	2,038	170
50	5,584	5,629	2,102	495	47	5,604	5,635	2,142	400
53	5,639	5,533	2,099	590	50	5,584	5,629	2,102	430
60	5,724	5,555	1,964	1,070	53	5,639	5,533	2,099	560
43	5,639	5,555	2,196	1,210	45	5,706	5,559	2,197	565
44	5,596	5,607	2,197	1,245	60	5,724	5,555	1,964	725
41	5,662	5,581	2,271	1,570	43	5,639	5,555	2,196	1,020
40	5,628	5,559	2,276	1,970	44	5,596	5,607	2,197	1,195
39	5,608	5,615	2,272	2,165	41	5,662	5,581	2,271	1,655
					40	5,628	5,559	2,276	2,110

Table 5.10 Source location results of MSLM-WV, TM, and methods in [17] and [18]

Event	MSLM-WV (m)			TM (m)			Refs. [17, 18] (m)		
	X	*Y*	*Z*	*X*	*Y*	*Z*	*X*	*Y*	*Z*
39	5,720.57	5,654.34	2,081.97	5,720.26	5,651.33	2,080.57	5,720	5,651	2,080
40	5,715.66	5,669.53	2,084.39	5,705.72	5,671.26	2,079.77	5,717	5,669	2,084
41	5,723.83	5,661.64	2,090.09	5,727.39	5,665.11	2,092.29	5,727	5,663	2,092
43	5,731.6	5,668.56	2,083.7	5,736.88	5,675.25	2,088.78	5,736	5,675	2,088

Table 5.11 Comparisons of location errors using different methods

Event	MSLM-WV (m)				TM (m)				Refs. [17, 18] (m)			
	X_{err}	Y_{err}	Z_{err}	D_{err}	X_{err}	Y_{err}	Z_{err}	D_{err}	X_{err}	Y_{err}	Z_{err}	D_{err}
39	4.43	5.66	6.03	9.38	4.74	8.67	7.43	12.37	5.00	9.00	8.00	13.04
40	9.34	9.53	3.61	13.83	19.28	11.26	8.23	23.80	8.00	9.00	4.00	12.69
41	1.17	1.64	2.09	2.90	2.39	5.11	4.29	7.09	2.00	3.00	4.00	5.39
43	6.60	8.56	4.30	11.63	11.88	15.25	0.78	19.35	11.00	15.00	0.00	18.60

Table 5.12 The characteristics and comparisons of the MSLM-WV and TM methods

Methods	Known parameters	Unknown parameters	advantages	disadvantages
TM	Coordinates of sensors; Arrivals of P-wave; Premeasured velocity	Source coordinates; Origin time of source	The minimum number of sensors is four	Induce locating error by the premeasured velocity; consume massive manpower and material for blasting test
MSLM-MV	Coordinates of sensors; Arrivals of P-wave	Source coordinates; Origin time of source; Average velocity	Weakening the locating error caused by premeasured velocity; Save manpower and material for blasting test	The minimum number of sensors is five

Table 5.13 Arrival times of MS/AE sources

Sensors	*A* (120, 260, 200)		*B* (202, −220, −80)		*C* (−150, 550, 520)	
	P	S	P	S	P	S
S_1	0.12764353	0.21376457	0.16455795	0.27770222	0.18497428	0.3130643
S_2	0.16620499	0.28055499	0.09158529	0.15130987	0.24726778	0.4209599
S_3	0.11630146	0.19411953	0.12636219	0.21154523	0.20574473	0.3490398
S_4	0.14084342	0.23662745	0.1489682	0.25069998	0.16484185	0.278194
S_5	0.13198361	0.2212818	0.17980035	0.30410284	0.18859451	0.3193348
S_6	0.23693611	0.40306488	0.1621189	0.27347767	0.30562814	0.522043

(continued)

Table 5.13 (continued)

Sensors	*D* (450, 336, 230)		*E* (−380, 480, 580)	
	P	S	P	S
S_1	0.0877421	0.1446533	0.2208269	0.375163
S_2	0.179976	0.3044071	0.2603757	0.4436635
S_3	0.1067876	0.177641	0.2406252	0.4094546
S_4	0.1974776	0.3347207	0.151591	0.2552427
S_5	0.1385451	0.2326466	0.2221509	0.3774561
S_6	0.2832176	0.4832267	0.3027593	0.517074

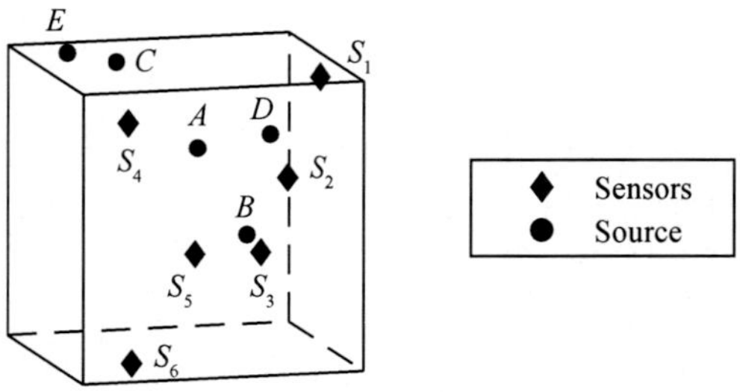

Fig. 5.13 Spatial position of sensors and MS/AE sources (unit: m)

Table 5.14 Location results and error of MS/AE source coordinates under different reading errors of arrivals

Sources	Locations (m)					
	E0			E1		
	X	*y*	*z*	*x*	*y*	*z*
A	120.00	260.00	200.00	120.02	259.98	199.96
B	202.00	−220.00	−80.00	202.04	−220.02	−79.99
C	−150.00	550.00	520.00	−149.95	549.99	520.01
D	450.00	336.00	230.00	450.00	336.06	230.01
E	−380.00	480.00	580.00	−380.03	479.97	579.94
Sources	Locations (m)					
	E2			E3		
	X	*y*	*z*	*x*	*y*	*z*
A	119.61	259.73	199.69	121.21	256.84	199.76
B	201.86	−220.14	−79.85	197.19	−217.28	−77.64
C	−150.38	550.81	520.54	−148.56	548.38	521.34
D	450.40	336.15	230.12	459.04	347.10	233.46
E	−379.32	479.77	579.35	−379.73	484.63	585.06

(continued)

Table 5.14 (continued)

Sources	Errors (m)					
	E0			E1		
	X	*y*	*z*	*x*	*y*	*z*
A	0.00	0.00	0.00	−0.02	0.02	0.04
B	0.00	0.00	0.00	−0.04	0.02	−0.01
C	0.00	0.00	0.00	−0.05	0.01	−0.01
D	0.00	0.00	0.00	−1.6	1.27	−0.42
E	0.00	0.00	0.00	0.28	−0.25	−2.89
Sources	Errors (m)					
	E2			E3		
	X	*y*	*z*	*x*	*y*	*z*
A	0.39	0.27	0.31	−1.21	3.16	0.24
B	0.14	0.14	−0.15	4.81	−2.72	−2.36
C	0.38	−0.81	−0.54	−1.44	1.62	−1.34
D	−0.40	−0.15	−0.12	−9.04	−11.10	−3.46
E	−0.68	0.23	0.65	−0.27	−4.63	−5.06

Table 5.15 Comparisons between authentic values and fitting results for the E0 condition

Sensor i ($j = 1$)	*A*		*B*		*C*	
	T	C	T	C	T	C
1	0.01483367	0.01483367	0.02021846	0.02021846	0.0238239	0.0238239
2	0.02049275	0.02049275	0.01098386	0.01098386	0.03758577	0.0375858
3	0.01347249	0.01347249	0.01467298	0.01467298	0.02795032	0.0279503
4	0.01659141	0.01659141	0.01776619	0.01776619	0.02026553	0.0202655
5	0.015391	0.015391	0.02286794	0.02286794	0.02450985	0.0245098
6	0.0350156	0.0350156	0.01981761	0.01981761	0.05425221	0.0542522
Sensor i ($j = 1$)	*D*		*E*			
	T	C	T	C		
1	0.0106557	0.0106557	0.0312364	0.0312364		
2	0.0228999	0.0228999	0.0410112	0.0410112		
3	0.012437	0.012437	0.0359202	0.0359202		
4	0.0262525	0.0262525	0.0181605	0.0181605		
5	0.0162719	0.0162719	0.0315365	0.0315365		
6	0.0474205	0.0474205	0.0533476	0.0533476		

1, 2, 3, 4, 5, and 6 indicating the values calculated using Eq. (5.26), and i is equal 1, 2, 3, 4, 5, and 6, while j is equal 1

As we all know, it is always difficult to read the accurate arrivals because of limitations of technologies, personal error, or machine error. In present work, the errors are considered as three levels within E1 [10^{-6}, 10^{-5}], E2 [10^{-5}, 10^{-4}], and E3 [10^{-4}, 10^{-3}], which are randomly generated by random function of Microsoft Excel. The calculated location results of the three levels were also listed in Table 5.14. Compared results between authentic values (T) and fitting results of MS/AE sources *A*, *B*, *C*, *D*, and *E* for error levels E0, E1, E2, and E3 are shown in Fig. 5.14. From Table 5.14 and Fig. 5.14, we can conclude that the location errors enlarged rapidity with the increase of arrival errors. The error E1 results in errors within 3 m, and the minimum and maximum are −0.28 and −2.98 m, respectively. The error E2 results in errors within 55 m, and the minimum and maximum are −3.99 and 54.08 m, respectively. The error E3 could induce errors more than 500 m, and the minimum and maximum are −1.62 and 530.39 m, respectively. Figure 5.14 also shows that the differences between fitted values and authentic values increased with the enlarging of arrival errors. The induced location errors were further investigated by ADEs between real MS/AE sources and fitted MS/AE sources in Fig. 5.15. It also clearly shows the ADEs enlarging exponentially with the increase of arrival errors, and the location errors for *A*, *B*, *C*, *D*, and *E* are increased from 0.19, 0.13, 0.28, 0.64, and 0.47 m to 82.49, 331.61, 656.54, 576.49, and 266.39 m, respectively.

5.5 Conclusions

Three iterative localization methods were developed to address the difficult problem of the location errors for microseismic monitoring system induced by wave velocity measurement deviation. After introducing the theory of the methods and validating them by numerical tests, experiments and blasting tests, some conclusions are drawn.

(1) For the MLM with narrowing the velocity interval on the basis of the TD method, the minimum velocity value and the maximum velocity value of the former localization process are taken as the lower limit and upper limit for the velocity interval, respectively, which is exactly the interval of the latter localization process. Similarly, the velocity interval can be optimized continuously, until the minimum and the maximum velocity differences between the adjacent localization processes are both less than the threshold value. At this time, it can be considered that the obtained velocity interval is the most suitable for the propagation media. Also, the localization results corresponding to this velocity interval are the most accurate among all the results of other localization processes. A source locating test was performed at a masonry structure building. Results show that the locating accuracy is improved by 5% compared to the one-step localization method without premeasured velocity. In addition, the locating results and errors of the TM with different premeasured velocity values were solved and discussed. It is shown that the MLM is obviously superior to the TM, where the locating accuracy is improved by 15% at least. It is concluded that the multi-step

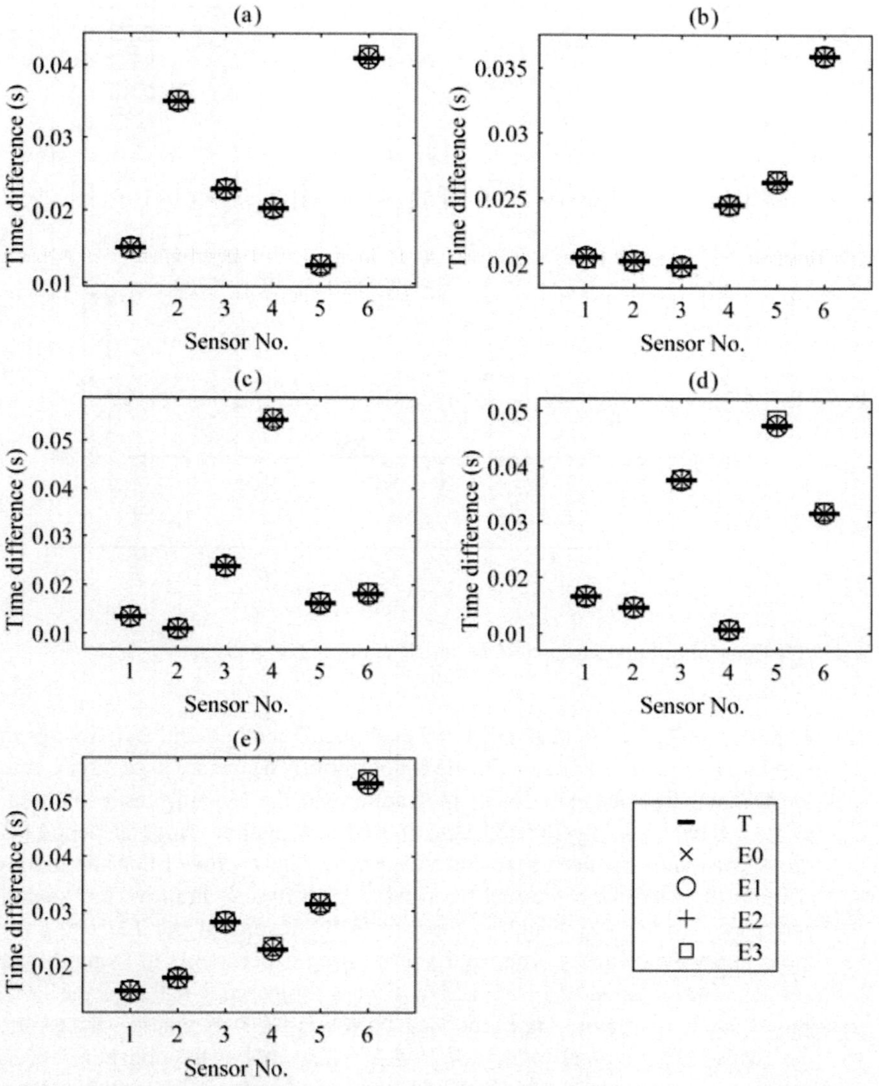

Fig. 5.14 Comparisons between authentic values and fitting results: **a** source *A*, **b** source *B*, **c** source *C*, **d** Source *D*, and **e** source *E*

can not only eliminate the locating error caused by the premeasured velocity, but also can improve the locating accuracy and efficiency for sources in the heterogeneous and complex media by optimizing the velocity interval, which is a beneficial complement for the localization theory of microseismic sources. Furthermore, it can provide some useful ideas for ensuring the stability and safety of complicated and focused structures, where the invisible cracks and potential damage may occur.

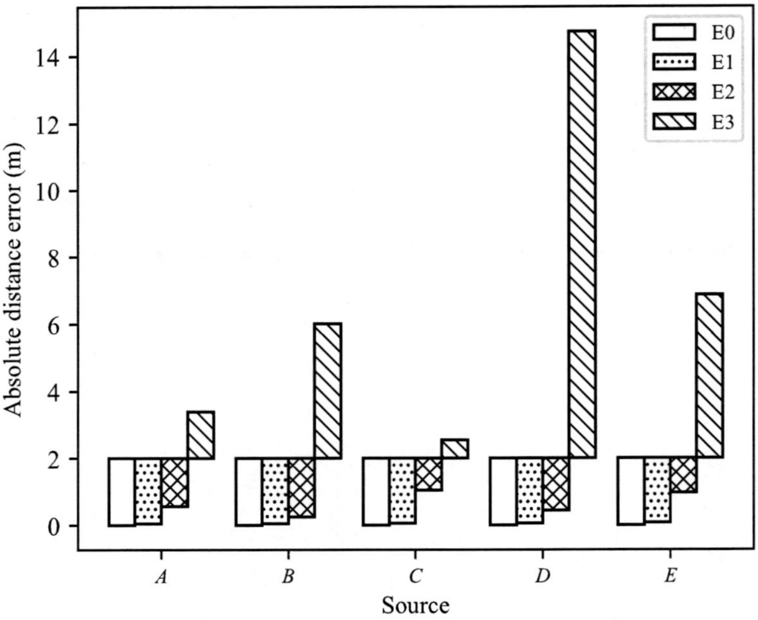

Fig. 5.15 Comparisons of ADE for different arrival errors

(2) The MSLM-MV is another method based on TD method and it optimizes the nonlinear objective function by LM algorithm. The pencil lead break tests, the thermal fracture experiment in granite, and the blasting tests (including published examples) were conducted to verify the method. Both the pencil lead break tests and the thermal fracture experiment have proved that the method without the need for a pre-measured wave velocity can improve the location accuracy significantly, which is superior to the results of the TM using pre-measured wave velocity. The results of blasting experiments in Dongguashan copper mine obtained by the MSLM-WV were compared with that by the TM. It shows that the average ADE of the MSLM-WV is 10.16 m, whereas the error of the TM is 17.55 m. Finally, the MSLM-WV is also applied to published data and it also gives superior results compared to the considered existing methods. It was proved that the MSLM-WV is reasonable and reliable with high location accuracy. Because it can accurately locate the MS/AE source coordinates without the need of a pre-measured velocity, the method can not only decrease the location errors induced by temporal and spatial errors of the wave velocity using the TM, but also determine the location of the MS/AE source in real time, which is a more efficient and effective methodology for detecting faults/breaking signals for both the healthy evaluation and disaster control than the most commonly used TM in mines.

(3) For the PSAFUVS which use PS waves arrivals for source localization, it can not only locate the MS/AE source for unknown velocity network but also determine

the PS wave real-time velocities for each event. Effects of arrival errors on location accuracy were investigated using PSAFUVS, and it shows location errors enlarged rapidity with the increase of arrival errors. The ADEs between real MS/AE sources and fitted MS/AE sources were further discussed, and results show the ADEs enlarging exponentially with the increase of arrival errors.

References

1. Tang L-Z, Pan C-L, Xie X-B (2003) Study on rockburst control in deep seated hard ore deposit [J]. Chin J Rock Mech Eng 22(7):l067-1071
2. Aki K, Lee W (1976) Determination of three-dimensional velocity anomalies under a seismic array using first P arrival times from local earthquakes: 1. A homogeneous initial model. J Geophys Res 81(23):4381–4399
3. Pavlis GL, Booker JR (1980) The mixed discrete-continuous inverse problem: application to the simultaneous determination of earthquake hypocenters and velocity structure. J Geophys Res: Solid Earth 85(B9):4801–4810
4. Tarantola A, Valette B (1982) Inverse problems= quest for information. J Geophys 50(1):159–170
5. Pujol J (1988) Comments on the joint determination of hypocenters and station corrections. Bull Seismol Soc Am 78(3):1179–1189
6. Dong L, Shu W, Han G, Li X, Wang J (2017) A multi-step source localization method with narrowing velocity interval of cyber-physical systems in buildings. IEEE Access 5:20207–20219. https://doi.org/10.1109/access.2017.2756855
7. Nelson GD, Vidale JE (1990) Earthquake locations by 3-D finite-difference travel times. Bull Seismol Soc Am 80(2):395–410
8. Nelder JA, Mead R (1965) A simplex method for function minimization. Comput J 7(4):308–313
9. Jordanov I, Georgieva A (2007) Neural network learning with global heuristic search. IEEE Trans Neural Netw 18(3):937–942
10. Levenberg K (1944) A method for the solution of certain problems in least squares. Q Appl Math 2:164–168
11. Marquardt DW (1963) An algorithm for least-squares estimation of nonlinear parameters. J Soc Ind Appl Math 11(2):431–441
12. Li X-B, Dong L-J, Gong F-Q, Zhou Z-L A location method for acoustic emission and microseismic sources. China Patent CN102129063B
13. Dong L-J, Li X-B, Tang L-Z, Gong F-Q (2011) Mathematical functions and parameters for microseismic source location without pre-measuring speed. Chin J Rock Mech Eng 30(10):2057–2067
14. Dong L-J, Li X-B, Tang L-Z (2013) Main influencing factors of microseismic source location accuracy. Sci Technol Rev 31(24):26–32
15. Dong L-J, Li X-B (2013) A microseismic/acoustic emission source location method using arrival times of PS waves for unknown velocity system. Int J Distrib Sens N. https://doi.org/Artn30748910.1155/2013/307489
16. Dong L-J, Sun D-Y, Li X-B, Du K (2017) Theoretical and experimental studies of localization methodology for AE and microseismic sources without pre-measured wave velocity in mines. IEEE Access 5:16818–16828. https://doi.org/10.1109/Access.2017.2743115

17. Brandstein MS, Adcock JE, Silverman HF (1997) A closed-form location estimator for use with room environment microphone arrays. IEEE Trans Speech Audio Process 5(1):45–50
18. Flanagan MP, Myers SC, Koper KD (2007) Regional travel-time uncertainty and seismic location improvement using a three-dimensional a priori velocity model. Bull Seismol Soc Am 97(3):804–825

Chapter 6
Collaborative Localization Method Using Analytical and Iterative Solutions

Although iterative localization methods can obtain a more accurate result than analytical methods, time-consuming calculation and divergent result would limit their application. In addition, there are some uncertainties for the data of arrivals in the practical mining process, which are recorded by the sensors of the microseismic monitoring system. Firstly, the construction environment is not only complex, but also influenced by the mining stress and geological structure [1, 2]. Thus, there is a lot of background noise in the monitoring area. Secondly, the sensors are easy to be triggered mistakenly by the interference signals in the monitoring area, hence external abnormal signals would be recorded [3]. All these problems could lead to the generation of the abnormal arrivals. Then, the calculation becomes more time-consuming, or even get more divergent results. It is difficult to identify the data of abnormal arrivals in the complex mining environment, which affects the locating accuracy greatly. Therefore, the abnormal arrivals recorded by sensors commonly in mining environment should be considered and removed carefully.

In this chapter, the collaborative localization method using analytical and iterative solutions (CLMAI) is introduced, which combined with the arrivals of multi-sensor and inversion of the real-time average wave velocity, to seek the optimal locating results. The analytical localization method is used to remove abnormal arrivals since it has a stable solution with the high precision when the input data are accurate. To weaken errors induced by the dynamic wave velocity, the iterative solution without the need of premeasured P-wave velocity is used to improve the locating accuracy since it can optimize results using the advantage of multiple sensors. The detailed description of the method is shown in Sect. 6.1. The CLMAI is verified through locating coordinates of blasts in Sect. 6.2 and microseismic events in Kaiyang phosphorous mine in Sect. 6.3.

L. Dong and X. Li, *Velocity-Free Localization Methodology for Acoustic and Microseismic Sources*, https://doi.org/10.1007/978-981-19-8610-9_6

6.1 Theory of the CLMAI

To eliminate the effects of abnormal arrivals, the analytical localization method is developed and used to remove abnormal arrivals since it has a stable solution with the high precision when the input data are accurate. To weaken errors induced by a dynamic wave velocity, the iterative solution is used to improve the locating accuracy since it can optimize results using the advantage of multiple sensors. The CLMAI was proposed, which combined with the arrivals of multi-sensor and inversion of the real-time average wave velocity, to seek the optimal locating results. Figure 6.1 shows the flowchart of the CLMAI, which includes two main processes: filtering the abnormal arrivals using the analytical solutions, and locating MS/AE using iterative localization method with clear arrivals.

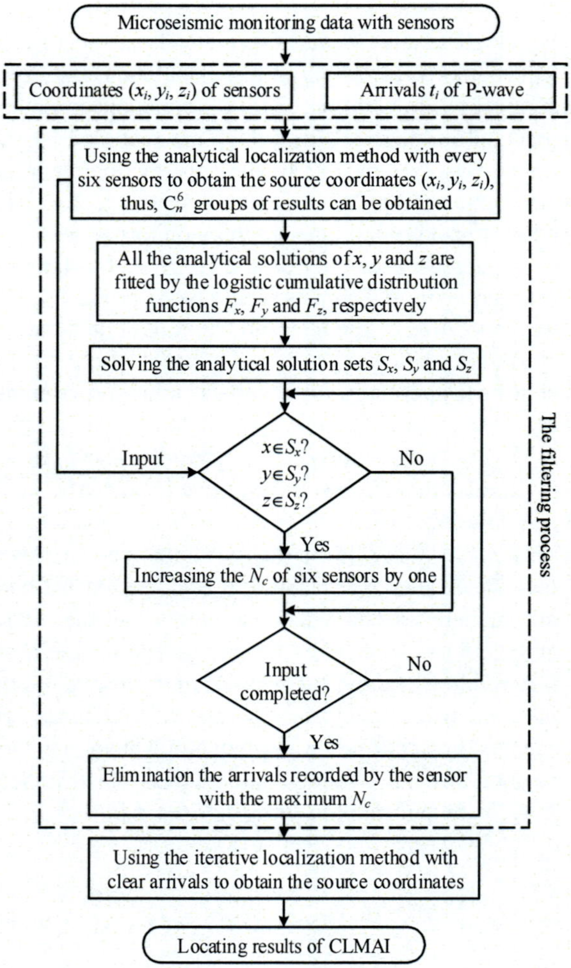

Fig. 6.1 The flowchart for the locating process of CLMAI

The specific steps of the CLMAI shown in Fig. 6.1 are detailed as follows:

(1) For a monitoring system with n sensors, the analytical localization method using six sensors is applied to obtain the source coordinates (x, y, z). Thus, C_n^6 groups of locating results can be obtained.
(2) The groups of locating results obtained in the first step are then fitted by the logistic cumulative distribution function (i.e. three logistic functions of F_X, F_Y, and F_Z can be obtained by fitting the results of x, y, and z, respectively).
(3) The solution sets S_x, S_y, and S_z can be calculated from F_X, F_Y, and F_Z according to the proposed method in Sect. 6.1.1.3. S_x, S_y, and S_z are used to characterize the abnormal locating results which deviate from the correct locating results.
(4) The locating result (x, y, z) is tagged as abnormal locating result if $x \in S_x$ or $y \in S_y$ or $z \in S_z$. . Then the count numbers (N_c) of the six sensors used to obtain the abnormal locating result are increased by 1, respectively. All of the C_n^6 groups of locating results are set as input to determine whether it is abnormal locating result. Thus, N_c is used to indicate the number of occurrences of the sensor in all of the abnormal locating results.
(5) If all of the C_n^6 groups of locating results have been determined (input completed), the arrival corresponding to the sensor with the maximum N_c is defined as abnormal arrival, as it has the most negative effect on the locating accuracy. The abnormal arrival is then eliminated to improve the location accuracy.
(6) After filtering the abnormal arrival, the iterative localization method is used to locate the source with the clear $n - 1$ arrivals to obtain the locating results with CLMAI.

6.1.1 Filtering the Abnormal Arrivals Using the Analytical Solutions

6.1.1.1 The Analytical Solutions

Leighton and Blake [4, 5] developed the USBM method to locate MS/AE. They supposed wave velocity value as a known parameter. However, the mining activities are intensive and usually carried out at numerous work sites in different stopes and levels at the same time, which indicates that the effect of dynamic disturbance cannot be ignored. The propagation paths are different between the known velocity measured by blasts and the events located at different coordinates. Due to the difference of rockmass structure (goafs, faults, tunnels, joint, fractures, etc.), the values of wave velocity will not be the same for different travel paths. In particular, the velocity increases in the stress concentration areas while decreases in the fracture areas of rockmass [6]. Therefore, it is inaccurate to use the average velocity premeasured through the blasts to locate the future events, where the velocity is different from the pre-measured velocity.

In this method, the wave velocity was taken as an unknown and real-time parameter. The coordinates of the MS/AE source is assumed as $P(x, y, z)$, while the coordinates of the six sensors are assumed as S_i (x_i, y_i, z_i) $(i = 1, 2, 3, 4, 5, 6)$. For an arbitrary event, the source coordinates satisfy the following equations:

$$(x_1 - x)^2 + (y_1 - y)^2 + (z_1 - z)^2 = v^2(t_1 - t_0)^2, \tag{6.1}$$

$$(x_2 - x)^2 + (y_2 - y)^2 + (z_2 - z)^2 = v^2(t_2 - t_0)^2, \tag{6.2}$$

$$(x_3 - x)^2 + (y_3 - y)^2 + (z_3 - z)^2 = v^2(t_3 - t_0)^2, \tag{6.3}$$

$$(x_4 - x)^2 + (y_4 - y)^2 + (z_4 - z)^2 = v^2(t_4 - t_0)^2, \tag{6.4}$$

$$(x_5 - x)^2 + (y_5 - y)^2 + (z_5 - z)^2 = v^2(t_5 - t_0)^2, \tag{6.5}$$

$$(x_6 - x)^2 + (y_6 - y)^2 + (z_6 - z)^2 = v^2(t_6 - t_0)^2, \tag{6.6}$$

where t_i $(i = 1, 2, 3, 4, 5, 6)$ is the arrival corresponding to the sensor S_i. t_0 is the origin time of the MS/AE event. The symbol v represents the average velocity of P-wave in the travel path.

By taking difference between the Eqs. (6.2)–(6.6) and the Eq. (6.1), it is feasible to obtain the Eq. (6.7).

$$2x(x_j - x_1) + 2y(y_j - y_1) + 2z(z_j - z_1) + 2v^2(t_j - t_1)t_0 + v^2\left(t_j^2 - t_2^2\right) = l_j \tag{6.7}$$

where $j = (2, 3, 4, 5, 6)$ and $l_j = \left(x_j^2 - x_1^2\right) + \left(y_j^2 - y_1^2\right) + \left(z_j^2 - z_1^2\right)$.

Obviously, the parameters including l_2, l_3, l_4, l_5, and l_6 are constants for the specific sensors network. By substituting V for v^2 and S for Vt_0, the Eq. (6.7) can be rewritten as:

$$2x(x_j - x_1) + 2y(y_j - y_1) + 2z(z_j - z_1) + 2(t_j - t_1)S + V(t_1^2 - t_j^2) = l_j. \tag{6.8}$$

The Eq. (6.8) is a five-variable linear equation, which can be solved according to the Cramer's rule.

We can get the source coordinates P (x, y, z), the wave velocity v, and the origin time t_0, which are shown as the Eqs. (6.9), (6.10), and (6.11).

$$x = \frac{D_z}{D}, \quad y = \frac{D_v}{D}, \quad z = \frac{D_z}{D} \tag{6.9}$$

$$v = \pm\sqrt{\frac{D_V}{D}} \tag{6.10}$$

where the negative value of v should be discard since it has no practical significance.

$$t_0 = \frac{D_S}{D_V} \tag{6.11}$$

where

$$D = \begin{vmatrix} 2x_{21} & 2y_{21} & 2z_{21} & 2t_{21} & t_1^2 - t_2^2 \\ 2x_{31} & 2y_{31} & 2z_{31} & 2t_{31} & t_1^2 - t_3^2 \\ 2x_{41} & 2y_{41} & 2z_{41} & 2t_{41} & t_1^2 - t_4^2 \\ 2x_{51} & 2y_{51} & 2z_{51} & 2t_{51} & t_1^2 - t_5^2 \\ 2x_{61} & 2y_{61} & 2z_{61} & 2t_{61} & t_1^2 - t_6^2 \end{vmatrix}; \ D_x = \begin{vmatrix} l_1 & 2y_{21} & 2z_{21} & 2t_{21} & t_1^2 - t_2^2 \\ l_2 & 2y_{31} & 2z_{31} & 2t_{31} & t_1^2 - t_3^2 \\ l_3 & 2y_{41} & 2z_{41} & 2t_{41} & t_1^2 - t_4^2 \\ l_4 & 2y_{51} & 2z_{51} & 2t_{51} & t_1^2 - t_5^2 \\ l_5 & 2y_{61} & 2z_{61} & 2t_{61} & t_1^2 - t_6^2 \end{vmatrix};$$

$$D_y = \begin{vmatrix} 2x_{21} & l_1 & 2z_{21} & 2t_{21} & t_1^2 - t_2^2 \\ 2x_{31} & l_2 & 2z_{31} & 2t_{31} & t_1^2 - t_3^2 \\ 2x_{41} & l_3 & 2z_{41} & 2t_{41} & t_1^2 - t_4^2 \\ 2x_{51} & l_4 & 2z_{51} & 2t_{51} & t_1^2 - t_5^2 \\ 2x_{61} & l_5 & 2z_{61} & 2t_{61} & t_1^2 - t_6^2 \end{vmatrix}; \ D_z = \begin{vmatrix} 2x_{21} & 2y_{21} & l_1 & 2t_{21} & t_1^2 - t_2^2 \\ 2x_{31} & 2y_{31} & l_2 & 2t_{31} & t_1^2 - t_3^2 \\ 2x_{41} & 2y_{41} & l_3 & 2t_{41} & t_1^2 - t_4^2 \\ 2x_{51} & 2y_{51} & l_4 & 2t_{51} & t_1^2 - t_5^2 \\ 2x_{61} & 2y_{61} & l_5 & 2t_{61} & t_1^2 - t_6^2 \end{vmatrix};$$

$$D_S = \begin{vmatrix} 2x_{21} & 2y_{21} & 2z_{21} & l_1 & t_1^2 - t_2^2 \\ 2x_{31} & 2y_{31} & 2z_{31} & l_2 & t_1^2 - t_3^2 \\ 2x_{41} & 2y_{41} & 2z_{41} & l_3 & t_1^2 - t_4^2 \\ 2x_{51} & 2y_{51} & 2z_{51} & l_4 & t_1^2 - t_5^2 \\ 2x_{61} & 2y_{61} & 2z_{61} & l_5 & t_1^2 - t_6^2 \end{vmatrix}; \ D_V = \begin{vmatrix} 2x_{21} & 2y_{21} & 2z_{21} & 2t_{21} & l_1 \\ 2x_{31} & 2y_{31} & 2z_{31} & 2t_{31} & l_2 \\ 2x_{41} & 2y_{41} & 2z_{41} & 2t_{41} & l_3 \\ 2x_{51} & 2y_{51} & 2z_{51} & 2t_{51} & l_4 \\ 2x_{61} & 2y_{61} & 2z_{61} & 2t_{61} & l_5 \end{vmatrix};$$

where x_{j1} represents the difference between x_j and x_1. Similarly, y_{j1}, z_{j1}, and t_{j1} represent the differences between y_j and y_1, z_j and z_1, as well as t_j and t_1, respectively.

The unknown parameters including the source coordinates (x, y, z), the average wave velocity v, and the origin time t_0 can be solved through the coordinates and arrivals of six different sensors. The solved source coordinates are unique for an event. The advantage of the analytical localization method lies in that the locating results can be solved quickly through the established explicit formulas. Besides, the analytical solutions are stable with high precision when the input data are accurate.

6.1.1.2 The Fitting of the Logistic Functions

To improve the accuracy of microseismic monitoring and the breadth of monitoring range, the number of sensors is usually greater than six in the practical mining engineering. However, the possibility for the generation of abnormal arrivals increases obviously as the increase of sensors in the meanwhile. For the microseismic monitoring system with more than six sensors, it is significant to locate the source coordinates and identify the abnormal arrivals. The filtering method for abnormal arrivals can be developed after obtaining the analytical solutions with more than six sensors.

The number of sensors is assumed as n ($n > 6$). By selecting six sensors from n sensors randomly, the C_n^6 groups of locating results can be obtained using the analytical localization method. Then, the logistic probability density function is applied to fit the coordinates for all the C_n^6 groups of analytical solutions. The abscissa corresponding to the maximum value of the logistic probability density function is exactly the coordinate to be solved. The logistic probability density function and the logistic cumulative distribution function are shown as the Eqs. (6.12) and (6.13), respectively.

$$f_L(x;\mu,s) = \frac{e^{-\frac{x-\mu}{s}}}{s\left(1+e^{-\frac{x-\mu}{s}}\right)^2} \tag{6.12}$$

$$F_L(x;\mu,s) = \frac{1}{1+e^{-\frac{x-\mu}{s}}} \tag{6.13}$$

where μ and σ are the average value and standard deviation of x; s is a scale parameter related to the standard deviation σ. In addition, the relationship between parameters σ and s is shown below:

$$s = \frac{\sqrt{3}\sigma}{\pi}. \tag{6.14}$$

As the C_n^6 groups of analytical solutions have been fitted through the logistic probability density function, the logistic cumulative distribution functions of F_x, F_y, and F_z can be obtained.

6.1.1.3 The Selection of the Control Values

The abnormal arrivals can be filtered by the F_x, F_y, and F_z as there is a great deviation between the normal locating results and the analytical locating results with abnormal arrivals. The solution sets S_x, S_y, and S_z can be obtained by adding a couple of control values (a, $1 - a$) to the F_x, F_y, and F_z, to determine whether the locating result is caused by the abnormal arrivals, where:

$$S_x = \{x|F_X(x) < a \cup F_X(x) > 1-a\} \tag{6.15}$$

$$S_y = \{y|F_Y(y) < a \cup F_Y(y) > 1-a\} \tag{6.16}$$

$$S_z = \{z|F_Z(z) < a \cup F_Z(z) > 1-a\} \tag{6.17}$$

The analytical locating result (x, y, z) is tagged as abnormal locating result if $x \in S_x$ or $y \in S_y$ or $z \in S_z$. Then the count numbers (N_c) of the six sensors used to obtain the abnormal locating result are increased by 1, respectively. For all of the C_n^6 groups of locating results, the arrival corresponding to the sensor with the maximum N_c is defined as abnormal arrival since it has the most negative effect on the locating accuracy.

It can be seen that the determination of the control values $(a, 1 - a)$ is the most critical factor affecting the correctness and accuracy of the filtering method. As shown in Fig. 6.2, the red area is S_x corresponding to the control values $(a, 1 - a)$. The determination of the control values $(a, 1 - a)$ should ensure that the filtered arrival with maximum N_c is the abnormal arrival of the event (i.e. the locating results using the abnormal arrivals should make up a large proportion of the red area). If the red area is too small (i.e. a is small), the number of the locating results in the red area are so small that there are many sensors with the same maximum N_c, thus it fails to filter the abnormal arrival. On the contrary, if the red area is too large (i.e. a is large), the arrivals corresponding to the sensor with the maximum N_c may not be the abnormal arrival (The reason is that there are a large number of correct locating results in the area). A synthetic test using 1,000 groups of locating examples with different couples of control values was carried out, to obtain the appropriate control values.

The synthetic test was carried out on account of the non-repeatability of real events in an opaque medium. It's difficult and time-consuming to determine whether the filtered arrivals are the abnormal arrivals while using the real events. Therefore, synthetic test, with controllable errors in their input data, i.e. the arrivals, is flexible

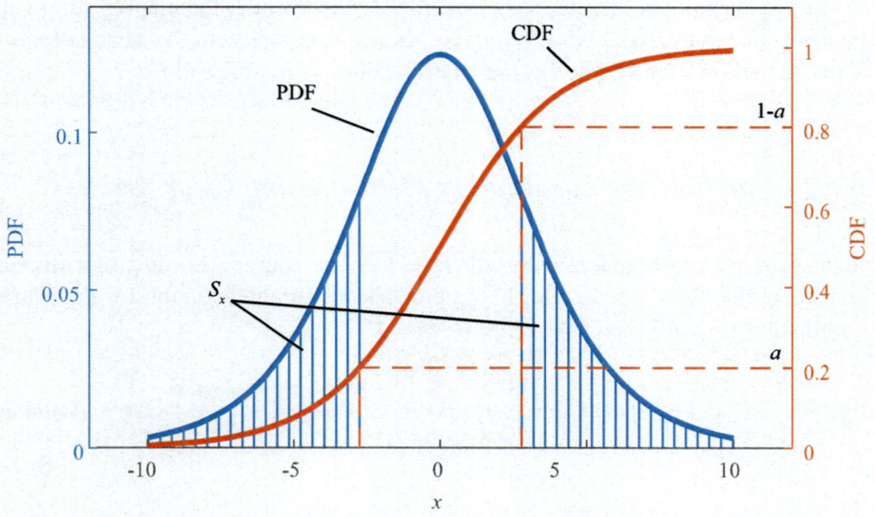

Fig. 6.2 The logistic probability density function (PDF) and corresponding cumulative distribution function (CDF)

Table 6.1 The statistical results of the 1,000 groups of examples

Item	Control values of the CDF						
	0.05, 0.95	0.1, 0.9	0.15, 0.85	0.2, 0.8	0.25, 0.75	0.3, 0.7	0.35, 0.65
Accuracy (%)	95.3	96.4	98.3	98.9	98.9	97.3	95.6
Time (s)	120.909	123.276	122.203	121.860	123.382	122.998	127.443

Note The CPU model of the computer used in the test is Intel(R) Core(TM) i5-6500 CPU @ 3.20 GHz

when comparing the accuracy of the filtering method using different control values. Specifically, we arrange a 1,000 m × 1,000 m × 1,000 m cubic array with 15 sensors. Meanwhile, the coordinates of the deployed sensors are: $A(0, 0, 0)$, $B(120, 360, 540)$, $C(450, 620, 870)$, $D(510, 630, 740)$, $E(960, 760, 130)$, $F(540, 460, 0)$, $G(730, 450, 630)$, $H(200, 300, 100)$, $I(300, 400, 500)$, $J(950, 20, 480)$, $K(560, 290, 780)$, $L(740, 860, 600)$, $M(400, 600, 800)$, $N(710, 630, 380)$, $O(340, 0, 0)$.

A thousand random sources were produced by the computer randomly and the arrivals corresponding to each sensor can be obtained by setting the velocity as 5,000 m/s. Besides, a random sensor S_{err} is selected and the arrivals of S_{err} are set as the abnormal arrivals (t_{err}) by adding an error of 1%. A random error of 0.01–0.05% is added to the arrivals of other sensors to simulate the small systemic picking errors. Seven couples of control values are selected to verify their filtering accuracies. If the filtered arrivals are exactly t_{err} we set at first, then it is proved that the abnormal arrivals have been filtered successfully. The statistical results of the 1,000 groups of examples are listed in Table 6.1.

It can be seen that the control values of 0.2 and 0.8 have the highest accuracy to filter the abnormal arrivals, which is 98.9%. Meanwhile, the elapsed time with control values of 0.2 and 0.8 is relatively shorter than that with the same accuracy. Thus, 0.2 and 0.8 are selected as the control values of F_x, F_y, and F_z.

6.1.2 The Iterative Localization Method Using Clear Arrivals

In this part, we adopt the following equations to locate sources coordinates with clear arrivals in the Cartesian coordinate system. The governing equation for the source coordinates is established as the Eq. (6.18).

$$t_{ij}^{\mathrm{mea}} - t_{ij}^{\mathrm{theor}} = \left(\frac{\partial t}{\partial x}\right)_{ij} \Delta x + \left(\frac{\partial t}{\partial y}\right)_{ij} \Delta y + \left(\frac{\partial t}{\partial z}\right)_{ij} \Delta z + \left(\frac{\partial t}{\partial v}\right)_{ij} \Delta v + \Delta t_{0i} + e_{ij} \tag{6.18}$$

where t_{ij}^{mea} and t_{ij}^{theor} are the measured and theoretical travel times from the ith event to the jth sensor, respectively. t_{0i} is the origin time of the ith event. Δ represents the perturbation of a parameter. e_{ij} denotes the higher-order terms of perturbations and data error. The deviation degree can be described using the difference between the measured travel time t_{ij}^{mea} and the theoretical travel time t_{ij}^{theor}.

Then, the objective function with the model of arrival time difference can be established (Eq. (6.19)), using the quadratic sum of differences between all the theoretical values and all the measured values. The locating accuracy is better when the deviation degree is smaller.

$$f(x, y, z, v) = \sum_{i,j=1}^{n} \left(t_{ij}^{\text{mea}} - t_{ij}^{\text{theor}}\right)^2 = \min \tag{6.19}$$

Since the Eq. (6.19) is a non-negative quadratic function about the independent variables *x*, *y*, *z*, and *v*, the accurate and stable source coordinates can be obtained all the time through minimizing the function value. The iterative localization method can be named as the TD method as the dependent variable is time difference. Compared to the TM used widely at present, TD method eliminates the locating error caused by the premeasured wave velocity, which is a significant advantage to improve the locating accuracy.

Combined with the arrivals of multi-sensor and inversion of the real-time average wave velocity, the CLMAI is proposed to eliminate the effects of abnormal arrivals and premeasured wave velocity. By taking the advantages of both analytical and iterative solutions, the CLMAI can not only remove abnormal arrivals by coupling the analytical solutions and the logistic probability density function, but also reduce the locating errors induced by premeasured wave velocity using the TD method. To clarify the differences between the TM, TD, and CLMAI, Table 6.2 lists the characteristics, advantages and disadvantages of three methods.

Table 6.2 The characteristics and comparisons of the TM, TD, and CLMAI methods

Methods	Known parameters	Unknown	Advantages	Disadvantages
TM	Coordinates of sensors; Arrivals of P-wave; Premeasured velocity	Source coordinates Origin time of source		The locating error induced by the premeasured velocity and abnormal arrivals
TD	Coordinates of sensors; Arrivals of P-wave;	Source coordinates; Origin time of source; Average velocity	Weakening the locating error caused by premeasured velocity	The locating error induced by abnormal arrivals
CLMAI	Coordinates of sensors; Arrivals of P-wave;	Source coordinates; Origin time of source; Average velocity	Eliminating the effects both of the pre-measured velocity and abnormal arrivals	

6.2 The Verification of the CLMAI by Blasts

The data of blasts in the Yongshaba mine are used to verify the proposed CLMAI. The Yongshaba mine is located in the Kaiyang County, Guizhou Province, China. The annual output of phosphate rock is more than 200 million tons. The surface elevation of Yongshaba mine is more than 1,500 m. The ore body of Yongshaba mine is controlled by ten obvious faults and the stability of stope roof is poor. Numerous underground goafs have formed due to the open stope mining method in the early mining stage, which result in the phenomenon of local stress concentration. The large-area rockmass instability or the rockburst may be induced at a time. Thus, a 32-channel microseismic monitoring system was established to avoid the destructive disasters. A total of 26 single-component sensors and two three-component sensors are distributed on the transport tunnels in 930, 1,080, and 1,120 levels. The natural frequency of the used sensors is 14 Hz $\pm$ 1 Hz and the response frequency is from 50 Hz to 5 kHz.

The blasts were carried out in the Yongshaba mine during the period from August 20th to August 22nd, 2013. There were eight blasts recorded by the IMS microseismic monitoring system. Figure 6.3 shows the authentic positions of eight blasts and 28 sensors. Table 6.3 lists the times and coordinates of eight blasts, as well as the number of triggered sensors corresponding to each test.

6.2.1 The Filtering of Abnormal Arrivals for Blasts

For the monitoring data, the arrival of P-wave corresponding to each triggered sensor is picked up using the improved Kurtosis method based on DWT and STA/LTA [8, 9]. Then, the arrivals and coordinates of triggered sensors are used to solve all the analytical solutions. The logistic probability density function is applied to fit the coordinates x, y, and z. The locating results and errors of eight blasts are listed in the Table 6.4. It is clear to see that the locating errors between the authentic coordinates and the locating results are great, where the maximum ADE reaches to 2,013.05 m. The locating results are unreasonable and unreliable. The reason is that the arrivals including abnormal ones are used, which affects the locating accuracy directly. Therefore, it is significant to identify and filter the abnormal arrivals for the localization of MS/AE in mines.

Based on the filtering method descripted in the Sect. 6.1.1, the logistic cumulative distribution function is combined with the analytical solutions to filter the abnormal locating results. Then, the locating results after the filtering of abnormal locating results can be obtained through the analytical localization method. Table 6.5 lists the locating results and errors using the analytical localization method after filtering abnormal locating results. Through the comparison between the Tables 6.4 and 6.5, it is obviously that the locating accuracy is improved significantly. For example, the ADE of No.5 reduces to 517.87 m from 2,013.05 m. The ADE of No.3 reduces to

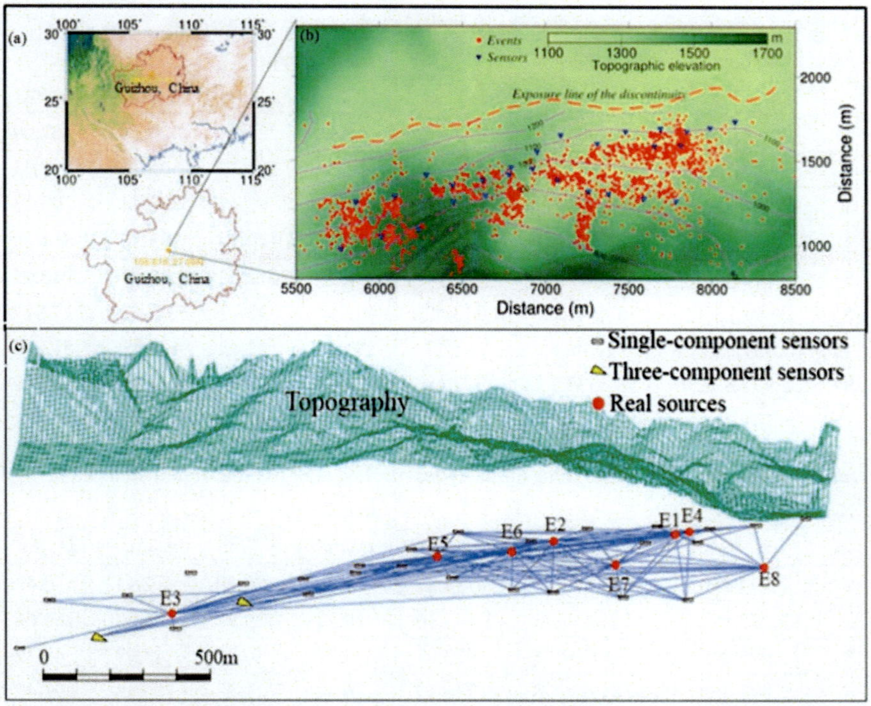

Fig. 6.3 **a** The location information of Yongshaba mine. **b** Map view of Yongshaba mine. **c** The positions of eight blasting tests and 28 sensors in Yongshaba mine (from [7]).

Table 6.3 Times, coordinates, and numbers of triggered sensors of the eight blasts

No	Time	Coordinate			Number of triggered sensors
		x/m	*y*/m	*z*/m	
1	Aug 20th,2013 15:25	2,997,760	381,683	1,107	11
2	Aug 20th,2013 15:38	2,997,405	381,653	1,099	19
3	Aug 21st,2013 15:05	2,996,224	381,194	1,014	14
4	Aug 21st,2013 15:26	2,997,777	381,684	1,107	19
5	Aug 22nd,2013 16:01	2,997,036	381,503	1,028	15
6	Aug 22nd,2013 16:17	2,997,278	381,590	1,053	17
7	Aug 22nd,2013 17:09	2,997,584	381,526	1,044	12
8	Aug 22nd,2013 17:20	2,998,029	381,442	1,017	13

45.32 m from 93.11 m. Figure 6.4 shows the fitting graphs of No.6 and No.8 using the logistic probability density function. The red parts represent the locating results before filtering, while the blue parts represent the locating results after filtering. It

Table 6.4 The locating results using the analytical method without filtering

No	Locating results			Locating errors			
	x/m	y/m	z/m	x_{err}/m	y_{err}/m	z_{err}/m	D_{err}/m
1	2,997,761.16	381,904.51	477.76	1.16	221.51	629.24	667.09
2	2,997,564.63	381,754.17	532.25	159.63	101.17	566.75	597.43
3	2,996,228.96	381,236.41	931.26	4.96	42.41	82.74	93.11
4	2,997,794.53	381,854.90	313.70	17.53	170.90	793.30	811.69
5	2,997,040.87	383,515.52	981.97	4.87	2012.53	46.03	2013.05
6	2,997,230.21	381,541.88	1,213.42	47.79	48.12	160.42	174.17
7	2,997,612.61	381,708.85	565.17	28.61	182.85	478.83	513.35
8	2,997,822.10	381,533.17	392.69	206.90	91.17	624.31	663.99

Table 6.5 The locating results using the analytical method after filtering abnormal locating results

No	Locating results			Locating errors			
	x/m	y/m	z/m	x_{err}/m	y_{err}/m	z_{err}/m	D_{err}/m
1	2,997,750.11	381,803.68	714.85	9.89	120.69	392.15	410.42
2	2,997,385.23	381,517.78	1,194.40	19.77	135.22	95.40	166.66
3	2,996,227.77	381,204.21	970.01	3.77	10.21	43.99	45.32
4	2,997,760.88	381,696.28	716.12	16.12	12.28	390.88	391.40
5	2,997,054.29	382,016.60	964.22	18.29	513.60	63.78	517.87
6	2,997,249.67	381,521.02	1,124.63	28.33	68.98	71.63	103.40
7	2,997,627.32	381,621.10	736.49	43.32	95.10	307.51	324.78
8	2,997,858.55	381,498.96	1,042.15	170.45	56.96	25.15	181.47

can be seen that the fitting precision of the curves after filtering is superior to the curves before filtering.

Therefore, it can be concluded that the locating accuracy can be improved effectively through the filtering of abnormal locating results. It is noticed that the locating errors of the analytical localization method are still large, because only six sensors are considered in each calculation process. As the complexity of rockmass structure in underground mining, the travel paths for P-wave are not straight, which caused large locating errors. Considering the multiple sensors, the iterative localization method can be applied to reduce the locating errors for its characteristic of seeking the optimal results in the whole range.

As described in the Sect. 6.1.1.3, the analytical solutions not only satisfy the inequalities including $F_x < 0.2$ or $F_x > 0.8$, $F_y < 0.2$ or $F_y > 0.8$, as well as $F_z < 0.2$ or $F_z > 0.8$, but also belong to the groups of locating results can be obtained. In these locating results, the sensor with the maximum N_c of each blasting test can be identified. The arrivals corresponding to the sensor with the maximum N_c are exactly the abnormal arrivals of each blasting test. Table 6.6 lists the count numbers

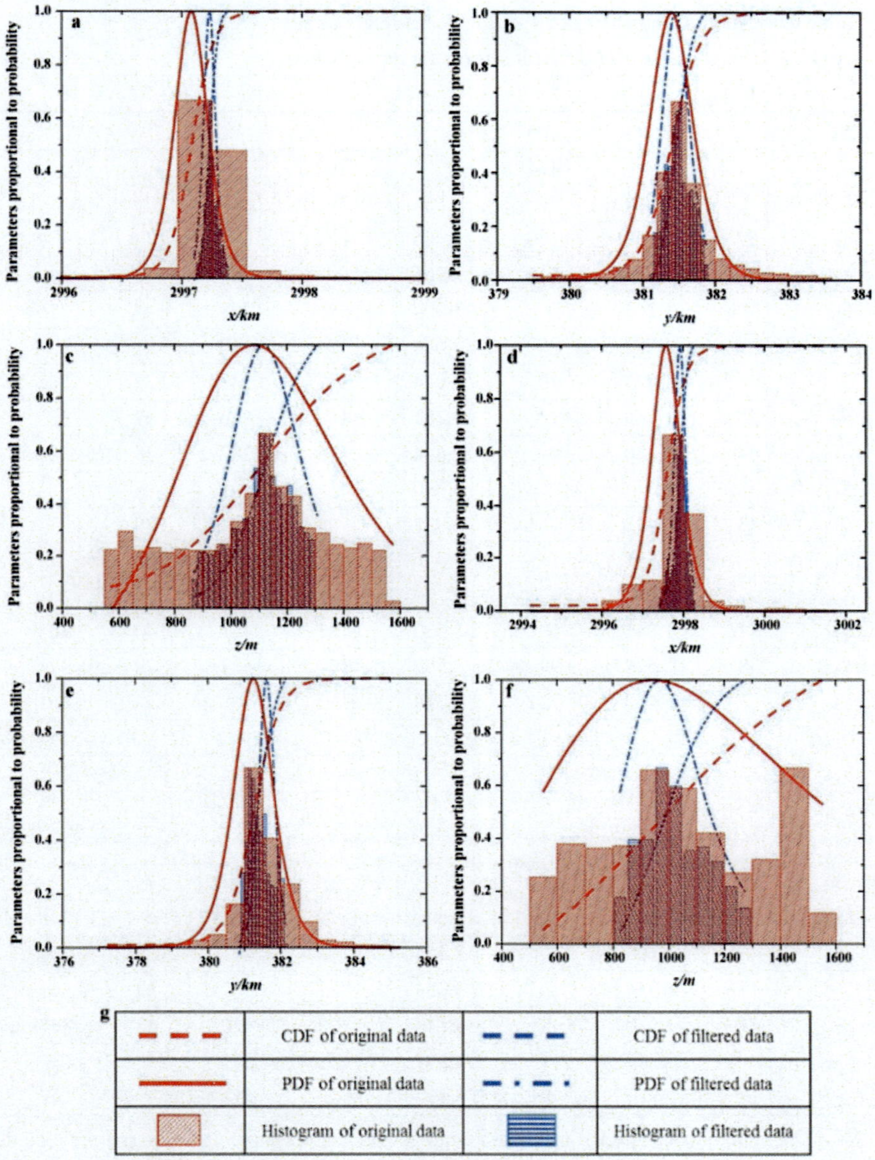

Fig. 6.4 The locating results of No.6 and No.8 through the fitting of logistic probability density function and logistic cumulative distribution function. **a–c** The results for *x*, *y*, and *z* of No.6. **d–f** The results for *x*, *y*, and *z* of No.8. **g** The legends of **a–f**

Table 6.6 The count numbers of 28 sensors for the eight blasts

No	Count numbers of different sensors for the eight blast							
	1	2	3	4	5	6	7	8
S_1			105					
S_2			134		100			
S_3		2,281	75	4,536	89	518		
S_4	57	2,147	140	3,406	84	402		439
S_5		1,910	141	3,256	99	448		
S_6		1,861		3,204	99	398	44	316
S_7		1,881		3,094	121	360	37	
S_8		1,875	**201**	2,974	141	343	39	388
S_9	31	1,769		2,781	142	296	47	380
S_{10}	44	1,730		2,644	125	282	48	367
S_{11}	38			3,080				
S_{12}	43	2,442		2,925			39	
S_{13}		1,811		3,446			24	
S_{14}		1,982				441		
S_{15}	54	1,841	103	3,530	101	451	40	368
S_{16}	38	1,528	99	3,579	131	514	29	371
S_{17}		1,460	72	**5,310**	123	568		**661**
S_{18}					**147**			
S_{19}			94					
S_{20}								
S_{21}			81					
S_{22}			94					
S_{23}								345
S_{24}	46	2,108		2,855		421	**59**	354
S_{25}	45	**4,314**		3,034		387	32	361
S_{26}	44	2,540		3,144	100	420	48	357
S_{27}		2,699	113	4,599	108	591		
S_{28}	**76**	2,879	78	5,197		**612**		441

Note The bold numbers are the maximum count numbers corresponding to the sensors recorded abnormal arrivals for each blasting test

of 28 sensors for the eight blasts. It is clear that the sensors with the maximum count numbers corresponding to the blasts No.1 to No.8 are S_{28}, S_{25}, S_8, S_{17}, S_{18}, S_{28}, S_{24}, and S_{17}, respectively. The arrivals recorded by these sensors are tagged as the abnormal arrivals. Therefore, the abnormal arrivals can be filtered to improve the locating accuracy.

6.2.2 The Validation for the Filtered Abnormal Arrivals

To further validate the filtered abnormal arrivals from above analytical method, the leave-one-out cross-validation is used to validate the filtering results. Firstly, the eight blasts are located by the TD method using the leave-one arrival for every locating process. Specifically, the calculated MS/AE source is represented as O. Thus, the distance d between every sensor and the source O, as well as the travel times t from the source O to each sensor can be solved. As for a blasting test with n triggered sensors, there are $n(n-1)$ groups of data for the distance d and the travel time t. Figure 6.5 shows the linear fitting graphs for the distance d and the travel time t of the eight blasts.

From the Fig. 6.5, it can be clearly seen that the arrivals of sensors S_{25}, S_8, S_{17}, S_{18}, and S_{17} for corresponding blasts No.2, No.3, No.4, No.5, and No.8 deviate from the fitting lines, which indicates that arrivals of sensors S_{25}, S_8, S_{17}, S_{18}, and S_{17} are abnormal arrivals. The results are consistent with the filtered results of Sect. 6.2.1 using the analytical solutions and the logistic probability density function.

For the blasts No.1 and No.7, it is hard to determine the sensors recorded abnormal arrivals from the d-t charts since the number of triggered sensors is relative scare, which means that the presence of abnormal arrivals has a great impact on the locating result of the TD method. Furthermore, the deviation points in d-t chart arise in the case of that the sensor recorded abnormal arrivals was eliminated. In this case, we can draw another kind of d-t chart as Fig. 6.6, where the color represents the sensor that was removed. It can be seen clearly from Fig. 6.6 that the S_{28} of No.1 and S_{10} of No.7 are the sensors recorded abnormal arrivals, which also meets the results of Sect. 6.2.1 using the proposed filtering method. The d-t chart of No.6 has no obvious deviation points, which means that there may be no abnormal arrival of No.6 or the error of abnormal arrival is too small to have an impact.

In summary, it is proved that the filtered sensors identified in Sect. 6.2.1 are exactly the sensors recorded abnormal arrivals. The proposed method for filtering abnormal arrivals is reasonable and feasible using the analytical solutions and the logistic probability density function.

6.2.3 The Locating Results Using the CLMAI and Discussion

The data of eight blasts were used to verify the effectiveness of proposed CLMAI. Table 6.7 lists the locating results and errors using the CLMAI. To compare the locating accuracy of the TM and TD methods, the locating results and errors of the TM and TD are also listed in Tables 6.8 and 6.9, respectively.

Results show that the average ADEs of the CLMAI, TM, and TD methods are 39.82 m, 147.47 m, and 69.87 m, respectively. The maximum ADEs of the CLMAI, TM, and TD methods are 65.87 m, 297.38 m, and 126.63 m, respectively. In the aspect of the average ADE or the maximum ADE, the locating error of the CLMAI reduces obviously. There is no doubt that the proposed CLMAI is superior to the TM and

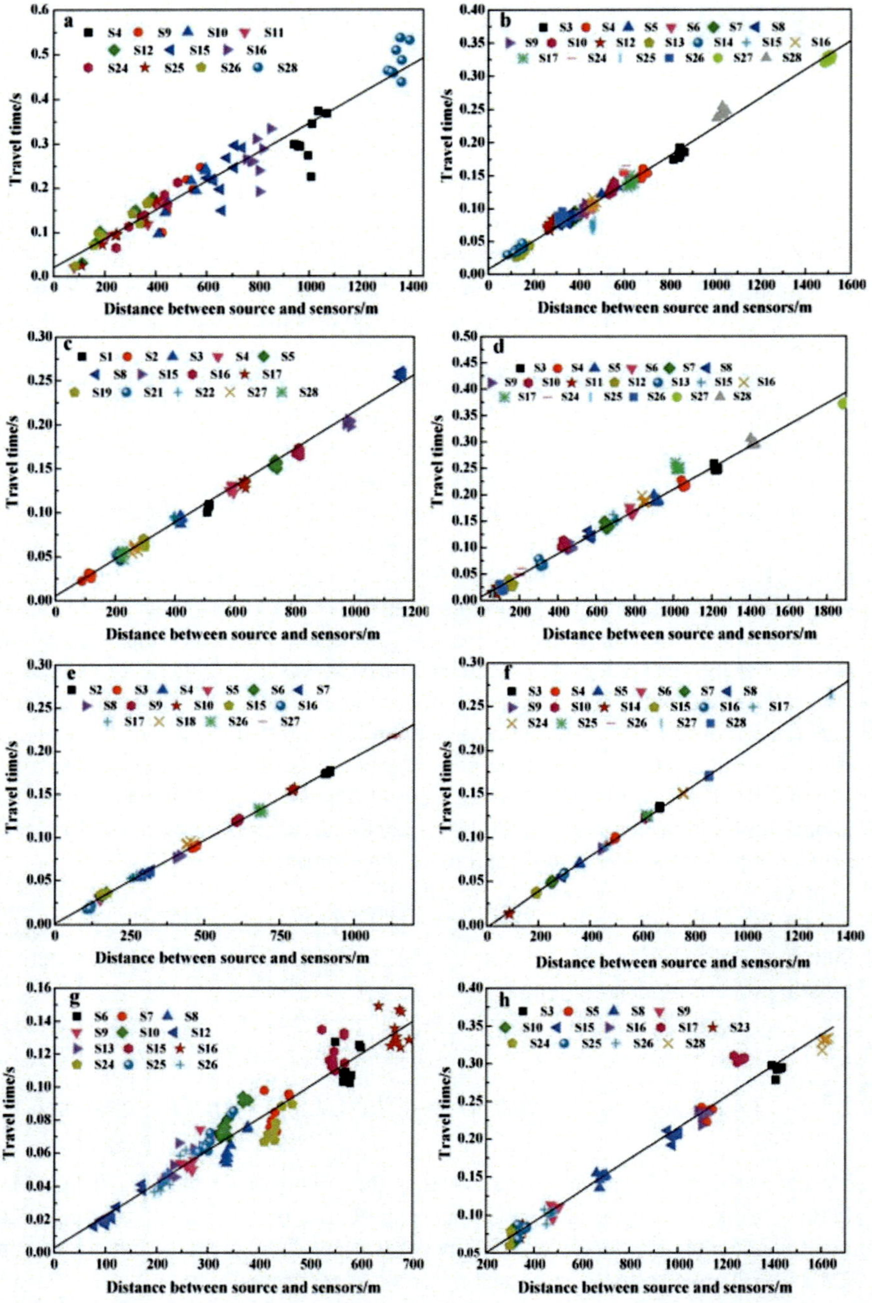

Fig. 6.5 The fitting graphs for the distance *d* and the travel time *t* of the eight blasts. **a–h** The linear fitting results of the blasts No.1 to No.8. The color represents the unfiltered sensor. The characteristics of abnormal arrivals indicate discrete points with larger distance between the fitting lines and points (For interpretation of the references to colour in this figure legend, the reader is referred to the web version of this article)

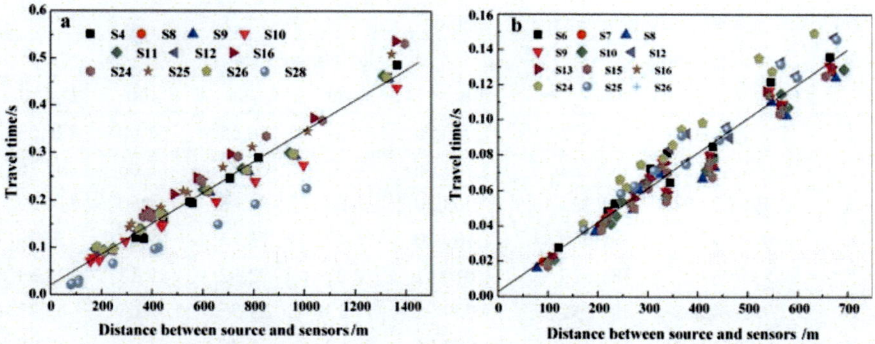

Fig. 6.6 The graphs **a** and **b** are the fitting graphs for the distance *d* and the travel time *t* of the blasts No.1 and No.7, respectively. The color in the figure represents the filtered sensor (For interpretation of the references to colour in this figure legend, the reader is referred to the web version of this article)

Table 6.7 Locating results using the CLMAI method

No	Locating results			Locating errors			
	x/m	y/m	z/m	x_{err}/m	y_{err}/m	z_{err}/m	D_{err}/m
1	2,997,760.88	381,645.90	1,119.49	0.88	37.10	12.49	39.15
2	2,997,370.70	381,629.61	1,058.48	35.30	23.39	40.52	58.61
3	2,996,228.42	381,193.99	948.28	4.42	0.01	65.72	65.87
4	2,997,794.69	381,660.58	1,109.92	17.69	23.42	2.92	29.50
5	2,997,034.72	381,514.83	1,029.39	2.28	11.83	1.39	12.12
6	2,997,260.36	381,542.21	1,064.24	17.64	47.79	11.24	52.17
7	2,997,584.77	381,503.16	1,069.62	0.77	22.84	25.62	34.33
8	2,998,041.09	381,421.84	1,029.95	12.09	20.16	12.95	26.84
Average value							39.82

Table 6.8 Locating results using the TM method

No	Locating results			Locating errors			
	x/m	y/m	z/m	x_{err}/m	y_{err}/m	z_{err}/m	D_{err}/m
1	2,997,787.50	381,647.52	1,103.11	27.50	35.48	3.89	45.06
2	2,997,401.47	381,672.33	961.53	3.53	19.33	137.47	138.87
3	2,996,264.64	381,116.99	1,092.55	40.64	77.01	78.55	117.27
4	2,997,764.13	381,679.28	1,065.29	12.87	4.72	41.71	43.90
5	2,997,054.39	381,640.87	773.06	18.39	137.87	254.94	290.41
6	2,997,238.55	381,480.99	1,212.02	39.45	109.01	159.02	196.79
7	2,997,556.20	381,494.89	1,071.75	27.75	31.11	27.75	50.10
8	2,998,066.18	381,292.51	1,271.37	37.18	149.49	254.37	297.38
Average value							147.47

Table 6.9 Locating results using the TD method

No	Locating results			Locating errors			
	x/m	*y*/m	*z*/m	x_{err}/m	y_{err}/m	z_{err}/m	D_{err}/m
1	2,997,751.62	381,572.35	1,046.00	8.38	110.65	61.00	126.63
2	2,997,400.23	381,583.54	1,167.06	4.77	69.46	68.06	97.36
3	2,996,223.97	381,224.40	903.20	0.03	30.40	110.80	114.89
4	2,997,795.43	381,660.84	1,109.80	18.43	23.16	2.80	29.73
5	2,997,038.27	381,525.52	1,010.87	2.27	22.52	17.13	28.39
6	2,997,248.65	381,551.37	1,029.91	17.13	29.35	38.63	53.73
7	2,997,613.93	381,499.64	1,099.05	29.93	26.36	55.05	67.98
8	2,998,029.58	381,435.72	1,056.71	0.58	6.28	39.71	40.21
Average value							69.87

TD methods, which can improve the locating accuracy significantly by eliminating the effects of premeasured wave velocity and abnormal arrivals.

It is noted that the CLMAI can improve the locating accuracy substantially, and decrease the average locating error from 147.47 m to 39.82 m. However, the location error around 40 m is still slightly larger, since the mining range is very large, where the span reaches 2,000 m and the vertical span is 900 m. However, the number of monitoring sensors is only 28 with 32 channels, which is too sparse for such a large mining area. It is suggested that the microseismic monitoring system of Yongshaba mine should be optimized, and the number of sensors should be increased to improve the layout density for the localization of MS/AE sources with higher accuracy.

6.3 A Case Study for Locating the Microseismic Sources in Kaiyang Mine

During the period from January to April 2014 in Kaiyang mine, there were 1,891 groups of microseismic data recorded by the IMS microseismic monitoring system. The TM, TD, and CLMAI methods are applied to locate all the 1,891 events. The arrival of P-wave corresponding to each triggered sensor is picked up using the improved Kurtosis method based on DWT and STA/LTA [10, 11]. As for the TM, the premeasured velocity is set as 4,500 m/s, which is equal to the average velocity value of the eight blasts.

The locating results of the TM, TD, and CLMAI methods are shown in the Fig. 6.7a, b, and c, respectively. The coordinates of microseismic sources are represented with points, where the colors represent the values of z-coordinate. Generally, the locating results distributed in the mining area are reasonable, which are named as effective results. Through the comparison of Fig. 6.7a-c, it can be seen that the effective results of the Fig. 6.7a are much less than the effective results of the Fig. 6.7b and

c. Thus, it can be concluded that the locating accuracy of the TD and CLMAI methods is superior to the TM. Also, the effective results of Fig. 6.8b is much smaller than the Fig. 6.7c, which means that the CLMAI has better locating accuracy compared to the TD method. Furthermore, the quantitative statistical analysis is used to evaluate the locating accuracy of these three methods, where the number of effective results is taken as the evaluation criterion. According to the Fig. 6.7d, the numbers of effective results for the TM, TD, and CLMAI are 955, 1,607, and 1,706, respectively. The number of the total events is 1,891. For the TM method, the effective result rate is merely 50.50%, because the premeasured fixed wave velocity is not suitable for the mines with dynamic wave velocity. The effective result rate for the TD method is 84.98%, while the effective result rate for the CLMAI method is 90.22%. It can be concluded that the proposed CLMAI eliminates the effects of premeasured wave velocity and abnormal arrivals, which is superior to the TM and TD methods. The effective and accurate CLMAI is applicable to the complex mining environment with multi-level and multi-stope.

Figure 6.8 shows the scatter diagram of P-wave velocity solved by the CLMAI method of the mining area for Yongshaba mine. From the Fig. 6.8, graphs a, b, and c, there is no obvious relationships between the wave velocity and the coordinates *x*, *y*, as well as *z*, where the points representing P-wave velocity of microseismic sources are discrete. Therefore, it is difficult to establish the wave velocity models in the complex mining environment with multi-level and multi-stope. As the Fig. 6.8d shown, the proportion for the wave velocity greater than 4,500 m/s, the average of eight blasts, is relatively large and reach to greater than 65%. The reason is that the continuous large-scale mining which caused the stress adjustment and stress concentration in the complex environment, then the wave velocity increases with different degrees. Also, the changes of multiple stope structures have an impact on the increase of wave velocity. On the contrary, the proportion for the wave velocity greater than 4,500 m/s is only about 35%. However, the value of the premeasured wave velocity in the TM is exactly equal to 4,500 m/s, which is unsuitable for the dynamic wave velocity in mines. As a result, the large locating error would be caused using the TM, as shown in Fig. 6.7, which only has availably located 955 from 1,891 events. While the proposed CLMAI can locate 1,706 from 1,891 events. Therefore, it is proved that the CLMAI is applicable to the complicated mining environment with abnormal arrivals, where the wave velocity changes dynamically.

6.4 Conclusions

Focusing on the problems that the abnormal arrivals and premeasured wave velocity in the complex deep mining environment have serious effects on the locating accuracy for MS/AE sources, the collaborative localization method CLMAI was proposed to improve the locating accuracy in this chapter. Firstly, the analytical solutions were resolved for the six sensors network in unknown velocity system. Then, the logistic density function was combined with the analytical solutions to determine the source

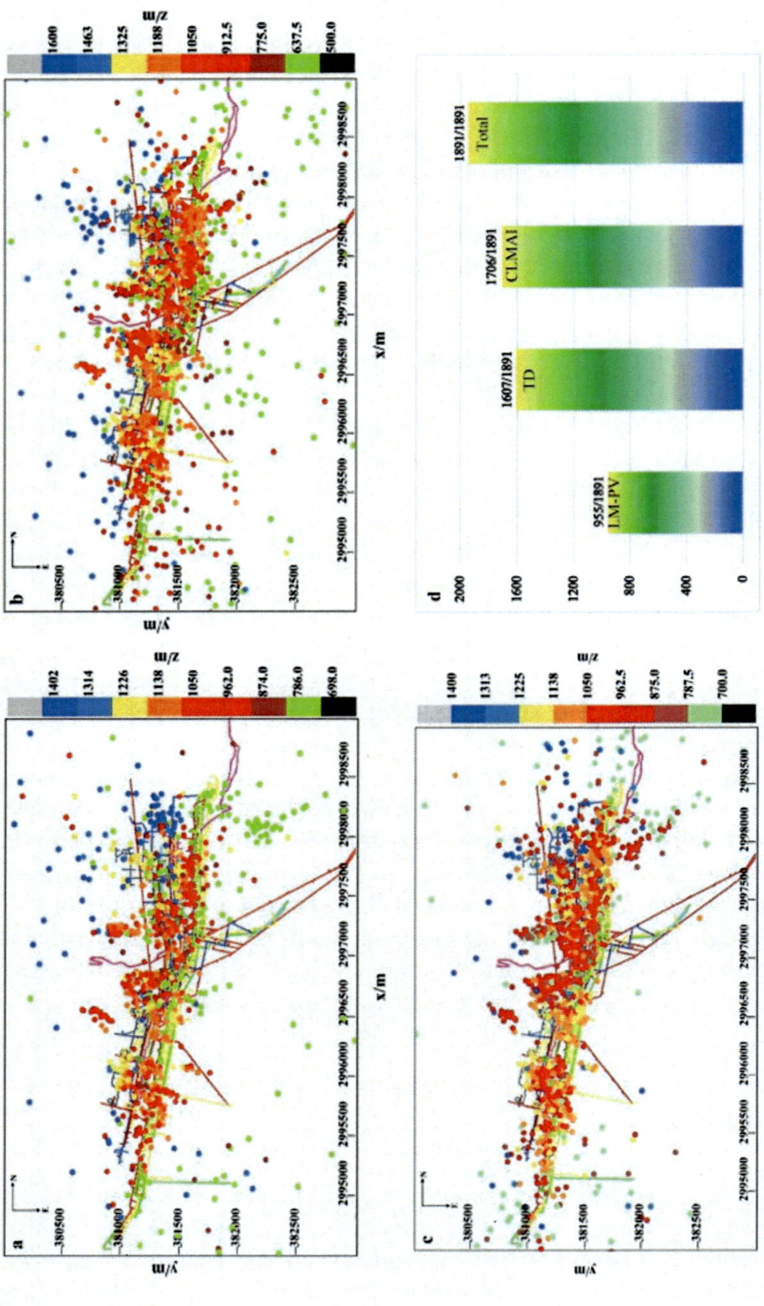

Fig. 6.7 The locating results of 1,891 events using three methods. **a** The locating results of the TM method. **b** The locating results of the TD method. **c** The locating results of the CLMAI method. **d** The number of the effective results for the three methods

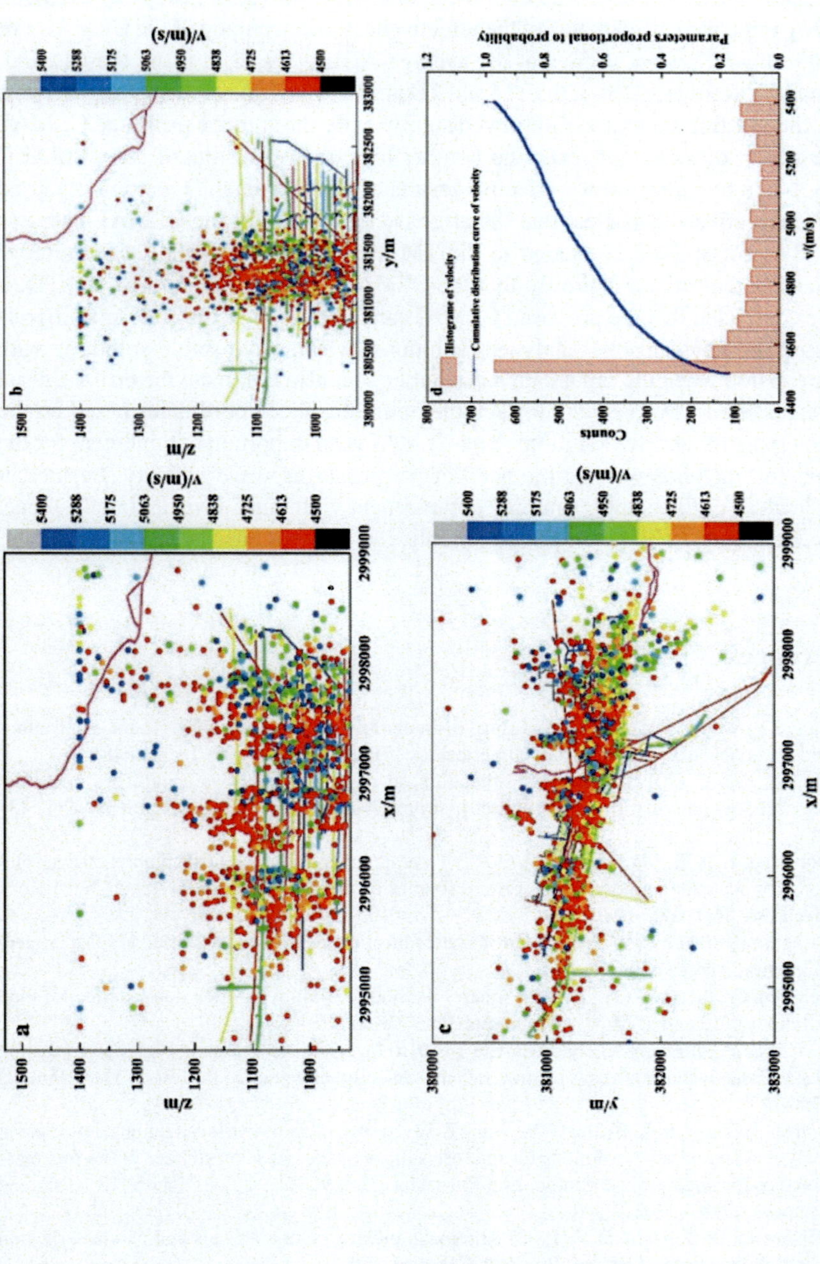

Fig. 6.8 The velocity structures of **a** x–z plane, **b** x–y plane, and **c** y–z plane, respectively. **d** The distribution of velocity for the 1,891 events, where the red parts represent the histogram of velocity and the blue curve represents the cumulative distribution curve of velocity (For interpretation of the references to colour in this figure legend, the reader is referred to the web version of this article)

coordinates for more than six sensors, as well as removing the abnormal arrivals using the advantage of the analytical solutions. Finally, by applying the iterative localization method with time differences after the filtering of abnormal arrivals, the temporal and spatial impact of premeasured wave velocity on the locating accuracy was eliminated effectively. The data of eight blasts and 1,891 groups of events in Kaiyang mine were used to verify the effectiveness and accuracy of the CLMAI. For the eight blasts, results show that the average ADE is 39.82 m using the CLMAI, whereas the average ADEs of the TM and TD are 147.47 m and 69.87 m, respectively. Thus, the locating accuracy is improved significantly through the proposed CLMAI. In the aspect of locating events, the number for effective results of the CLMAI is 1,706, while the numbers for effective results of the TM and TD are only 955 and 1,607, respectively. It is clear that the effective result rate for the CLMAI increases by 5.28% and 39.76% compared to the TM and TD. Through the localization of the blasts and events, it is proved that the CLMAI is superior to the TM and TD. It can be concluded that the proposed CLMAI can not only filter the abnormal arrivals by using the advantage of analytical solutions, which have stable solutions with high precision when the input data are accurate, but also eliminate the errors caused by premeasured wave velocity to seek the optimal source coordinates based on the characteristic of iterative solutions. The CLMAI is an important complement for the localization methods without the need of premeasuring wave velocity. Besides, it is applicable for the complex mining environment with multi-level and multi-stope, where the abnormal arrivals exist commonly.

References

1. Jia P, Zhu W (2015) Mechanism of zonal disintegration around deep underground excavations under triaxial stress–Insight from numerical test. Tunn Undergr Space Technol 48:1–10
2. Zhu W, Wei C (2011) Numerical simulation on mining-induced water inrushes related to geologic structures using a damage-based hydromechanical model. Environ Earth Sci 62(1):43–54
3. Dong L-J, Li X-B, Ma J, Tang L-Z (2017) Three-dimensional analytical comprehensive solutions for acoustic emission/microseismic sources of unknown velocity system. Chin J Rock Mech Eng 36(1):186–197
4. Leighton F, Blake W (1970) Rock noise source location techniques, vol 7432. US Department of Interior, Bureau of Mines
5. Leighton F, Duvall W (1972) Least squares method for improving rock noise source location techniques. Bureau of Mines, Washington, DC (USA)
6. Lu C-P, Liu G-J, Zhang N, Zhao T-B, Liu Y (2016) Inversion of stress field evolution consisting of static and dynamic stresses by microseismic velocity tomography. Int J Rock Mech Min Sci 87:8–22
7. Dong L-J, Zou W, Li X-B, Shu W-W, Wang Z-W (2019) Collaborative localization method using analytical and iterative solutions for microseismic/acoustic emission sources in the rockmass structure for underground mining. Eng Fract Mech 210:95–112. https://doi.org/10.1016/j.engfracmech.2018.01.032
8. Gelchinsky B, Shtivelman V (1983) Automatic picking of first arrivals and parameterization of traveltime curves. Geophys Prospect 31(6):915–928

9. Li X-B, Shang X-Y, Wang Z-W, Dong L-J, Weng L (2016) Identifying P-phase arrivals with noise: an improved Kurtosis method based on DWT and STA/LTA. J Appl Geophys 133:50–61
10. Dong L-J, Shu W-W, Li X-B, Han G-J, Zou W (2017) Three dimensional comprehensive analytical solutions for locating sources of sensor networks in unknown velocity mining system. IEEE Access 5:11337–11351
11. Li X-B, Wang Z-W, Dong L-J (2016) Locating single-point sources from arrival times containing large picking errors (LPEs): the virtual field optimization method (VFOM). Sci Rep 6(1):1–12

Chapter 7
Velocity-Free Localization Methods for the Complex Structures Based on Non-straight Wave Travel Paths

The methods introduced in Chaps. 4–6 do not take into account the effect of geometric irregularity and still assume the wave path between the source and the AE sensor is a straight path. However, geometric irregularities (such as lugs and caves) in the complex material structure would significantly affect the propagation paths of the traveling wave [1]. Therefore, some localization methods considering complex media conditions were investigated to improve the location accuracy recently. Gollob et al. [2] proposed the FastWay method considering the geometry. It used the Dijkstra's algorithm [3] to search the fastest wave travel path between the source and the sensor to estimate the position of the source. Park et al. [4] identified the impact location at the artificial training data set of the impulse response function. It was applied to the airplane wing and aluminum fuselage and achieved good results. Baxter et al. [1] proposed Delta T source location for AE, which was applicable source location in complex geometric structures. It utilized a large quantity of artificial sources and differences in times of arrival information to train the data map of ΔT manually to determine source location. This approach was further assessed in the performance and the robustness by Eaton et al. [5]. Hensman et al. [6] improved this technique which only requires less training data by introducing probabilistic interpretation. Both using Dijkstra's algorithm and training data require a considerable amount of time.

In this chapter, some localization methods considering the complex structures are introduced. In Sect. 7.1, an A* location method without premeasured velocity (ALM) is descripted assuming that the empty area is known and the sources are on a plane. Considering an unknown empty area, localization method SUEA is introduced in Sect. 7.2. Then, a 3-D localization method for the hole-containing structure, called VFH, is introduced in Sect. 7.3. The validating tests are demonstrated and the results are discussed in Sect. 7.4.

L. Dong and X. Li, *Velocity-Free Localization Methodology for Acoustic and Microseismic Sources*, https://doi.org/10.1007/978-981-19-8610-9_7

7.1 A* Localization Method Without Premeasured Velocity

ALM uses the improved A* search algorithm and the arrivals to determine the location of the AE source. There are some algorithms similar to the A* algorithm, including ant colony algorithm [7], rapidly-exploring random tree algorithm [8] and the artificial potential field method [9]. Compared to these methods, A* search algorithm can find the global optimal path and has strong robustness. The ALM can be divided into four parts. Firstly, meshing the objects to be tested and representing the shape of the object with 0 and 1. Secondly, collecting the arrival information of the P wave since the P wave is first received by the sensor (Fig. 7.1) and is easy to separate [10]. Thirdly, using A* search algorithm to search for the fastest wave path between each sensor and each grid node. Finally, identifying the location of AE source according to the principle of least square method. Figure 7.2 shows the flowchart of the localization process of the ALM. Each part of the ALM is clarified in this section.

7.1.1 Initializing the Text Environment

The geometric shape and specific location of irregular empty area are confirmed in the location area. According to the situation of irregular empty space and the accuracy requirement of location, the size of each square cell is determined in the grid. In general, the amount of calculation and the accuracy of localization would increase as the grid is continuously partitioned. However, when the grid has been partitioned enough, the accuracy of localization will no longer change obviously if partitioning continues. The zero matrix **M** of the same size as the grid node is established, and the **M** value corresponding to the empty area node is marked as 1. The grid nodes form a set *P*, which serves as the starting node for searching the fastest waveform path. It is assumed that the propagation velocity of P wave in the non-empty area is

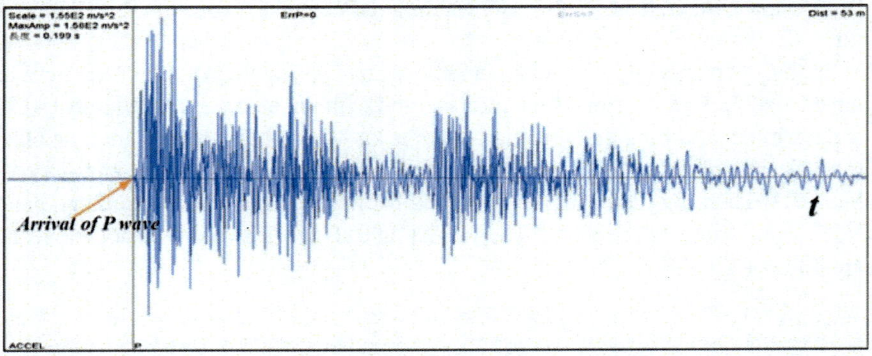

Fig. 7.1 Waveform signal received by the sensor (from [11])

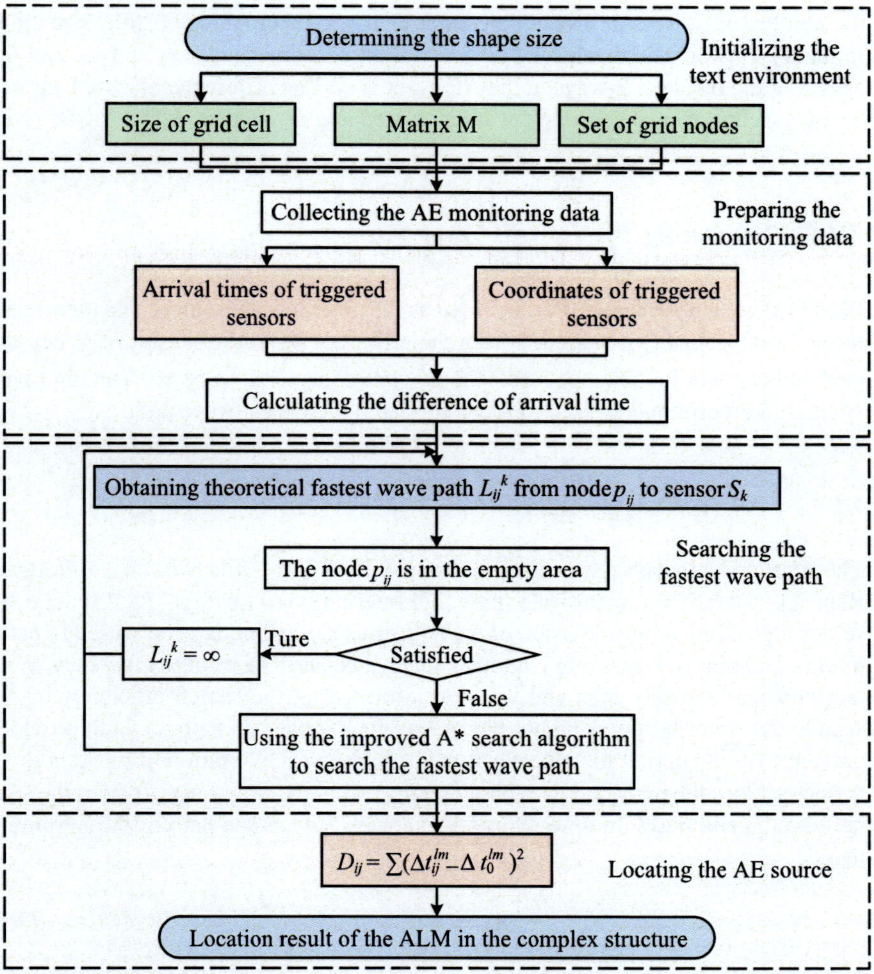

Fig. 7.2 The flowchart for the locating process of the ALM

expressed as C. To reduce the location error, the velocity is brought into equation as an unknown, which does not need to be premeasured in the ALM. The unknown source coordinates and the initial time are set as (x_0, y_0) and t_0.

7.1.2 Collecting Arrivals

There are m sensors fixed at different positions on the object. They were located on the grid nodes and their coordinates were known. Since there are four unknowns (wave velocity C, coordinate x_0, coordinate y_0, initial time t_0) in this 2-D model, the number of sensors m should be greater than or equal to 4 to construct equations in theory. As for the kth sensor S_k, the location coordinates are x_k and y_k, and the arrival of the received P-wave in the AE event is t_0^k. The difference of actual arrivals between sensor S_l and sensor S_m was calculated and denoted as Δt_0^{lm}.

7.1.3 Searching the Fastest Wave Path

Each grid node p_{ij} in the set P is regarded as the potential AE source. The theoretical fastest wave path L_{ij}^k is searched from the grid node p_{ij} to the sensor S_k. When the grid node p_{ij} was in the empty area, the L_{ij}^k was considered to be ∞. The improved A* search algorithm was used in ALM to search the fastest wave path.

7.1.3.1 A* Search Algorithm

The A* search algorithm was first proposed by Peter Hart, Nils Nilsson and Bertram Raphael of the Stanford Institute (now SRI International) in 1968 [12, 13]. The A* search algorithm, as an extension of the Dijkstra's algorithm, is widely used in path-finding and graph traversal in computer science because of its high efficiency. With the increase of empty area and location information, the search efficiency of A* search algorithm becomes higher because of its heuristic function, which provides guidance for the optimum path selection [14, 15]. In each iteration of its main loop, A* determines the extended path based on the estimated function $f(n)$ from the start node to the end node. In other words, A* always selects the path of the minimum length

$$f(n) = g(n) + h(n) \tag{7.1}$$

where n represents the current extended node. $g(n)$ represents the minimum length of the searching path from the starting node to the node n. $h(n)$ is a heuristic function. It uses the Euclidean distance to estimate the fastest path length between the current node and the end node. The A* search algorithm stops searching when the end node is the current expansion node or no next node can be expanded.

In the actual application of the node grid, the traditional A* search algorithm generally uses the center node of the cell and selects the next node from the eight nodes of the adjacent layer, as shown in Fig. 7.4a. Figure 7.3 shows that a 2-D plane with empty areas was divided into 20 × 20 grids. The wave path as shown in Fig. 7.3 can be obtained by using the traditional A* search algorithm. It seems unreasonable at position (1) and position (2). The path searched at the position (1) is not the optimal path because it is not close to the empty space. In order to reduce the errors caused by searching the path with the center node, the grid can be densified. However, it will increase the amount of computation several times. The path at position (2) is

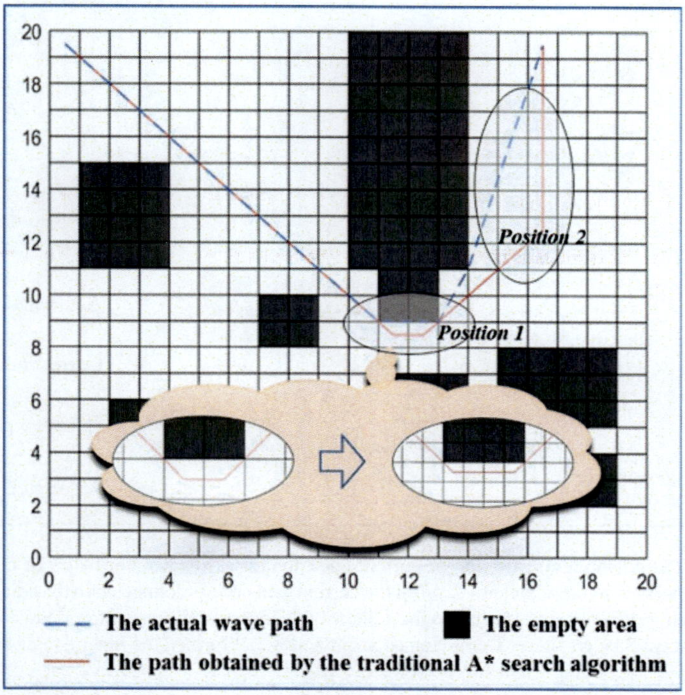

Fig. 7.3 The unreasonable positions in the path obtained by the traditional A* search algorithm (from [11])

obviously not the fastest path, which is caused by the limitations of the traditional A* search algorithm. One node extends to eight nodes of an adjacent layer, which means that there are only eight directions (measured counterclockwise from the positive half of the x-axis) to select: 0°, 45°, 90°, 135°, 180°, 225°, 270°, and 315°.

7.1.3.2 The Improved A* Search Algorithm

To search the wave path more effectively, the A* search algorithm was improved. The grid nodes were applied to search and the nodes were established effective connection with more adjacent layer nodes. The use of grid nodes makes the path of the search close to the edge of the empty area, which is more in line with the actual situation. The relationship between the node and more adjacent layer nodes is extended, so that the more directions of the local path, the smaller the angle between any two adjacent directions, and the more accurate the searched path. As can be seen in Fig. 7.4b–d, the node is connected to the adjacent two layers (16 nodes), three layers (32 nodes), and four layers (48 nodes). The relationship between the node $z(k)$ and the number of layers k is expressed as:

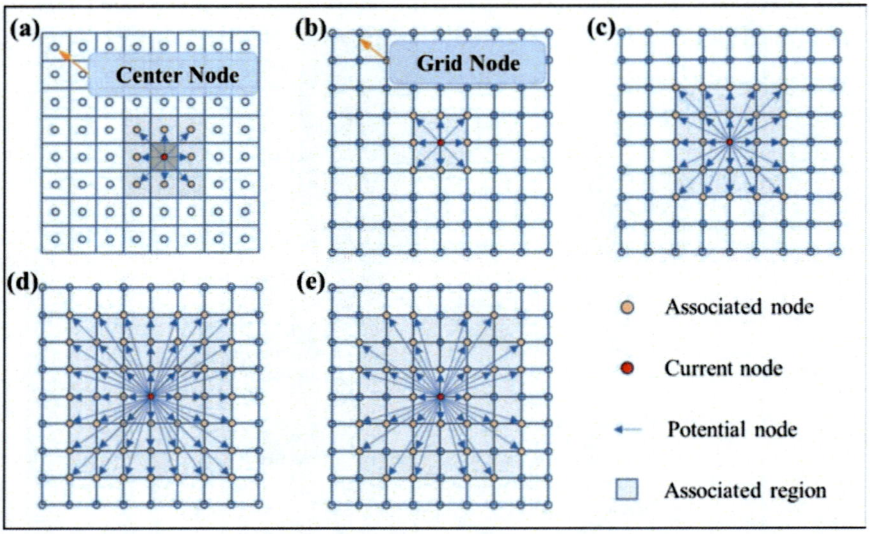

Fig. 7.4 The potential direction of the current node and the associated nodes: **a** The current center node is connected to the adjacent 1 layer; **b** the current grid node is connected to the adjacent 1 layer; **c** the current grid node is connected to the adjacent two layers; **d** the current grid node is connected to the adjacent 3 layers; and **e** Removing adjacent nodes that formed the repetitive directions (from [11])

$$z(k) = (2k + 1)^2 - 1. \tag{7.2}$$

The more layers the nodes are connected, the smaller the relative error between the search path and the actual path is. However, the amount of calculation increases since the linear relationship between the amount of calculation and the connected nodes. It is necessary to calculate the maximum theoretical relative error between the search path length and the actual path to determine the optimal number of association layers and nodes. One-eighth of the graph is taken for research due to the symmetry of the node expansion. In the Fig. 7.5a, the number of layers is assumed as k. The angle β is formed by the extension direction of the node and the the positive half of the x-axis. β_i is the $(i + 1)$th extension direction (counted counterclockwise from the positive half of the x-axis), which is equal to arctan(ia/ka). The angle α is the subtraction of two adjacent angle β, then α_i is equal to $(\beta_i - \beta_{i-1})$. According to the geometric relationship of Fig. 7.5a, the angle α can be expressed as

$$\alpha_i = \arctan\frac{i}{k} - \arctan\frac{i-1}{k} \quad (\mathrm{i} = 1, 2, \ldots, k). \tag{7.3}$$

When $i = 1$, α_i takes the maximum value. The direction formed by the current node and the internal node can be regarded as the increase of the direction connected by the current node and the outermost node. Therefore, α_1 is the global maximum. As shown in the Fig. 7.5b, the number of layers is assumed as k. In the infinite sector

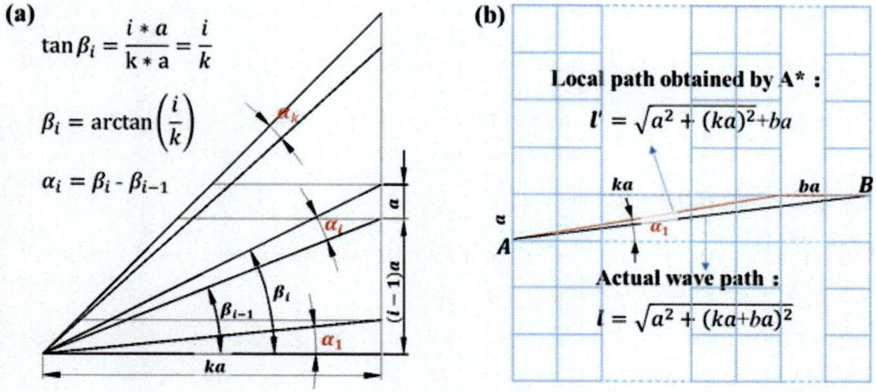

Fig. 7.5 Localization paths: **a** angle α and directions β; **b** actual wave path l and local path l' (from [11])

region with angle α_1, l is the actual path from the node A to the node B, which is equal to $\sqrt{a^2 + (ka + ba)^2}$. The search path l' obtained by the improved A* search algorithm is equal to $\sqrt{a^2 + (ka)^2} + ba$. $y(k)$ is the maximum theoretical relative error of the distance between l and l'. Then, the relationship between k and $y(k)$ can be obtained and is shown in Eq. (7.4).

$$\begin{aligned} y(k) &= \max\left(\frac{\sqrt{a^2 + (ka)^2} + ba}{\sqrt{a^2 + (ka + ba)^2}} - 1\right) \\ &= \sqrt{1 + \frac{1}{\sqrt{1 + k^2}} \cdot \frac{2}{2k + \frac{1+k^2}{b} + b}} - 1 \\ &= \sqrt{2k^2 + 2 - 2k\sqrt{k^2 + 1}} - 1 \end{aligned} \tag{7.4}$$

In the process of outward expansion, some directions repeat with the previous direction, which are not necessary to consider. Therefore, these connections between nodes that formed the repetitive directions can be removed to reduce the amount of calculations, as shown in Fig. 7.4e. The relationship between the node and the number of layers is corrected by fitting

$$z_c(k) = 0.61 \times (2k + 1)^2 + 1.2. \tag{7.5}$$

Figure 7.6 shows the relationships between the k and the $y(k)$, between the k and the $z(k)$, as well as between the k and the $z_c(k)$ in the process of searching for the fastest wave path in the grid. When the number of layers to be expanded reaches four layers, the relative error of the theoretical path is less than 0.8%, which can meet the location requirements and have a relatively small amount of calculation. If the k is extended to 5 layers, the relative error of the theoretical path is about 0.5%, while

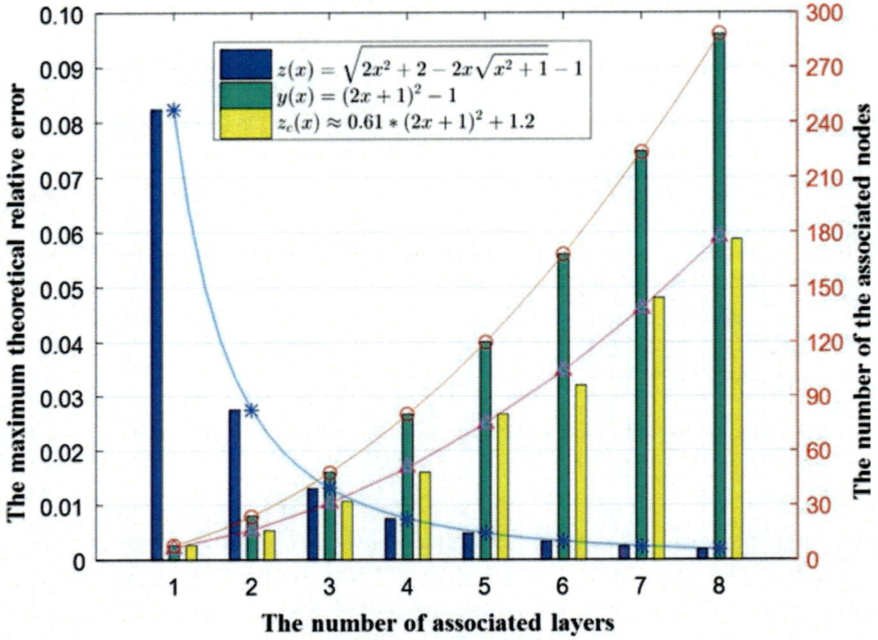

Fig. 7.6 The number of layers, the maximum theoretical relative error, and the number of the associated nodes (from [11])

the amount of calculation is about 67% more than that of the 4 layers. It means that the amount of calculation is greatly increased, but the relative error of the theoretical path cannot be significantly reduced. Therefore, the improved A* search algorithm in this paper considers a node to establish contact with nodes in the associated four layers. In order to determine the specific location of the path, the A* (or the Dijkstra) needs to perform a reverse search after obtaining the length of the fastest path. An array is added to the algorithm to record the coordinates of the previous node at the array position corresponding to the current node. It can avoid the reverse search and increase the efficiency of calculation.

7.1.4 Locating AE Source

The theoretical travel time t_{ij}^{k} is equal to the fastest wave path L_{ij}^{k} divided by the P wave velocity C. As for the source excited at point P_{ij}, the time difference of sensor S_l and sensor S_m is equal to the difference of theoretical travel time Δt_{ij}^{lm}.

According to the principle of least square method, the D_{ij} is introduced to describe the degree of deviation of the point P_{ij} from the unknown AE source by using Δt_{0}^{lm} and Δt_{ij}^{lm}, expressed as

$$D_{ij} = \sum \left(\Delta t_{ij}^{lm} - \Delta t_0^{lm}\right)^2. \tag{7.6}$$

When the grid node p_{ij} in the set P falls into the empty area, $D_{ij} = \infty$. The summation is performed over all the possible sensor pairs. Each grid node p_{ij} will get the corresponding D_{ij} value. The larger the value of D_{ij}, the greater the deviation between the point p_{ij} and the unknown source. Therefore, the position with the minimum D_{ij} value can be considered as the location of the MS/AE source.

7.2 Localization Method for Structures Containing Unknown Empty Areas

The AE source location method for structures containing unknown empty areas (SUEA) can be divided into two steps, including the determination of unknown empty areas and the localization of AE sources. The detailed flowchart of SUEA is shown in Fig. 7.7.

7.2.1 Determination of Unknown Empty Areas

7.2.1.1 On-Site Data Acquisition

Acoustic emission sensors are arranged at different positions in the interesting target area. The arrival time of signal from the active AE source to each sensor is collected on site. The position of the active source is known. The time recorded by the cth sensor S_c is set as t^c and recorded by the rth sensor S_r is t^r. Then the difference between the actual arrivals of the signal received by the S_c and S_r is seen in Eq. (7.7).

$$\Delta t^{cr} = t^c - t^r. \tag{7.7}$$

7.2.1.2 Establishing the Model Data of the Empty Area

Different empty area models would be constructed for the target area. The target area can be meshed according to a certain proportion, and grid intersections are obtained. Each grid intersection can be regarded as a sample point, and a set of sample points are obtained.

Firstly, a sample point is taken as the starting point and all available values of the empty area size are traversed. The empty area sizes take values that are integer multiples of the grid cell size and less than or equal to the size of the target area. In this study, the shapes of the empty area are divided into three categories: square,

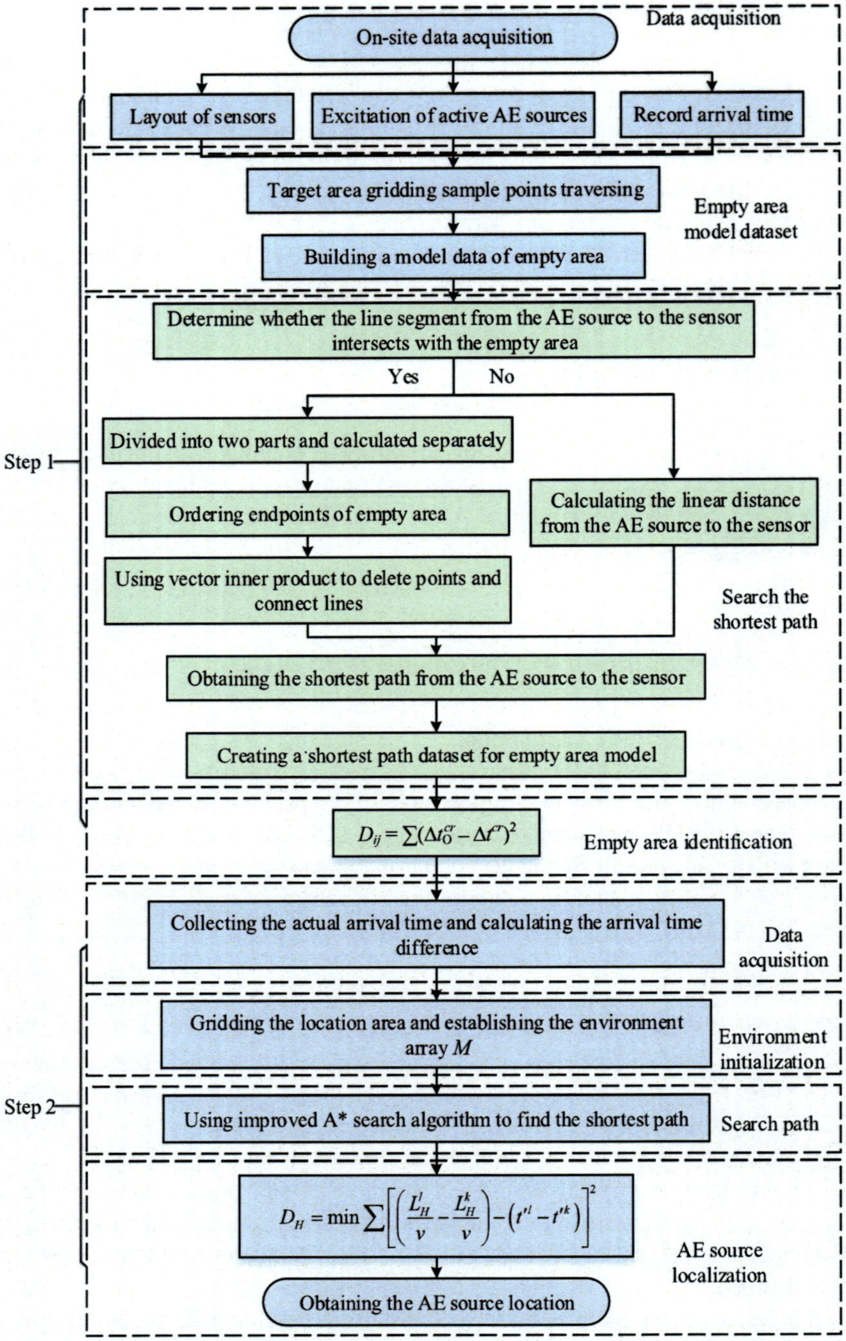

Fig. 7.7 Flowchart of SUEA

oval, and horseshoe, which are the most common void shapes in the underground. Then, all the shapes of empty areas are traversed to obtain all the empty area models in the target region.

7.2.1.3 Calculating the Shortest Time Path from Each Active AE Source to Each Sensor in Each Model

(1) Determining whether the line segment from the AE source to the sensor intersects with the empty area
For convenient application, the circular shape in this study is replaced by an external cut dodecagon, and the horseshoe shape is replaced by an external cut eight-sided shape, as shown in Fig. 7.8. The empty area and the AE source O divide the plane into five parts (except the empty area, in the Fig. 7.8). Method description takes the square hole as an example, and the rest are similar. The directed path from the AE source O to the sensor S is defined as the vector $\overrightarrow{OS}$. The vertices of the square hole are labeled as a, b, c, and d. The vectors $\overrightarrow{Oa}$, $\overrightarrow{Ob}$, $\overrightarrow{Oc}$, $\overrightarrow{Od}$ are made with O as the starting point and a, b, c, d as the ending points; X_{Sa} is obtained by calculating the value of the vector outer product of $\overrightarrow{OS}$ (x_{OS}, y_{OS}) and $\overrightarrow{Oa}$ (x_{Oa}, y_{Oa}) (Eq. 7.8).

$$X_{Sa} = \overrightarrow{OS} \times \overrightarrow{Oa} = x_{OS}y_{Oa} - x_{Oa}y_{OS} \tag{7.8}$$

X_{Sb}, X_{Sc}, and X_{Sd} are all calculated and recorded in the same way, respectively. If X_{Sa}, X_{Sb}, X_{Sc}, and X_{Sd} are all non-positive or all non-negative, it means that the sensor S is in region II or IV. If not, the sensor S would be in region I or III or V. Then, the vector inner products Y_{Sa} of the vector $\overrightarrow{OS}$ with $\overrightarrow{Oa}$ is calculated (Eq. 7.9).

$$Y_{Sa} = \overrightarrow{OS} \cdot \overrightarrow{Oa} = x_{OS}x_{Oa} + y_{OS}y_{Oa} \tag{7.9}$$

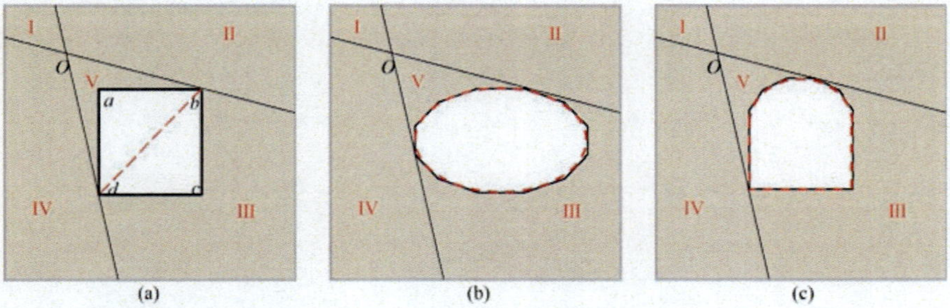

Fig. 7.8 A plane divided into five parts by the source O and the **a** square, **b** ellipse, and **c** horseshoe empty area (from [16])

Y_{sb}, Y_{sc} and Y_{sd} are all calculated and recorded in the same way, respectively. If Y_{Sa}, Y_{Sb}, Y_{Sc} and Y_{Sd} are all negative, it means that the sensor S is in region I. If not, the sensor S would be in region III or V. In the circumstances, the point b and point d can be determined since they correspond to the maximum and minimum values of the vector outer products X (X_{Sa}, X_{Sb}, X_{Sc}, or X_{Sd}). Then, the AE source O and the two points b and d form a triangle. The center of gravity method is then used to confirm whether the sensor S is inside the triangle. If so, it means that the sensor S is in the region V since the sensor S cannot be arranged in the empty area; if not, the sensor S is in region III.

If the sensor S is in region I or II or IV or V, the shortest time path between the AE source O and the sensor S is the length of the line segment of OS, as shown in Fig. 7.9a. If the sensor S is in region III, it is the intersection problem considered in the following. The shortest time path of acoustic wave between the AE source O and the sensor S is then calculated by bypassing the empty area, as shown in Fig. 7.9b. Then, the distance of the shortest time path would require further analysis, which is illustrated in steps (2), (3) in detail.

(2) Ordering endpoints of empty area

As shown in Fig. 7.9b, the OS divides the empty area into two parts. The endpoints of two parts can be divided according to the value of the corresponding vector outer product. The endpoints in Part I have a positive value and that in Part II have the negative value, as shown in Fig. 7.10a. As for the endpoint with the corresponding value of 0, it can be ignored since it is on the connection line between point O and point S. Then, these two parts would be dealt with separately.

For Part I, the projection value P_{Oa} can be obtained by projecting the vectors $\overrightarrow{Oa}$ in the direction of the vector $\overrightarrow{OS}$ (Eq. 7.10).

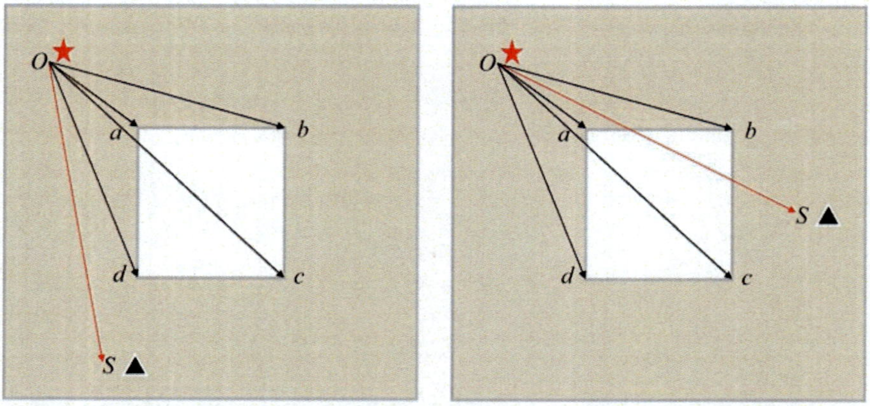

(a) Connecting lines out of the empty areas between the source and sensor

(b) Connecting line between the source and sensor intersects with empty area

Fig. 7.9 Schematic diagram of AE source, sensor, and empty area (from [16])

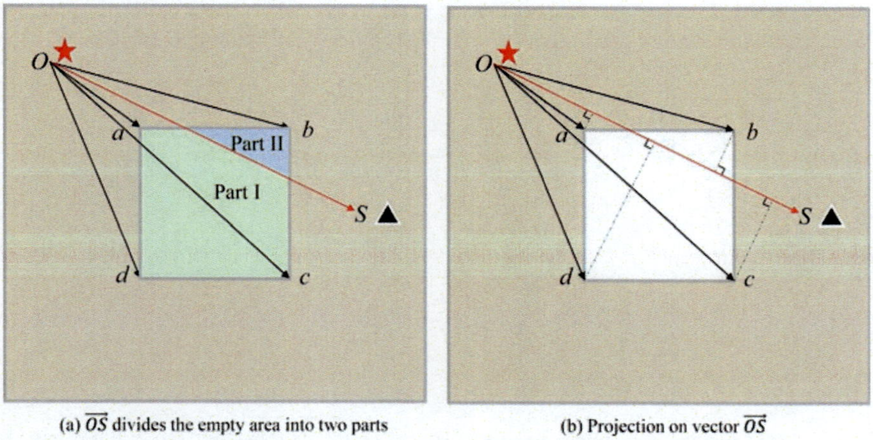

Fig. 7.10 Dividing the empty area into two parts and projecting on the vector $\overrightarrow{OS}$ (from [16])

$$P_{Oa} = \frac{\overrightarrow{OS} \cdot \overrightarrow{Oa}}{\left|\overrightarrow{OS}\right|} \tag{7.10}$$

P_{Ob} and P_{Oc} can also be calculated and recorded in the same way in Fig. 7.10b. The vector projections of each part are sorted by size. According to the length of the projection, the order of the first curve segment can be obtained as *O-a-d-c-S*. Similarly, the order of another curve segment is *O-b-S*.

(3) Deleting unnecessary points

The vector angle between the vector $\overrightarrow{OS}$ and the vector $\overrightarrow{Oa}$ is calculated by Eq. (7.11).

$$\cos\varphi_a = \frac{\overrightarrow{OS} \cdot \overrightarrow{Oa}}{\left|\overrightarrow{OS}\right|\left|\overrightarrow{Oa}\right|} \tag{7.11}$$

where φ_a is the angle between the vector $\overrightarrow{OS}$ and the vector $\overrightarrow{Oa}$. The same is done for the rest of the endpoints. Then, the endpoint that corresponds to the vector with the largest angle to the vector $\overrightarrow{OS}$ would be marked as endpoint 1. For example, as shown in Fig. 7.11a, the endpoint *d* is marked as endpoint 1. Then the vector angle of vector $\overrightarrow{OS}$ and vector $\overrightarrow{Sa}$ is calculated, and the rest of the endpoints are done in the same way. Another endpoint that corresponds to the vector with the largest angle with vector $\overrightarrow{OS}$ can be obtained and marked as endpoint 2. As shown in Fig. 7.11a, the endpoint *c* is then marked as 2. To ensure that the resulting path is the shortest time path for wave propagation, all endpoints between point *O* and endpoint 1 are deleted, and all endpoints between point *S* and endpoint 2 are deleted. Then, the first curve segment would become

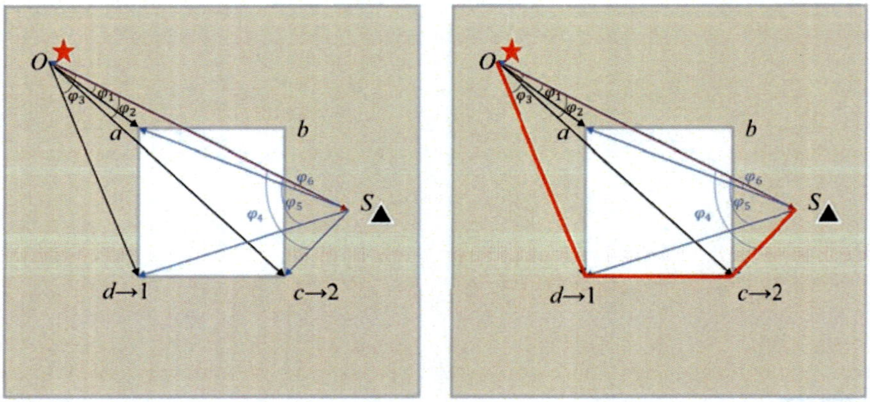

(a) Calculate the included angle (b) Find the maximum angle and delete the remaining points

Fig. 7.11 Deleting points by comparing the included angle size (from [16])

as *O-d-c-S*, as shown in Fig. 7.11b. Another curve segment would be processed in the same way, as shown in Fig. 7.12.

(4) Calculating the shortest path distance and building a dataset
The shortest time path is obtained by comparing the distances of two paths of the two parts. Assuming that there are m sensors and n AE sources, a single empty area model can get $m \times n$ the shortest time paths. The data of the shortest time path for this empty area model is then established.

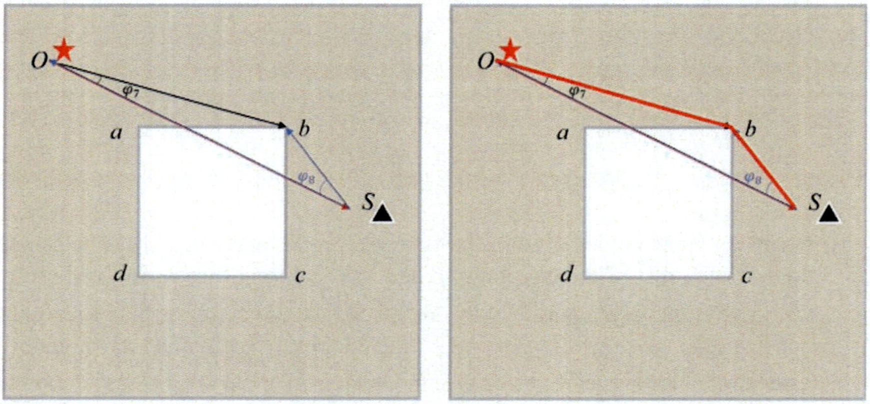

(a) Calculate the included angle (b) Find the maximum angle and delete the remaining points

Fig. 7.12 The shortest path on the other side (from [16])

7.2.1.4 Identifying Empty Areas

In the empty area model data, the shortest time path between the AE source and the sensor is known. The propagation velocity of the AE source signal in the non-empty region is set as v. The value of the difference of the shortest time path between the signal excited by the AE source O to the AE sensor S_c and to the AE sensor S_r is ΔL_O^{cr}. Then, the difference of the theoretical travel time of the signal received by the AE sensor S_c and the AE sensor S_r is seen in Eq. (7.12).

$$\Delta t_O^{cr} = \frac{\Delta L_O^{cr}}{v} \tag{7.12}$$

The deviation is calculated by Eq. (7.13).

$$D = \sum \left(\Delta t_O^{cr} - \Delta t'^{cr}\right)^2 \tag{7.13}$$

The smaller the value of D, the smaller the degree of deviation between the selected empty area model and the actual empty area. When D takes $D_{\min}$, the corresponding empty area model is the identified empty area.

7.2.2 *Localization of AE Sources*

Based on the identification of the empty area in Sect. 7.2.1, AE sources are located in the target area.

7.2.2.1 Data Acquisition

Receiving the acoustic signal (P-wave signal) generated by the unknown actual AE source through the AE sensor, the actual time is recorded as t'^k at which the kth AE sensor S_k receives the acoustic signal.

7.2.2.2 Calculation of Theoretical Values

The target area is gridded at a certain scale to get n grid intersections. Each grid intersection is treated as a node, and a set of n nodes is obtained. An array **M** is created, and the elements in **M** correspond to the nodes one by one. If a node falls in a solid region in the target area, its corresponding element in the **M** matrix is set to 0, representing a passable position for that node. If a node falls in an empty area in the target region, its corresponding element in the **M** matrix is set to 1, which means that the node position is not passable.

All the nodes in the set are traversed, and each node H is regarded as the potential AE source location. The following processing is performed: The improved A* shortest path search algorithm is used to track the shortest time path of signal propagation between H and the kth AE sensor S_k, denoted as L_H^k; If H is in the empty area, then $L_H^k = \infty$; Compared with the A* search algorithm, the improved A* search algorithm adapts grid points instead of center points to search the paths which are more in line with the actual boundary of the empty region. Besides, the improved A* search algorithm can search more accurate paths by using more connection nodes.

7.2.2.3 Locating Calculation

D_H is introduced to describe the degree of the deviation of each node H from the actual AE source location is estimated. The node with the smallest deviation from the actual AE source position among all nodes is determined, and the coordinates of this node are used as the positioning coordinates of the actual AE source. D_H is calculated as shown in Eq. (7.14).

$$D_H = \min \sum \left[\left(\frac{L_H^l}{v} - \frac{L_H^k}{v} \right) - \left(t'^l - t'^k \right) \right]^2 \tag{7.14}$$

A smaller value of D_H indicates that the deviation of node H from the actual AE source location is the smaller, and thus the node corresponding to the smallest D_H is regarded as the position of the actual AE source.

7.3 Localization Method for the Hole-Containing Structure

The VFH uses the improved A* search algorithm and the location method without pre-measured velocity to estimate the location of the MS/AE source. It can be divided into four parts: first, dividing the mesh of the object which needs locating and represents the shape of the object with 0 and 1; second, acquiring the P-wave arrivals signal generated by the MS/AE event received by the sensor; then, using the A* search algorithm to search for the fastest waveform path between each sensor and each grid point; finally, the minimum deviation amount D is introduced to calculate the position of the MS/AE source. Figure 7.13 shows a flow chart of the location process of the VFH. This section describes each part of the method.

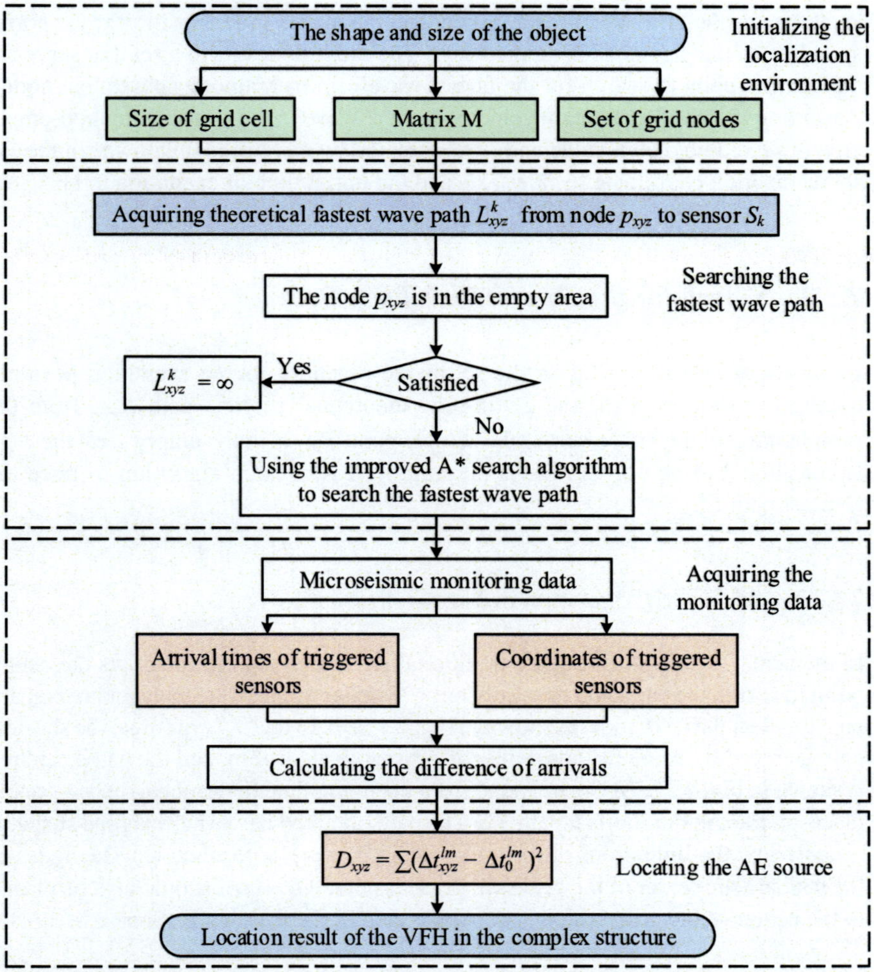

Fig. 7.13 The flowchart for the locating process of the VFH. S_k: the kth sensor receiving the signal; D_{xyz}: the deviation degree of the point P_{xyz} from the unknown MS/AE source P_0; Δt_0^{lm}: the actual time difference of the two different sensors S_l and S_m; Δt_{xyz}^{lm}: the travel time difference between two different stations l and m

7.3.1 Determine the Initial Environment

The geometry and specific location of the empty area is determined in the location area. Moreover, the size of the unit square grid is determined according to the condition of the empty area and the accuracy requirement of the position. In general, the denser the grid is, the higher the location accuracy is, while the calculation will be multiplied. However, the location accuracy would not be obviously improved by partitioning when the grid is sufficiently densely divided. A zero matrix M with the same size as the grid node is established, and the matrix index position x, y, z corresponding to the grid node is marked as 1. The mesh nodes form a set that serves as the starting point to search for the fastest waveform path among subsequent nodes. Assuming that the propagation velocity of the P wave in the surrounding non-empty area is a fixed unknown value and is represented by C. For the unknown source P_0, set its position coordinate to x_0, y_0, z_0, and the initial time of excitation to t_0.

7.3.2 Search for the Fastest Waveform Path

Using every grid point P_{xyz} in the set as the potential source excitation position, tracing the shortest path, and getting the theoretical shortest path L_{xyz}^k from the sample point to the kth sensor. If the sample point P_{xyz} is in the empty area, the L_{xyz}^k is considered to be ∞. VFH uses the improved A* search algorithm to trace the shortest path L_{xyz}^k.

7.3.2.1 A* Search Algorithm in 3-D Application

In the actual 3-D application, the traditional A* search algorithm adopts the center point to search the path, and generally only 26 nodes of the adjacent layer are considered to select the next node, as shown in Fig. 7.14a, b. In the "L" type area, the shortest path is traced by using the conventional A* search algorithm, and the path searched is shown in Fig. 7.14c. It can be found from the figure that there are two unreasonable places in tracing the shortest path. (1) The path obtained has a sharp jagged shape. It is caused by the limitations of the traditional A* search algorithm. (2) The nodes of the path are the center of the grid, which means that the sensor should also be placed in the center of the grid, which is not applicable in the real application.

7.3.2.2 Improved A* Search Algorithm

To trace the shortest path more effectively, the A* search algorithm is improved to search by grid points, as shown in Fig. 7.15. This allows the nodes of the searched path to be on the mesh node, which also means that the sensor can be attached to the surface node of the object. It is more in line with the actual situation. To avoid a sharp jagged shape of the searched path, an effective connection between the node and nodes around it is established. In the traditional A* search algorithm, one node expands to 26 nodes in an adjacent layer, as shown in Fig. 7.15a. This means that the current node can only expand around 26 directions. The more nodes in adjacent layers connected with the current node, the more directions available for the path node to select, which leads to the increasing accuracy of the searched path. Because of the symmetry of the directions between the nodes, only one eighth of the total

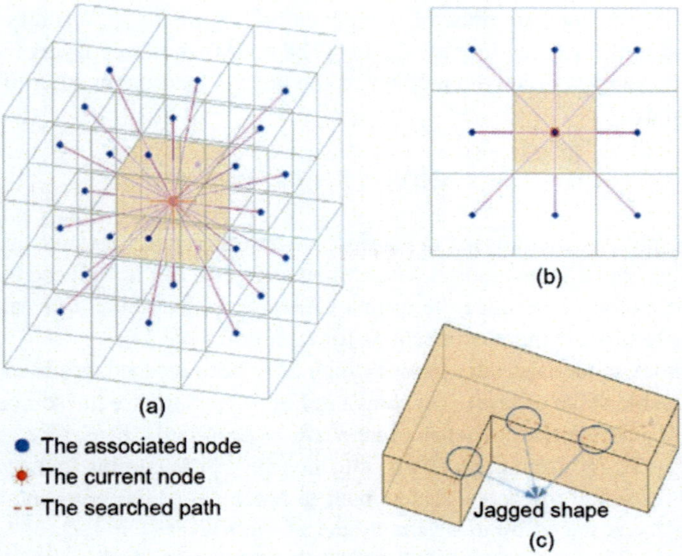

Fig. 7.14 The traditional A* search algorithm. **a** The current node is connected to the associated 26 nodes; **b** the main view of (**a**); **c** the path searched by the traditional A* search algorithm (from [17])

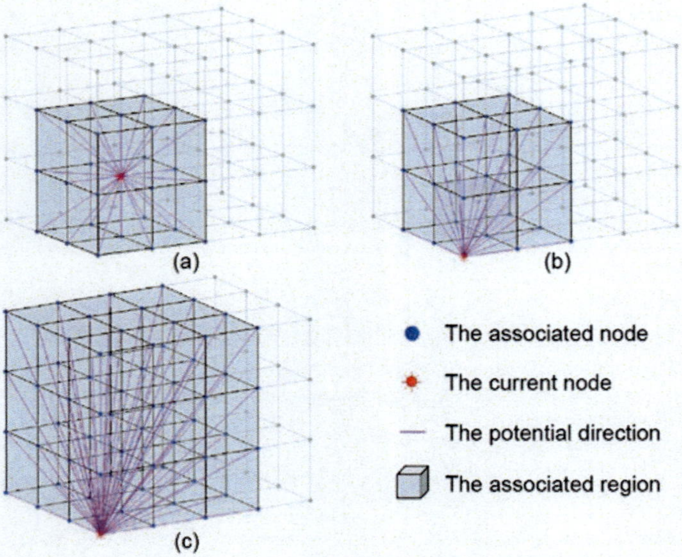

Fig. 7.15 The potential direction between the current grid node and the associated grid nodes. **a** The current node is connected to the adjacent one layer. **b** The current node is connected to the adjacent two layers. **c** The current node is connected to the adjacent three layers (from [17])

number of directions are drawn for explanation. Figure 7.15b, c shows the grid is connected to the surrounding two layers (124 nodes) and three layers (342 nodes). The relationship between the number of the node Z_i and the number of layers i is expressed as

$$z(i) = (2i + 1)^3 - 1. \quad (7.15)$$

In the process of outward expansion, certain directions formed by nodes are repeated, so they can be ignored. The computation cost can be reduced by removing these grid points. To reduce the error of the search, the amount of calculation is increased to provide more connections for each grid node.

The block model and strip model which have been meshed, are used to explore the reasonable value of layer i, as shown in Fig. 7.16. Suppose the source is excited at point O, and the wave reaches point K ($K = A, B, C, D, E, F, G, H, A', B', C', D', E', F', G', H'$) from point O, forming the path L_K. According to the geometrical relation, recording the actual fastest path distance D_R of the wave into Table 7.1. The path L_K is traced by using the model of the i layer (i = 1, 2, 3, 4, 5, 6, ...), respectively. The relative error E between the searched path distances D_{Si} and D_R can be expressed as:

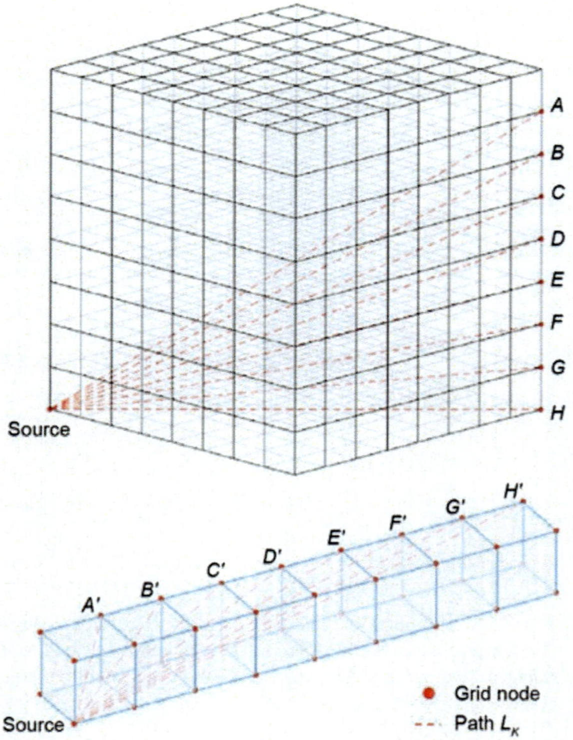

Fig. 7.16 The ideal paths from the source node to neighbor nodes K ($K = A, B, \cdots, H'$) (from [17])

Table 7.1 Relative error between the distance of actual path and the distance of searched path

Path symbol	D_R	Relative error E (%) of searching path using the model of layer i							
		$i=1$	$i=2$	$i=3$	$i=4$	$i=5$	$i=6$	$i=7$	$i=8$
L_A	11.31	1.76	0.65	0.35	0.20	0.12	0.07	0.03	0
L_B	11.36	3.24	0.92	0.30	0	0	0	0	0
L_C	11.49	4.31	0.73	0.09	0.09	0.02	0.02	0.02	0
L_D	11.70	4.88	0	0	0	0	0	0	0
L_E	12.00	4.81	1.01	0.11	0.11	0.02	0.02	0.02	0
L_F	12.37	4.01	1.40	0.49	0	0	0	0	0
L_G	12.81	2.41	1.10	0.63	0.38	0.23	0.13	0.06	0
L_H	13.30	0	0	0	0	0	0	0	0
L'_A	1.73	0	0	0	0	0	0	0	0
L'_B	2.45	11.54	0	0	0	0	0	0	0
L'_C	3.32	12.53	4.01	0	0	0	0	0	0
L'_D	4.24	11.54	4.88	1.74	0	0	0	0	0
L'_E	5.20	10.31	4.88	2.32	0.89	0	0	0	0
L'_F	6.16	9.21	4.62	2.47	1.27	0.51	0	0	0
L'_G	7.14	8.27	4.31	2.45	1.42	0.77	0.32	0	0
L'_H	8.12	7.48	4.01	2.37	1.46	0.89	0.50	0.21	0
$E_{i-\max}$	–	12.53	4.88	2.47	1.46	0.89	0.50	0.21	0

$$E_i = \left(\frac{D_{Si}}{D_R} - 1\right) \times 100\%. \quad (7.16)$$

The relative error E of obtained path is recorded in Table 7.1, and the maximum path error $E_{i\text{-max}}$ of each model is selected. The time complexity O of the improved A* search algorithm is positively related to the number of expanded nodes Z, i.e. $O_i \propto Z_i$. Thus, the number of nodes Z can be used to approximate the calculation amount O. Figure 7.17 shows the relationship between the number of layers i and $E_{i\text{-max}}$ and the number of associated nodes Z_i, respectively.

Through different testing, we found that the relative error of the theoretical path is less than 3%, when the layer number of topologies reaches three. The relative error is influenced by the size of sensors or the system error. This means the algorithm can basically meet the location requirement with relatively small growth rate of calculation amount. Therefore, this article would utilize the A* search algorithm to establish connection between a node and the surrounding three layers. In order to determine the specific location of the path, after determining the length of the shortest path by A* search algorithm (or the Dijkstra algorithm), a reverse search is also needed. Therefore, this article adds the function of recording the coordinates of the previous node at the corresponding position of the current node when searching

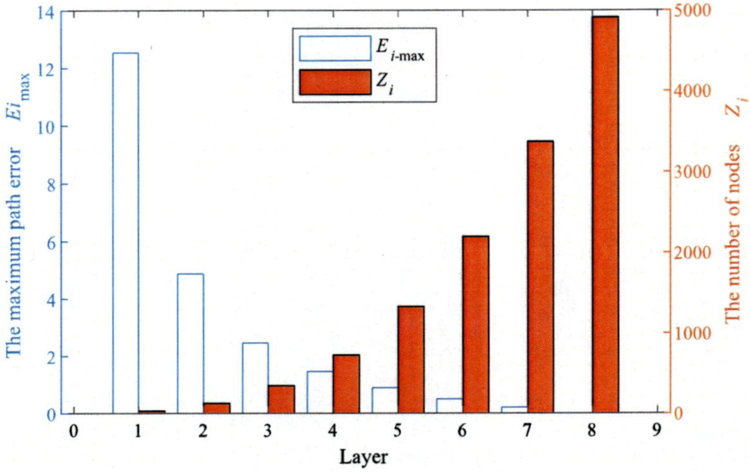

Fig. 7.17 The number of layers, relative errors, and the number of the associated nodes

for a path. It is also helpful to avoid reverse search and enhance computational efficiency.

7.3.3 Collect Data of Arrivals

Suppose that m sensors are installed at different positions of the specimen to capture the arrivals of the MS/AE. Each sensor is located on a grid point so the location is known. For the 3-D model, there are five unknowns (wave velocity of P wave, coordinates of MS/AE source x_0, y_0, z_0, and initial time of excitation t_0), so sensor number m should be an integer greater than or equal to five. For the kth sensor S_k receiving the signal, the coordinates are recorded as x_k, y_k, z_k, and the first arrival time of the received P wave signal is t_0^k. Calculate the actual time difference of the two different sensors S_l and S_m, denoted by Δt_0^{lm}.

7.3.4 Source Location

For the source generated by the point P_{xyz}, the theoretical travel time t_{xyz}^k corresponds to the shortest propagation path L_{xyz}^k between the source and the kth sensor divided by the velocity C. The time difference between two different stations l and m corresponds to the difference in travel time Δt_{xyz}^{lm}.

According to the square of the difference between Δt_0^{lm} and Δt_{xyz}^{lm}, D_{xyz} is introduced to describe the deviation degree of the point P_{xyz} from the unknown MS/AE source P_0, which is expressed as

$$D_{xyz} = \sum \left(\Delta t_{xyz}^{lm} - \Delta t_0^{lm}\right)^2. \tag{7.17}$$

Among them, when the sources fall into the empty area, $D_{xyz} = \infty$.

Each grid point will get the corresponding D_{xyz} value. The deviation degree of the point P_{xyz} from the unknown source P0 increases with the growth of the value of D_{xyz}. Therefore, the coordinate corresponding to the minimum D_{xyz} value can be considered as the coordinate of the MS/AE source location.

7.4 Verification and Discussions

7.4.1 Verification for ALM

Virtual localization tests and lead-breaking tests were carried out to evaluate the performance of the proposed ALM method. Virtual localization was used to compare the ALM with the FastWay [2], whereas the lead-breaking test was conducted to compare four localization methods, including: the ALM, the FastWay, the TD, and the TM [18].

7.4.1.1 Virtual Localization Tests

The center node of the cell was applied to connect the adjacent center nodes when the FastWay was applied for searching, whereas the grid node was applied to the ALM. Thus, the sensors should be located at the center nodes when using the FastWay, however the sensors should be located at the grid nodes when using the ALM in the same grid. The difference in sensor arrangement of the two methods would cause trouble in evaluating the performance of the two methods. Virtual localization test was specially designed to accommodate the difference in sensor arrangement of the two methods. It was utilized to compare the performance between the ALM and the FastWay.

There were three cavities in a square plane of 20 m × 20 m, as shown in Fig. 7.18. 400 grid cells and 441 grid nodes were obtained by dividing the 2-D plane into 20 × 20 grid cells. Four sensors were arranged on the grid nodes for localization of ALM. The coordinates of the four sensors were S_A^1 (1, 20), S_A^2 (19, 20), S_A^3 (2, 3), S_A^4 (18, 2). Generally, the P-wave velocity of granite is from 4, 000 to 6, 000 m/s. The propagation velocity of the wave in the medium was assumed to be 5, 000 m/s, and the occurrence time of the AE event was 0 s as we identified the moment of the event occurrence as the initial moment. In fact, it is no problem that the propagation speed of waves and the occurrence time of events are assumed to other values. The coordinates of the four sensors S_A^1, S_A^2, S_A^3, and S_A^4 were translated by vector (0.5,

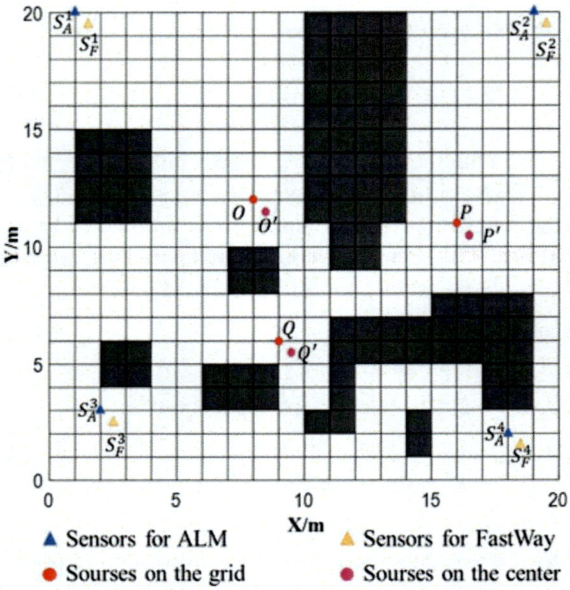

Fig. 7.18 Distribution of sensors and AE sources (from [11])

−0.5) to obtain the coordinates of the center nodes of the four cells. The other four sensors were arranged at these 4 center nodes for the FastWay with coordinates: S_F^1 (1.5, 19.5), S_F^2 (19.5, 19.5), S_F^3 (2.5, 2.5), and S_F^4 (18.5, 1.5). The AE sources O (8, 12), P (16, 11), and Q (9, 6) were assumed to be on the grid nodes. Three other new AE sources were assumed on the center nodes of the grid cell. The coordinates of the new AE sources were O' (8.5, 7.5), P' (16.5, 10.5), and Q' (9.5, 5.5). The spatial distribution of sensors and AE sources were shown in Fig. 7.18. The distance from each AE source to each sensor was calculated by geometry and listed in Table 7.2. The arrivals of P-wave can be obtained according to the assumed medium velocity and the distance in Table 7.2.

Table 7.2 The distances of fastest path between assumed sources and sensors (unit: m)

Sensor	Sources					
	O	O'	P	P'	Q	Q'
S_A^1	11.31	12.02	20.47	20.67	16.28	16.98
S_F^1	10.61	11.31	19.77	19.97	15.59	16.30
S_A^2	18.77	18.07	9.49	9.82	17.53	17.48
S_F^2	18.59	17.88	9.19	9.49	17.35	17.30
S_A^3	11.18	11.07	15.70	16.33	7.63	8.01
S_F^3	11.07	10.95	15.53	16.15	7.46	7.84
S_A^4	18.72	18.01	10.66	9.95	12.41	11.82
S_F^4	18.89	18.18	10.82	10.12	12.78	12.19

The ALM and the FastWay were used to locate six AE sources when using real velocity (the error of velocity is 0%). The location results and errors of the two methods were recorded in Table 7.4. When the ALM was used for localization, the location errors of three AE sources at grid nodes were 0 m, and those at center nodes were 0.7 m. When using the FastWay at real velocity, the location errors of three AE sources at grid nodes were 0.7 m, and those at the center nodes showed that errors of two from them were 0 m and one was 1 m. In practical monitoring, the wave velocity would be greatly influenced by environmental factors such as the pressure. It was assumed that the velocity had a floating error of ±5%, ±10%, ±15%, and ±20%, respectively. The FastWay was used to calculate the coordinates of AE source with the velocity error. The absolute errors of X-coordinates E_x, the absolute errors of Y-coordinates E_y, and the absolute location errors E_l of each calculated AE source were recorded in Tables 7.3, 7.4, and 7.5. The maximum absolute location errors E_l of the FastWay was 3.0 m in Tables 7.3, 7.4, and 7.5. The location error increased with the velocity error increasing. It was obvious that the velocity error would induce great error to the location result of the FastWay.

The results show that the location errors of FastWay and ALM are small when the true velocity of the object is known. However, the location accuracy of ALM is higher than that of FastWay with real wave velocity. The reason is that the fastest path searched by ALM is more accurate than the path obtained by FastWay. Tables 7.3, 7.4, and 7.5 show that the location error of FastWay increases as the floating error of velocity increases. In order to get the satisfactory location result with the FastWay, the error of the wave velocity must be within 5%, which is demanding. It is difficult to accurately determine the material velocity, which is not completely uniform. Li et al. [19] performed multiple wave velocity tests on different locations of the same rock block with no obvious defects. He found that the difference in the average velocity of the measurement would reach more than 10%. Moreover, the generation of AE events implies the change in the environment of the object to be tested, which would

Table 7.3 The location errors of source O and source O'

Method	Error of the wave velocity (%)	Source O			Source O'		
		E_x (m)	E_y (m)	E_l (m)	E_x (m)	E_y (m)	E_l (m)
ALM		0	0	0.0	0.5	0.5	0.7
FastWay	0	0.5	0.5	0.7	1	0	1.0
	5	−0.5	0.5	0.7	0	0	0.0
	−5	1.5	0.5	1.6	1	0	1.0
	10	−1.5	1.5	2.1	−1	1	1.4
	−10	1.5	0.5	1.6	2	1	2.2
	15	−2.5	1.5	2.9	−2	1	2.2
	−15	1.5	0.5	1.6	2	1	2.2
	20	−2.5	1.5	2.9	−2	1	2.2
	−20	2.5	1.5	2.9	2	1	2.2

Table 7.4 The location errors of source P and source P'

Method	Error of the wave velocity (%)	Source P			Source P'		
		E_x (m)	E_y (m)	E_l (m)	E_x (m)	E_y (m)	E_l (m)
ALM		0	0	0.0	0.5	−0.5	0.7
FastWay	0	0.5	−0.5	0.7	0	0	0.0
	5	0.5	0.5	0.7	0	0	0.0
	−5	0.5	−0.5	0.7	0	0	0.0
	10	−0.5	0.5	0.7	0	1	1.0
	−10	0.5	−0.5	0.7	0	−1	1.0
	15	−0.5	1.5	1.6	0	2	2.0
	−15	1.5	−0.5	1.6	1	−1	1.4
	20	−0.5	2.5	2.5	0	3	3.0
	−20	1.5	−1.5	2.1	1	−1	1.4

Table 7.5 The location errors of source Q and source Q'

Method	Error of the wave velocity (%)	Source Q			Source Q'		
		E_x (m)	E_y (m)	E_l (m)	E_x (m)	E_y (m)	E_l (m)
ALM		0	0	0.0	0.5	−0.5	0.7
FastWay	0	0.5	−0.5	0.7	0	0	0.0
	5	0.5	−0.5	0.7	0	0	0.0
	−5	−0.5	−0.5	0.7	0	0	0.0
	10	0.5	−0.5	0.7	1	1	1.4
	−10	−0.5	−0.5	0.7	0	0	0.0
	15	1.5	1.5	2.1	1	1	1.4
	−15	−0.5	−0.5	0.7	0	0	0.0
	20	1.5	1.5	2.1	2	1	2.2
	−20	−0.5	−0.5	0.7	−1	0	1.0

cause the change of wave velocity in the monitoring process. The location accuracy of the ALM is not affected by the given velocity because the ALM can accurately fit the relationship between the coordinates of the sensors and the time difference. In conclusion, ALM has more obvious advantages than FastWay.

7.4.1.2 Lead-Breaking Tests

The lead-breaking test was conducted to evaluate the performance of the ALM. As shown in Fig. 7.19a, a complex geometric specimen was made of 15 cm thick marble, with a total size of 60 cm × 60 cm. Three empty areas, circular, rectangular and triangular, were cut out in the specimen. Figure 7.19b shows the shape and size

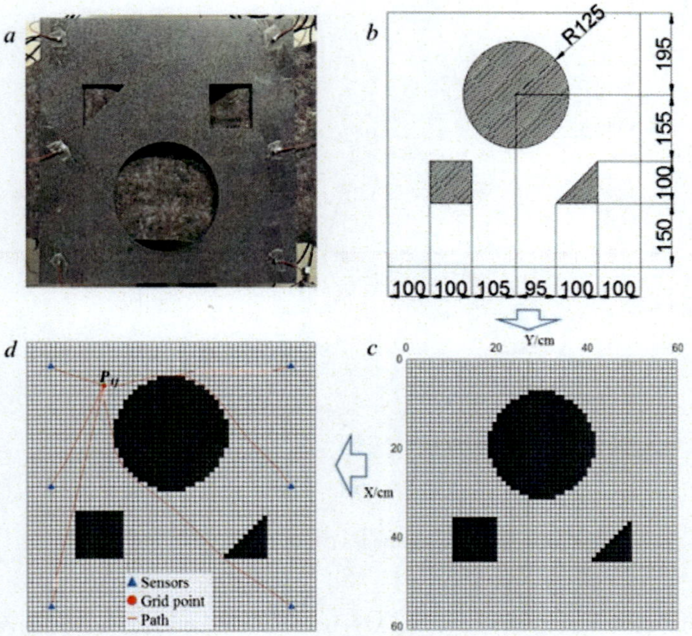

Fig. 7.19 The process of localization with the ALM. **a** The complex geometric; **b** the size and the shape of the specimen; **c** adjustment of the coordinate system; and **d** the paths searched by the improved A* search algorithm from a node to six sensors (from [11])

of the specimen. The coupling agent was added between the sensor and the specimen to provide better acoustic coupling. A layer of sound insulation cotton was placed under the specimen to reduce the impact of noise. Figure 7.19d shows the relative locations of six AE sensors on the specimen. The coordinates (unit: cm) of the sensor are as follows: S_1 (5, 5), S_2 (30, 5), S_3 (55, 5), S_4 (5, 55), S_5 (30, 55), S_6 (55, 55). The resonance frequency of P-wave of the granite specimens is usually less than 100 kHz.

The resonance frequency of P-wave of the granite specimens falls in the range from 1×10^4 Hz to 5×10^5 Hz. To avoid the distortion of the waveform signal, the sampling frequency needs to be greater than ten times of the resonant frequency for high frequency sampling in the tests. Therefore, the sampling rate is kept at 5 MHz to collect AE data. The threshold of 30 dB was chosen for AE data acquisition since the maximum amplitude of background noise is about 30 dB. The specimen was divided into 60×60 grids. The coordinate system in the specimen was adjusted to correspond with the environment matrix **M** (in Fig. 7.19c). The path from each grid node p_{ij} to each sensor can be searched by the improved A* search algorithm (Fig. 7.19d).

When the threshold was supposed to be set as θ dB, the time difference used for localization was the time difference $\Delta t'_{12}$ in where the instantaneous intensity of P wave pulses was θ dB instead of the actual arrival-time difference Δt_{12}. Figure 7.20

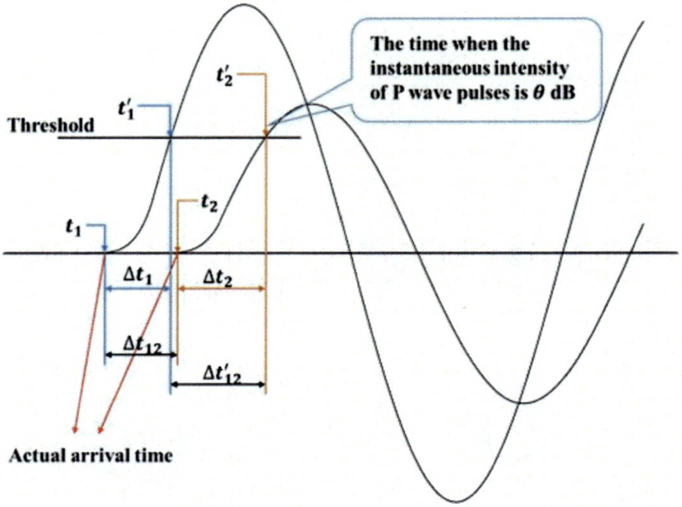

Fig. 7.20 Waveforms with different amplitudes (from [11])

shows that the time difference (Δt_1, Δt_2) between the actual P-wave arrival time (t_1, t_2) and the time (t_1', t_2') where the instantaneous intensity of P wave pulses is θ dB.

The intensity of P wave pulses is related to the amplitude. The Δt_1 and Δt_2 would change with the change of the P wave amplitude. The relationship between $\Delta t_{12}'$ and Δt_{12} is as follows

$$\Delta t_{12}' = t_2' - t_1' = (t_2 + \Delta t_2) - (\Delta t_2) = \Delta t_{12} + (\Delta t_2 - \Delta t_1) \tag{7.18}$$

When difference of the distances from AE source to each sensor is large, the amplitude of P wave at each sensor position is also quite different. At this time, the value of $(\Delta t_2 - \Delta t_1)$ cannot be ignored. If the time difference $\Delta t_{12}'$ is still used for localization, it will result in significant errors. Therefore, attention should be paid to the selection of lead-breaking points on specimens to verify the effectiveness of the algorithm. The amplitude of the P waves received by each sensor should be as close as possible to avoid the interference caused by the pickup error. If the test can meet the requirements of collecting the full waveform data, then more accurate arrivals can be picked up by AIC and other methods [20, 21].

A series of points were selected as the AE sources on the specimens. Two lead-breaking tests were carried out at each AE source. The coordinates of AE source were recorded in Table 7.6. The arrivals of P waves t_i ($i = 1, 2, \ldots, 6$) were also recorded in Table 7.6, which were received by six sensors. The locations of AE source were calculated by using ALM, FastWay, TD, and TM respectively. The coordinates of location results were recorded in Table 7.6. The absolute location errors E_l of four methods were also obtained in Table 7.6. Figure 7.21 shows the visualization of the location results. It can be seen that four methods all have obtained satisfactory results with the average error less than 3 cm. The comparison of Table 7.6 directly shows

Table 7.6 Location results and errors of four methods (Unit: cm)

The coordinates of AE source	ALM			FastWay			TD			TM		
	X	*Y*	*D*	*X*	*Y*	*D*	*X*	*Y*	*D*	*X*	*Y*	*D*
(15, 15)	14	15	1.0	13	13	2.8	15.7	12.7	2.4	15.0	10.0	5.0
	14	15	1.0	14	13	2.2	15.7	12.7	2.4	15.4	9.5	5.5
(25, 15)	27	16	2.2	28	14	3.2	26.6	13.1	2.5	29.2	11.7	5.3
	26	16	1.4	26	15	1.0	27.2	13.8	2.5	26.9	13.3	2.6
(30, 15)	29	15	1.0	29	15	1.0	29.3	12.3	2.8	30.9	14.2	1.2
	30	15	0.0	30	15	0.0	29.3	12.3	2.8	29.4	12.7	2.4
(35.25)	34	25	1.0	34	25	1.0	34.5	24.7	0.6	34.5	24.5	0.7
	34	24	1.4	34	25	1.0	33.8	24.5	13	31.0	25.6	4.1
(40.25)	40	24	1.0	39	24	1.4	39.0	24.7	1.0	40.6	24.5	0.8
	39	24	1.4	39	25	1.0	38.7	24.9	1.3	40.6	24.5	0.8
(45.25)	45	24	1.0	44	25	1.0	42.8	25.9	2.3	43.6	25.6	1.5
	45	24	1.0	44	25	1.0	45.2	27.9	2.9	44.7	27.9	2.9
(50, 25)	48	25	2.0	48	25	2.0	50.3	27.5	2.5	48.3	25.9	1.9
	49	26	1.4	49	26	1.4	48.4	25.9	1.8	49.4	27.8	2.8
(35, 35)	34	35	1.0	34	35	1.0	34.9	34.8	0.3	34.7	34.9	0.3
	34	35	1.0	34	35	1.0	33.9	37.2	2.5	33.9	37.0	2.3
(40, 35)	39	35	1.0	38	35	2.0	38.2	34.6	1.8	39.8	34.3	0.7
	39	35	1.0	38	35	2.0	42.0	33.2	2.7	39.8	34.3	0.7
(45, 35)	43	35	2.0	42	34	3.2	43.7	35.6	1.5	43.2	35.4	1.8
	43	35	2.0	43	34	2.2	43.7	35.6	1.5	43.2	35.4	1.8
(50, 35)	49	35	1.0	47	35	3.0	46.9	35.2	3.1	49.1	35.8	1.2
	48	35	2.0	47	35	3.0	51.8	34.7	1.8	47.3	35.1	2.7
(35.40)	34	38	2.2	34	39	1.4	32.6	38.3	3.0	34.5	41.5	1.6
	33	39	2.2	33	40	2.0	33.8	39.4	1.4	33.8	39.5	1.3
(25, 45)	27	44	2.2	28	47	3.6	28.4	48.4	4.8	28.4	48.4	4.8
	26	43	2.2	28	46	3.2	28.9	47.4	4.6	28.0	47.6	4.0
(30, 45)	30	44	1.0	30	47	2.0	33.9	47.1	4.5	31.8	47.4	3.0
	30	45	0.0	29	47	2.2	33.3	48.1	4.6	35.0	50.8	7.6
Average error			1.4			1.8			2.4			2.6

that the location accuracy of the ALM is higher than that of the FastWay, the TD, and the TM. It is embodied in two aspects: (1) The maximum location error of the ALM is 2.2 cm, whereas the maximum positioning errors of the FastWay, the TD, and the TM are 3.6 cm, 4.6 cm, and 7.6 cm respectively; (2) The average location error of the ALM is 1.4 cm, whereas the FastWay, TD, and the TM are 1.9 cm, 2.4 cm, and 2.6 cm, respectively. Compared with FastWay, the average location accuracy of

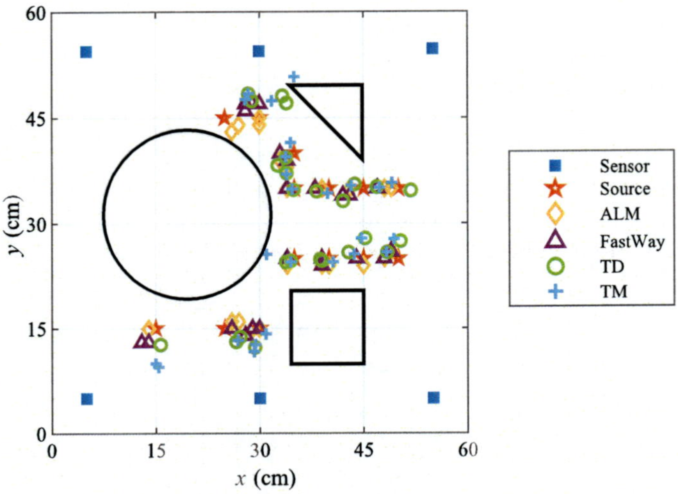

Fig. 7.21 Location results of AE sources using four methods on a complex specimen

ALM has managed a 26% improvement. Moreover, ALM can further densify the mesh to improve the location accuracy of the AE source.

Further analysis of the Fig. 7.21 reveals that the location accuracy of the four methods is higher when the AE sources are at the center of the sensor array. The location results far from the center of the sensor array are affected by the picking error of the arrival-time. Pearson M R et al. [22] pointed out that lowering the threshold level can also improve the location accuracy. Figure 7.20 can explain the viewpoint of Pearson M R, etc. The value of $(\Delta t_2 - \Delta t_1)$ can be reduced by lowering the threshold value, which can improve the location accuracy.

7.4.2 Verification for SUEA

To verify the practical feasibility of this method for complex structural empty area identification and localization, AE tests were carried out on five square rock slabs containing empty areas of different shapes and sizes. The AE monitoring equipment that used was the AMSY-6 multi-channel device (composed of parallel measurement channels) from Vallen, Germany. The system has 32 channels for AE signal acquisition, with a sampling frequency of 10 MHz and a preamplifier gain of 34 dB. It can be used to identify, measure, and locate the AE sources of materials. It records information on the temporal and spatial changes of micro-level damage of the test object in the case of rupture, corrosion, leakage, partial discharge, friction, wear, as well as impact. The system is equipped with the dual channel AE signal processor ASIP-2, which has a resolution of up to 40 MHz sampling rate at an 18-bit dynamic range and a high sensitivity. The acquisition of AE signals is carried out simultaneously during the test, and the acquisition and analysis of AE characteristic parameters and waveforms are carried out. The type of AE sensor adopted is VS45-H. The VS45-H is a passive piezoelectric AE sensor with a wide frequency response. Its frequency response is characterized by a peak at 280 kHz and can be used in the frequency range from 20 to 450 kHz. It is a broad band response AE sensor covering the range of low frequency and standard frequency. The main chassis adopts military grade VME bus structure with multiple transmission modes. The transmission speed can reach 5 Gbps, which can better meet the measurement and analysis requirements of this test.

The common rock slabs with less anisotropy were used in the test. Five square rock slabs with dimensions of 60 cm × 60 cm and thickness of about 1 cm were used. On the rock slab, 20 cm × 20 cm square holes, 20 cm × 40 cm rectangular holes, $R = 10$ cm circular holes, $a = 20$ cm, $b = 10$ cm oval holes and horseshoe-shaped holes consisting of 20 cm square sides and 10 cm radius semicircle were machined and cut respectively. The parallelism, flatness, and perpendicularity of the rock sample were in accordance with the test protocol. The specimen, AE sensor, preamplifier, AE detector and computer were connected in an orderly manner. An appropriate amount of Vaseline was applied between the sensor and the specimen to obtain better acoustic coupling, and a layer of acoustic bubble paper was placed underneath the specimen to reduce the effect of noise. Figure 7.22 shows the schematic diagram of the test setup.

In order to identify and locate the empty area, a total of 28 sensors were arranged around each specimen in this experiment. Their relative positions are shown in Fig. 7.22. The coordinate system is also shown in Fig. 7.22 and the coordinates of the sensor positions are shown in Table 7.7.

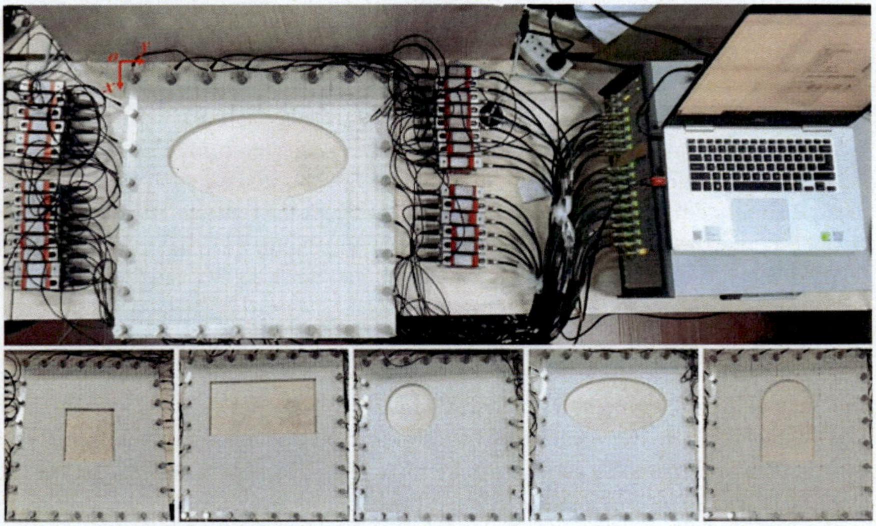

Fig. 7.22 Experimental setup diagram (from [16])

Table 7.7 Sensors position coordinates

No.	x (cm)	y (cm)	No.	x (cm)	y (cm)
1	2	2	15	58	58
2	2	10	16	58	50
3	2	18	17	58	42
4	2	26	18	58	34
5	2	34	19	58	26
6	2	42	20	58	18
7	2	50	21	58	10
8	2	58	22	58	2
9	10	58	23	50	2
10	18	58	24	42	2
11	26	58	25	34	2
12	34	58	26	26	2
13	42	58	27	18	2
14	50	58	28	10	2

Lead break tests were conducted next to each sensor before starting the experiment. It was determined that the amplitude of the lead-breaking signal received by each sensor from the adjacent source was above 95 dB to ensure the coupling quality of the sensor and the specimen. Considering reducing the influence of environmental noise and improving the reliability of signal acquisition, the threshold was set at 35 dB according to the monitoring results of environmental background noise.

The experimental procedure was divided into two parts: empty area identification and AE source localization. (1) Empty area identification: During the experiment, the pulse signal from the sensor was used as the AE source. The pulse emission time interval was 1 s. There were 28 channels, and each channel emitted four pulses. The AE acquisition system recorded the initial arrival time of each sensor. (2) AE source localization: The points next to the empty area were selected for the lead break. This is because the edge stress of the empty area is high and it is easy to crack, which is used to simulate the fracture signal. In order to avoid the adverse effects of human errors on the experiment during lead breaking, a diameter of 0.5 mm and hardness of HB pencil core with a special pencil (which can ensure a 30° angle between the core and the surface of the rock slab) was used.

In order to explore the problem of AE source localization in the target area containing unknown empty area, a newly AE source localization method is proposed and the validity of the method is verified by experiments. For each lead break point, SUEA, and TD these two location methods were used for locating and error analysis, and the location results are shown in Tables 7.8, 7.9, 7.10, 7.11, and 7.12. In order to display the location results more visually, the location results of the five specimens are plotted as shown in Fig. 7.23. As can be seen from the results, the new location method got satisfactory results on all five specimens. The location errors of SUEA

Table 7.8 Location results of Specimen 1

No.	Actual lead break point (cm)		SUEA location result (cm)		SUEA location error (cm)	TD location result (cm)		TD location error (cm)
	x	*y*	*x*	*y*		*x*	*y*	
1	18	18	17	17	1.4	19.3	18.4	1.3
2	18	26	17	26	1.0	5.5	28.8	12.8
3	18	34	17	34	1.0	12.4	35.9	6.0
4	18	42	18	42	0	19.4	41.2	1.6
5	26	42	25	42	1.0	24.0	45.7	4.2
6	34	42	34	42	0	34.7	44.7	2.8
7	42	42	43	43	1.4	40.3	39.5	3.0
8	42	34	44	35	2.2	44.0	35.2	2.3
9	42	26	44	24	2.8	45.0	23.9	3.7
10	42	18	42	18	0	41.9	18.9	0.9
11	34	18	34	17	1.0	34.9	15.5	2.6
12	26	18	25	17	1.4	25.4	15.6	2.5
Average error					1.1			3.6

Table 7.9 Location results of specimen 2

No.	Actual lead break point (cm)		SUEA location result (cm)		SUEA location error (cm)	TD location results (cm)		TD location error (cm)
	x	*y*	*x*	*y*		*x*	*y*	
1	8	8	8	8	0	10.1	8.6	2.2
2	8	16	7	16	1.0	10.8	18.2	3.6
3	8	24	7	24	1.0	0	22.0	8.2
4	8	30	8	30	0	0	32.2	8.3
5	8	36	8	36	0	0	39.0	8.6
6	8	44	2	43	6.1	3.2	43.7	4.8
7	8	52	8	52	0	10.3	51.7	2.3
8	16	52	15	54	2.2	14.3	60.0	8.2
9	24	52	24	54	2.0	28.4	60.0	9.1
10	32	52	32	54	2.0	34.2	47.7	4.8
11	32	44	32	45	1.0	39.0	42.7	7.1
12	32	36	32	37	1.0	40.7	38.1	8.9
13	32	30	32	30	0	44.7	31.8	12.8
14	32	24	32	25	1.0	40.9	23.2	9.0

(continued)

Table 7.9 (continued)

No.	Actual lead break point (cm)		SUEA location result (cm)		SUEA location error (cm)	TD location results (cm)		TD location error (cm)
	x	*y*	*x*	*y*		*x*	*y*	
15	32	16	32	14	2.0	39.1	14.7	7.2
16	32	8	32	7	1.0	35.5	13.5	6.5
17	24	8	23	6	2.2	28.0	9.5	4.2
18	16	8	15	6	2.2	14.5	0	8.1
Average error					1.4			6.9

Table 7.10 Location results of specimen 3

No.	Actual lead break point (cm)		SUEA location result (cm)		SUEA location error (cm)	TD location results (cm)		TD location error (cm)
	x	*y*	*x*	*y*		*x*	*y*	
1	8	8	7	8	1.0	9.2	9.6	2.0
2	8	16	8	15	1.0	8.3	16.2	0.3
3	8	24	8	24	0	8.4	25.9	1.9
4	8	32	9	31	1.4	7.7	33.4	1.4
5	16	32	15	33	1.4	15.2	34.7	2.8
6	24	32	23	33	1.4	25.0	33.5	1.8
7	32	32	31	32	1.0	32.4	32.2	0.5
8	32	24	32	24	0	33.7	25.1	2.0
9	32	16	32	16	0	33.6	16.6	1.7
10	32	8	32	8	0	33.0	7.7	1.0
11	24	8	24	8	0	25.6	8.3	1.6
12	16	8	16	9	1.0	16.8	8.5	1.0
Average error					0.7			1.5

Table 7.11 Location results of Specimen 4

No.	Actual lead break point (cm)		SUEA location result (cm)		SUEA location error (cm)	TD location results (cm)		TD location error (cm)
	x	*y*	*x*	*y*		*x*	*y*	
1	8	8	8	8	0	8.8	8.8	1.1
2	8	16	7	15	0	11.0	17.2	3.2
3	8	24	6	23	3.2	0	21.2	8.5
4	8	30	5	29	3.0	0	31.1	8.1
5	8	36	8	35	0	10.4	36.1	2.4

(continued)

Table 7.11 (continued)

No.	Actual lead break point (cm)		SUEA location result (cm)		SUEA location error (cm)	TD location results (cm)		TD location error (cm)
	x	*y*	*x*	*y*		*x*	*y*	
6	8	44	8	43	0	11.2	43.1	3.3
7	8	52	8	52	0	10.4	50.3	3.0
8	16	52	16	51	0	17.3	50.6	1.9
9	24	52	24	52	0	26.2	50.0	3.0
10	32	52	32	52	1.0	33.1	48.9	3.3
11	32	44	32	44	0	36.3	43.2	4.3
12	32	36	31	36	1.0	39.8	39.0	8.4
13	32	30	32	30	1.0	38.4	30.4	6.4
14	32	24	31	24	1.0	37.9	21.4	6.4
15	32	16	32	16	1.0	36.8	15.0	4.9
16	32	8	32	9	0	34.0	12.0	4.5
17	24	8	23	9	1.4	25.9	9.2	2.3
18	16	8	16	8	0	16.6	8.9	1.0
Average error					0.7			4.2

Table 7.12 Location results of Specimen 5

No.	Actual lead break point (cm)		SUEA location result (cm)		SUEA location error (cm)	TD location results (cm)		TD location error (cm)
	x	*y*	*x*	*y*		*x*	*y*	
1	8	18	8	18	0	11.3	18.7	3.3
2	8	26	8	26	0	9.5	25.9	1.5
3	8	34	8	34	0	10.1	34.9	2.2
4	8	42	9	41	1.4	11.9	40.6	4.1
5	16	42	17	41	1.4	17.6	42.3	1.7
6	24	42	24	41	1.0	22.4	46.7	5.0
7	32	42	31	41	1.4	33.6	44.4	2.8
8	42	42	42	42	0	40.4	40	2.5
9	42	34	43	35	1.4	43.5	35.9	2.5
10	42	26	43	25	1.4	44.7	24.4	3.2
11	42	18	42	18	0	40.6	19.4	2.0
12	32	18	32	18	0	33.1	15.7	2.5
13	24	18	23	17	1.4	22.5	12.9	5.3
14	16	18	16	18	0	17.0	16.5	1.8
Average error					0.7			2.9

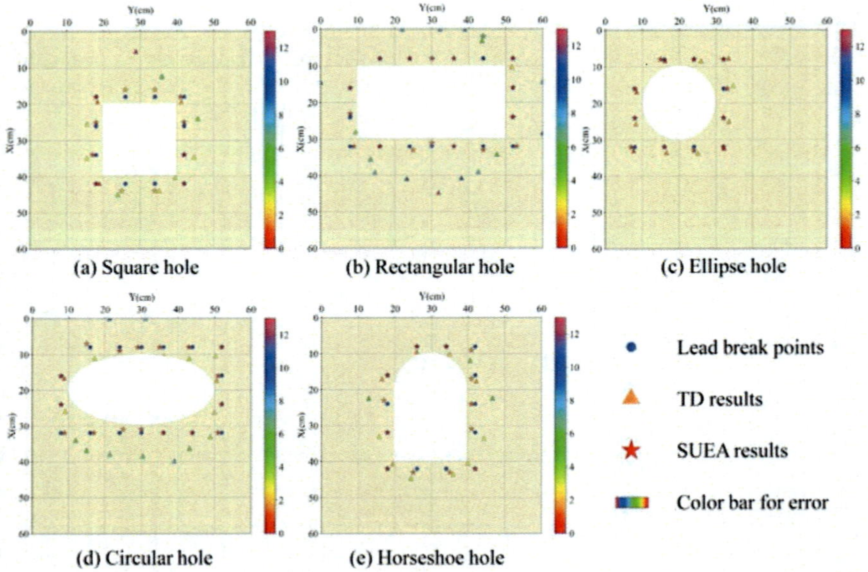

Fig. 7.23 SUEA and TD location results (from [16])

are less than or equal to 2.0 cm (less than the diameter of the sensor) for almost all of source points in five tests, while that of TD for most source points are larger than 2.0 cm.

According to the new AE source location method, the largest location error is 2.8 cm and the smallest error is 0 cm in the 12 lead breaking points of Specimen 1. Among the 18 lead breaking points of Specimen 2, the largest location error is 6.1 cm, and the smallest error is 0 cm. The average location error of Specimen 2 is 1.4 cm, which is the largest average location error among five specimens. Among the 12 lead break points of Specimen 3, the largest location error is 1.4 cm and the smallest error is 0 cm. Specimen 4 has the largest location error of 3.2 cm and the smallest error of 0 cm among the 18 lead break points. At last Specimen 5 has the largest location error of 1.4 cm and the smallest error of 0 cm among the 14 lead breaks points.

As a control, the TD location method was chosen to locate the lead break points. The reason for choosing TD is that this method can get a high-accuracy location result and does not require a pre-measured velocity [23]. Therefore, the location results of TD was chosen for comparison. TD location results show that in Specimen 1, the largest location error is 12.8 cm and the smallest error is 0.9 cm. In Specimen 2, the largest location error is 12.8 cm and the smallest error is 2.2 cm, which is the largest average location error among the five specimens. In Specimen 3, the largest location error was 2.8 cm and the smallest error was 0.3 cm, and Specimen 3 was also the one with the smallest average location error among the five specimens. In Specimen 4, the largest location error was 8.5 cm and the smallest error was 1.0 cm. In Specimen 5, the largest location error was 5.3 cm and the smallest error was 1.5 cm.

For further analysis, the average location errors of SUEA for specimens 1–5 were calculated and the corresponding values are 1.1, 1.4, 0.7, 0.7, and 0.7 cm. The average location errors of TD location method are 3.6, 6.9, 1.5, 4.2, and 2.9 cm. Because the traditional method cannot bypass the empty area, the location result has a large deviation from the actual situation. Compared with the TD location method, the new method showed a significant reduction in the average location error on the five specimens. The average location error of the new method is only 22% of that of the traditional location method.

This indicates that the accuracy of the new location method has been improved significantly. This is because the new method takes into account that the presence of potential unknown empty areas, which can significantly affect the location accuracy of the conventional location method. The new method could identify the potential unknown empty area and then locate the AE source, which can effectively solve the problem of locating the AE source in complex areas containing unknown empty areas. The results of the empty area identification are shown in Table 7.13 and Fig. 7.24, which are in line with the actual situation.

In the actual mining, cavern, tunnel, and other environments, the strata will encounter complex empty areas such as roadways and quarries. Using AE technology to monitor and locate the areas with cavities is a hot issue at this stage. There are two main types of problems with the currently existing AE localization methods: one is that the empty area is not taken into account, and the travel path of the wave between the source and the AE sensor is assumed to be a straight-line path; the other is that localization can only be performed on the basis of a known empty area. As we all know, acoustic waves propagate much faster in solids than in air. Therefore, waves that bypass the empty area will reach the sensor location faster and be recorded by the sensor in the structure with complex empty areas. In this case, for the first type of methods, the assumption that the travel path of the wave between the AE source and the AE sensor is a straight line will no longer meet the actual situation. Besides, in

Table 7.13 Empty area identification results

No.	Actual empty area location (cm)	Identification result (cm)
1	20, 20, 20, 20	18, 18, 26, 24
2	10, 10, 20, 40	10, 6, 22, 48
3	10, 10, 20, 20	6, 10, 26, 22
4	10, 10, 20, 40	10, 8, 22, 44
5	10, 20, 20, 20	8, 18, 24, 22

Notes The first two numbers represent the coordinates of the endpoints of the empty area near the origin of the coordinates (circles and ellipses are based on their external tangent squares); and the last two numbers represent the two side lengths of the empty area in the x and y directions

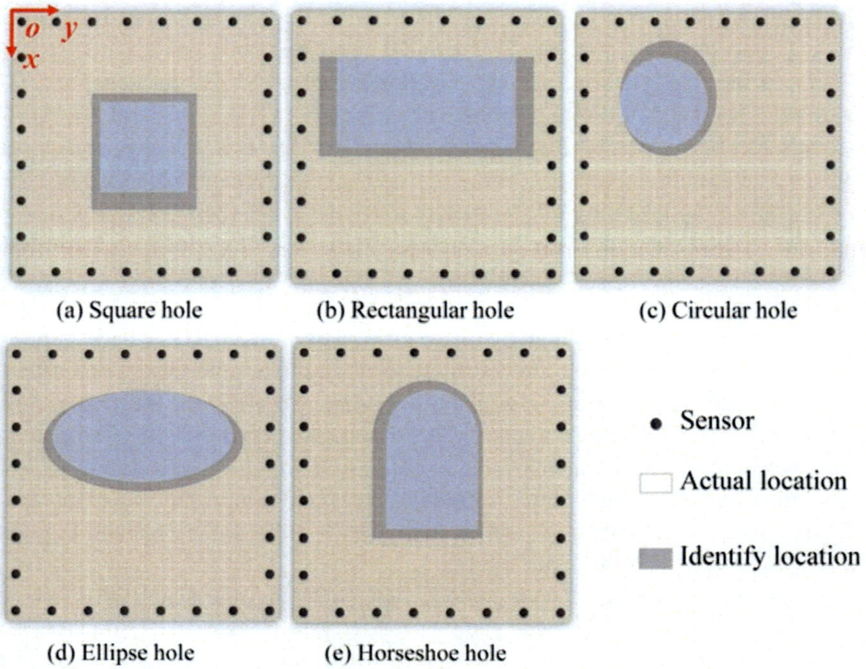

Fig. 7.24 Five test pieces empty area identification results (from [16])

the actual engineering environment, it is impossible to know all the empty positions in each area to be located. So the application of the second type of methods is very limited. The newly proposed localization method can realize the precise localization of AE sources under the premise that the location of the empty in the target area is unknown.

This method takes into account the fact that the empty location will be encountered in the actual engineering environment where it is difficult to identify, making the accuracy of AE source localization in complex environments much higher and improving the application range. However, this method also has the following limitations:

- The sensor size is too large, which affects the wave velocity. The pulse signal emitted by any sensor is used as the AE source, and the remaining 27 sensors receive the AE signal. The shortest time paths of wave propagation from the AE source to 27 sensors are calculated respectively. Combined with the arrival times collected by the AE acquisition system, it can be seen that the wave velocity is relatively small at a distance far from the AE source and relatively large at a distance close to the AE source. The reason for this may be that when a sensor is excited and receives a pulse signal, it transmits and receives the signal on a plane rather than a single central point. Therefore, when a sensor emits a pulse signal, the area of the receiving sensor that is the closest to the emitting sensor is the first to receive the signal. So the shortest time path of the actual acoustic wave is the shorter than the distance from the point of pulse emission to the center of the sensor. In the path search dataset with empty areas, the theoretical path is larger than the actual path. The wave velocity $v = s/t$, when s is larger, the incremental Δs caused by the AE signal being a surface source instead of a point source has a smaller or negligible effect on the wave velocity; when s is smaller, the effect of Δs on the wave velocity is larger.

- The ray density is too small. Using the pulse signal excited by the sensor as the active AE source, the distance between two adjacent sensors is 8 cm by dividing the grid. The sensors are not closely arranged, resulting in a small ray density. This causes large errors when identifying small empty areas. If the number of sensors is increased or the number of AE sources is increased (e.g., more lead break points are used as active AE sources), the discrimination effect will be better, and thus the location accuracy will be higher.
- Traversing all shapes is obviously not feasible due to computational and time costs.

7.4.3 Verification for VFH

The lead-breaking test was carried out on the hollow column mortar structures to evaluate the performance of the proposed method. A cubic concrete test specimen with the size of 10 cm × 10 cm × 10 cm was cut out into a cylindrical empty space with the size of ϕ6 cm × 10 cm, as shown in Fig. 7.25a. There are 106 cubic grid blocks of the same size to form the model when the sides of cube are 1 mm in length. To speed up the calculation, the test piece is divided into 25 × 25 × 25 cubes with the size of 4 mm in length, as shown in Fig. 7.25b.

The cubes of the empty area are marked as not passed, and the other cubes as pass. Six MS/AE sensors were attached to the test piece, and a coupling agent was added between the sensor and the test piece for better acoustic coupling. The position of the sensor is located on the grid node (Fig. 7.25c), and the coordinate position is shown in Table 7.14.

To facilitate the calculation, the model coordinates were transformed so that the index of the node matrix **M** is in one-to-one correspondence with the coordinate positions. The data acquisition uses a 40 dB threshold and a 5 MHz sampling rate. The lead-breaking tests were carried out at different positions of the test piece, and each position (shown in Table 7.15) was performed twice. The arrivals of the P wave generated by MS/AE event are recorded by the sensors (Table 7.15). Searching the path L_S between each potential source and the sensor and calculating the distance of the path, are shown in Fig. 7.25d. Then, according to the arrivals recorded by the sensors, the deviation value D_{xyz} of each potential source is calculated. The coordinates (x, y, z) of source location on the specimen corresponding to the minimum D_{xyz} is determined and is converted to the position coordinates. The location results of VFH are shown in Table 7.16. Meanwhile, the results of TD are performed and shown in Table 7.16.

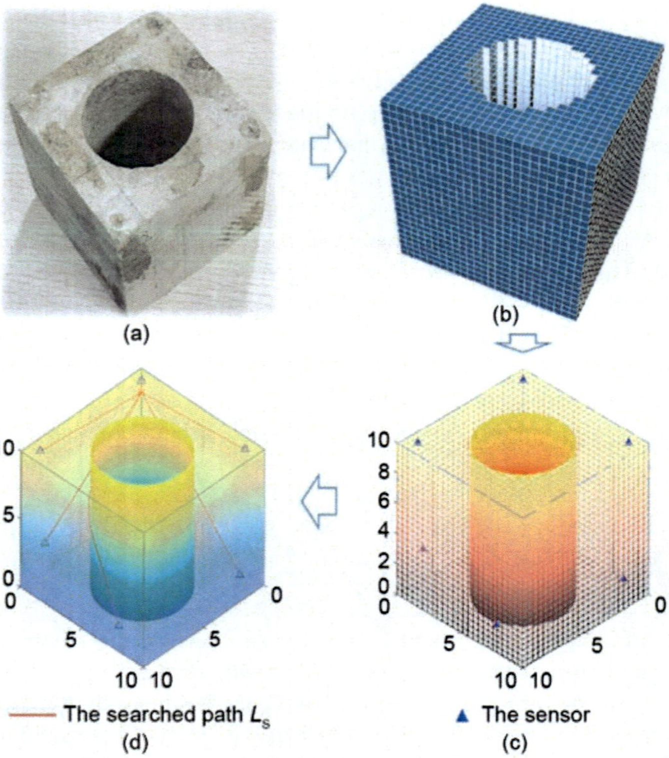

Fig. 7.25 The process of localization with the VFH. **a** The specimen; **b** modelling, and meshing the specimen; **c** determining the coordinate of sensors (unit: cm); **d** the path searched by the improved A* search algorithm from node to sensor (unit: cm) (from [17])

Table 7.14 The coordinates of the sensors

Number of sensors	The corresponding coordinates (cm)		
	x	y	z
1	0.8	0.8	10.0
2	9.2	0.8	10.0
3	0.8	9.2	10.0
4	8.0	0	2.0
5	10.0	8.0	2.0
6	2.0	10.0	2.0

The localization results between VFH and TD with the actual lead-breaking points, and the error records are shown in Table 7.16. As seen from the table, the maximum location error of the TD is 4.5 cm, which is much larger than the location error of the VFH. For the convenience of observation. Figure 7.26 shows the visualization of the location results and location errors of the two methods. In Fig. 7.26a, the

Table 7.15 Coordinates of the sources and Arrivals received by each sensor

Source		Arrivals of P-wave received by each sensor (s)					
Number	Coordinate (cm)	t_0^1	t_0^2	t_0^3	t_0^4	t_0^5	t_0^6
1	(2, 2, 10)	5.7511340	5.7511397	5.7511543	5.7511710	5.7511860	5.7511773
2	(2, 2, 10)	77.1878437	77.1878503	77.1878667	77.1878797	77.1878967	77.1878850
3	(4, 2, 10)	142.0323807	142.0323873	142.0324037	142.0324063	142.0324170	142.0324177
4	(4, 2, 10)	149.8894003	149.8894053	149.8894243	149.8894240	149.8894370	149.8894363
5	(6, 2, 10)	160.1324277	160.1324220	160.1324540	160.1324433	160.1324497	160.1324707
6	(6, 2, 10)	168.7844300	168.7844217	168.7844593	168.7844437	168.7844507	168.7844717
7	(8, 2, 10)	200.8480153	200.8480110	200.8480553	200.8480400	200.8480437	200.8480630
8	(8, 2, 10)	208.4753257	208.4753217	208.4753647	208.4753503	208.4753540	208.4753733
9	(2, 4, 10)	224.7515927	224.7516070	224.7516027	224.7516263	224.7516340	224.7516227
10	(2, 4, 10)	231.2690027	231.2690200	231.2690117	231.2690357	231.2690430	231.2690317
11	(8, 4, 10)	239.7299927	239.7299827	239.7300150	239.7300070	239.7300053	239.7300173
12	(8, 4, 10)	248.6253717	248.6253610	248.6253917	248.6253857	248.6253840	248.6253960
13	(2, 6, 10)	260.6217517	260.6217653	260.6217460	260.6217897	260.6217783	260.6217707
14	(2, 6, 10)	266.9614870	266.9615013	266.9614827	266.9615220	266.9615147	266.9615070
15	(8, 6, 10)	276.0698310	276.0698170	276.0698303	276.0698380	276.0698317	276.0698413
16	(8, 6, 10)	281.5980240	281.5980103	281.5980230	281.5980313	281.5980247	281.5980350
17	(2, 8, 10)	299.8790950	299.8791300	299.8790753	299.8791293	299.8791120	299.8791033
18	(2, 8, 10)	306.1023487	306.1023673	306.1023300	306.1023800	306.1023667	306.1023577
19	(4, 8, 10)	321.6901460	321.6901560	321.6901283	321.6901680	321.6901533	321.6901503
20	(4, 8, 10)	328.3538907	328.3539017	328.3538733	328.3539133	328.3538983	328.3538950

(continued)

Table 7.15 (continued)

Source		Arrivals of P-wave received by each sensor (s)					
Number	Coordinate (cm)	t_0^1	t_0^2	t_0^3	t_0^4	t_0^5	t_0^6
21	(6, 8, 10)	343.9337353	343.9337263	343.9337130	343.9337407	343.9337287	343.9337303
22	(6, 8, 10)	350.3266527	350.3266437	350.3266300	350.3266577	350.3266463	350.3266480
23	(8, 8, 10)	358.3117887	358.3117633	358.3117593	358.3117800	358.3117677	358.3117773
24	(8, 8, 10)	372.6762447	372.6762160	372.6762150	372.6762370	372.6762237	372.6762337
25	(2, 0, 8)	469.6502823	469.6502983	469.6503073	469.6503127	469.6503263	469.6503197
26	(2, 0, 8)	476.5324683	476.5324800	476.5324933	476.5324997	476.5325127	476.5325060
27	(2, 0, 6)	487.2123353	487.2123510	487.2123573	487.2123530	487.2123720	487.2123633
28	(2, 0, 6)	496.1957583	496.1957730	496.1957803	496.1957763	496.1957940	496.1957867
29	(2, 0, 4)	516.2830867	516.2830983	516.2831053	516.2830917	516.2831123	516.2831053
30	(2, 0, 4)	539.7597230	539.7597350	539.7597420	539.7597293	539.7597497	539.7597423
31	(4, 0, 8)	552.4292987	552.4293053	552.4293257	552.4293203	552.4293320	552.4293340
32	(4, 0, 8)	558.2833970	558.2834027	558.2834203	558.2834177	558.2834297	558.2834323
33	(4, 0, 4)	604.1152323	604.1152360	604.1152503	604.1152267	604.1152480	604.1152500
34	(4, 0, 4)	611.2469390	611.2469420	611.2469570	611.2469327	611.2469543	611.2469567
35	(6, 0, 8)	646.9034513	646.9034447	646.9034767	646.9034630	646.9034737	646.9034853
36	(6, 0, 8)	662.0699583	662.0699517	662.0699853	662.0699660	662.0699800	662.0699917
37	(6, 0, 6)	676.0523813	676.0523767	676.0524093	676.0523790	676.0523973	676.0524090
38	(6, 0, 6)	685.3599257	685.3599203	685.3599547	685.3599233	685.3599417	685.3599537
39	(10, 2, 8)	758.6056710	758.6056637	758.6057070	758.6056820	758.6056927	758.6057077

(continued)

Table 7.15 (continued)

Source		Arrivals of P-wave received by each sensor (s)					
Number	Coordinate (cm)	t_0^1	t_0^2	t_0^3	t_0^4	t_0^5	t_0^6
40	(10, 2, 8)	767.6932820	767.6932720	767.6933153	767.6932900	767.6932983	767.6933147
41	(10, 2, 6)	800.9608797	800.9608573	800.9609003	800.9608623	800.9608720	800.9608920
42	(10, 2, 6)	816.6084260	816.6084123	816.6084450	816.6084170	816.6084257	816.6084480
43	(10, 2, 4)	855.6525857	855.6525677	855.6525953	855.6525590	855.6525700	855.6525893
44	(10, 2, 4)	865.6914593	865.6914370	865.6914673	865.6914290	865.6914400	865.6914623
45	(10, 4, 8)	915.7629410	915.7629163	915.7629477	915.7629333	915.7629320	915.7629483
46	(10, 4, 8)	924.0719167	924.0718900	924.0719230	924.0719073	924.0719060	924.0719223
47	(10, 4, 4)	945.0755100	945.0754873	945.0755170	945.0754827	945.0754813	945.0755040
48	(10, 4, 4)	960.0038163	960.0037973	960.0038203	960.0037930	960.0037917	960.0038147
49	(10, 6, 8)	981.8797490	981.8797173	981.8797347	981.8797343	981.8797250	981.8797393
50	(10, 6, 8)	991.4208480	991.4208160	991.4208330	991.4208317	991.4208210	991.4208380
51	(10, 6, 6)	1006.533625	1006.533603	1006.533618	1006.533611	1006.533597	1006.533618
52	(10, 6, 6)	1027.621851	1027.621829	1027.621848	1027.621836	1027.621823	1027.621844
53	(10, 6, 4)	1038.115605	1038.115585	1038.115600	1038.115579	1038.115567	1038.115579
54	(10, 6, 4)	1046.868003	1046.867983	1046.867999	1046.867977	1046.867965	1046.867977
55	(8, 10, 8)	1151.870872	1151.870859	1151.870851	1151.870873	1151.870851	1151.870864
56	(8, 10, 8)	1162.700972	1162.700959	1162.700951	1162.700973	1162.700950	1162.700962
57	(8, 10, 6)	1174.590833	1174.590821	1174.590812	1174.590828	1174.590804	1174.590813
58	(8, 10, 6)	1180.248350	1180.248338	1180.248329	1180.248345	1180.248320	1180.248331

(continued)

Table 7.15 (continued)

Source		Arrivals of P-wave received by each sensor (s)					
Number	Coordinate (cm)	t_0^1	t_0^2	t_0^3	t_0^4	t_0^5	t_0^6
59	(8, 10, 4)	1188.623414	1188.623402	1188.623395	1188.623402	1188.623376	1188.623386
60	(8, 10, 4)	1195.139257	1195.139247	1195.139238	1195.139240	1195.139219	1195.139230
61	(6, 10, 8)	1202.616754	1202.616753	1202.616733	1202.616764	1202.616743	1202.616749
62	(6, 10, 8)	1211.729176	1211.729172	1211.729154	1211.729185	1211.729164	1211.729166
63	(6, 10, 4)	1225.341341	1225.341338	1225.341322	1225.341339	1225.341314	1225.341312
64	(6, 10, 4)	1230.283246	1230.283243	1230.283227	1230.283244	1230.283219	1230.283217
65	(4, 10, 8)	1238.183980	1238.183987	1238.183958	1238.183998	1238.183979	1238.183974
66	(4, 10, 8)	1246.735645	1246.735654	1246.735623	1246.735664	1246.735645	1246.735639
67	(3.2, 2.8, 8)	1343.326448	1343.326458	1343.326465	1343.326470	1343.326492	1343.326476
68	(3.2, 2.8, 8)	1351.281429	1351.281441	1351.281449	1351.281451	1351.281475	1351.281457
69	(3.2, 2.8, 6)	1369.293569	1369.293578	1369.293581	1369.293578	1369.293606	1369.293587
70	(3.2, 2.8, 6)	1377.155674	1377.155684	1377.155687	1377.155685	1377.155711	1377.155694
71	(3.2, 2.8, 4)	1390.602172	1390.602179	1390.602181	1390.602169	1390.602194	1390.602180
72	(3.2, 2.8, 4)	1398.425528	1398.425535	1398.425538	1398.425528	1398.425550	1398.425539
73	(7.2, 6.8, 8)	1427.161920	1427.161888	1427.161894	1427.161904	1427.161892	1427.161900
74	(7.2, 6.8, 8)	1435.785064	1435.785038	1435.785044	1435.785053	1435.785042	1435.785050
75	(7.2, 6.8, 6)	1447.024657	1447.024636	1447.024641	1447.024642	1447.024629	1447.024638
76	(7.2, 6.8, 6)	1488.583798	1488.583775	1488.583776	1488.583781	1488.583768	1488.583777
77	(7.2, 6.8, 4)	1529.275322	1529.275301	1529.275301	1529.275288	1529.275283	1529.275294
78	(7.2, 6.8, 4)	1541.953987	1541.953968	1541.953968	1541.953964	1541.953949	1541.953961

Table 7.16 Location result and error of the lead-breaking points

Number	VFH							TD						
	Location result (cm)			Location Error (cm)				Location result (cm)			Location Error (cm)			
	x	y	z	Δx	Δy	Δz	E	x	y	z	Δx	Δy	Δz	E
1	3.6	2.4	10.0	1.6	0.4	0	1.6	3.7	1.9	10.0	1.7	−0.1	0	1.7
2	3.6	2.4	10.0	1.6	0.4	0	1.6	3.7	1.7	10.0	1.7	−0.3	0	1.7
3	4.0	1.2	10.0	0	−0.8	0	0.8	4.8	1.2	9.2	0.8	−0.8	−0.8	1.4
4	4.4	1.2	10.0	0.4	−0.8	0	0.9	5.2	1.1	9.4	1.2	−0.9	−0.6	1.6
5	6.0	2.4	9.2	0	0.4	−0.8	0.9	5.8	1.1	10.0	−0.2	−0.9	0	0.9
6	6.4	2.4	9.2	0.4	0.4	−0.8	1.0	7.2	1.7	10.0	1.2	−0.3	0	1.3
7	6.8	2.8	10.0	−1.2	0.8	0	1.4	6.5	0.8	9.1	−1.5	−1.2	−0.9	2.1
8	6.8	2.8	10.0	−1.2	0.8	0	1.4	5.8	0.8	9.2	−2.2	−1.2	−0.8	2.6
9	2.4	3.6	10.0	0.4	−0.4	0	0.6	2.9	4.0	9.8	0.9	0	−0.2	0.9
10	2.0	4.0	10.0	0	0	0	0	1.9	4.2	9.1	−0.1	0.2	−0.9	0.9
11	7.2	2.8	10.0	−0.8	−1.2	0	1.4	7.2	4.0	9.1	−0.8	0	−0.9	1.2
12	7.2	3.2	10.0	−0.8	−0.8	0	1.1	7.1	4.1	9.0	−0.9	0.1	−1.0	1.4
13	2.4	6.0	10.0	0.4	0	0	0.4	2.8	5.8	10.0	0.8	−0.2	0	0.8
14	2.0	5.6	10.0	0	−0.4	0	0.4	3.1	5.8	10.0	1.1	−0.2	0	1.1
15	8.0	5.2	10.0	0	−0.8	0	0.8	7.6	5.4	8.6	−0.4	−0.6	−1.4	1.6
16	8.0	5.6	10.0	0	−0.4	0	0.4	7.3	5.6	9.2	−0.7	−0.4	−0.8	1.2
17	2.4	7.6	8.8	0.4	−0.4	−1.2	1.3	0.5	7.9	8.2	−1.5	−0.1	−1.8	2.3
18	3.2	7.2	10.0	1.2	−0.8	0	1.4	2.1	7.6	10.0	0.1	−0.4	0	0.4
19	3.6	8.0	10.0	−0.4	0	0	0.4	5.0	8.8	9.6	1.0	0.8	−0.4	1.3

(continued)

Table 7.16 (continued)

Number	VFH							TD						
	Location result (cm)			Location Error (cm)				Location result (cm)			Location Error (cm)			
	x	y	z	Δx	Δy	Δz	E	x	y	z	Δx	Δy	Δz	E
20	3.6	8.0	10.0	−0.4	0	0	0.4	5.0	8.8	9.6	1.0	0.8	−0.4	1.3
21	6.0	7.6	9.6	0	−0.4	−0.4	0.6	7.3	10.1	10.0	1.3	2.1	0	2.5
22	6.0	7.6	9.6	0	−0.4	−0.4	0.6	7.3	10.1	10.0	1.3	2.1	0	2.5
23	7.6	7.6	9.6	−0.4	−0.4	−0.4	0.7	8.8	9.2	10.0	0.8	1.2	0	1.4
24	7.6	7.6	10.0	−0.4	−0.4	0	0.6	10.1	9.1	10.0	2.1	1.1	0	2.4
25	2.8	1.2	10.0	0.8	1.2	2.0	2.5	3.4	2.3	10.0	1.4	2.3	2.0	3.3
26	3.6	1.6	10.0	1.6	1.6	2.0	3.0	3.4	2.3	10.0	1.4	2.3	2.0	3.3
27	2.8	1.6	7.2	0.8	1.6	1.2	2.2	3.0	1.7	6.7	1.0	1.7	0.7	2.0
28	2.8	1.2	7.6	0.8	1.2	1.6	2.2	0.9	0	5.3	−1.1	0	−0.7	1.3
29	2.4	0.4	5.2	0.4	0.4	1.2	1.3	2.7	1.7	5.0	0.7	1.7	1.0	2.1
30	2.8	1.2	5.6	0.8	1.2	1.6	2.2	3.3	2.0	5.7	1.3	2.0	1.7	2.9
31	4.4	0.8	9.2	0.4	0.8	1.2	1.5	6.2	3.2	9.3	2.2	3.2	1.3	4.1
32	4.4	0.8	9.6	0.4	0.8	1.6	1.8	4.2	1.5	9.6	0.2	1.5	1.6	2.2
33	4.0	0	4.8	0	0	0.8	0.8	3.5	1.4	5.0	−0.5	1.4	1.0	1.8
34	4.4	0	4.8	0.4	0	0.8	0.9	3.4	−0.1	3.6	−0.6	−0.1	−0.4	0.7
35	6.0	1.6	8.8	0	1.6	0.8	1.8	6.4	1.2	8.6	0.4	1.2	0.6	1.4
36	6.0	1.2	8.0	0	1.2	0	1.2	5.7	−0.1	7.0	−0.3	−0.1	−1.0	1.0

(continued)

Table 7.16 (continued)

Number	VFH							TD						
	Location result (cm)			Location Error (cm)				Location result (cm)			Location Error (cm)			
	x	y	z	Δx	Δy	Δz	E	x	y	z	Δx	Δy	Δz	E
37	5.6	1.2	6.4	−0.4	1.2	0.4	1.3	8.2	−0.1	5.9	2.2	−0.1	−0.1	2.2
38	6.0	0.8	6.4	0	0.8	0.4	0.9	8.2	−0.1	5.9	2.2	−0.1	−0.1	2.2
39	6.4	1.6	8.0	−3.6	−0.4	0	3.6	6.5	−0.1	7.6	−3.5	−2.1	−0.4	4.1
40	6.8	2.8	8.4	−3.2	0.8	0.4	3.3	7.1	−0.1	9.0	−2.9	−2.1	1.0	3.7
41	8.4	2.8	6.4	−1.6	0.8	0.4	1.8	10.1	4.7	9.6	0.1	2.7	3.6	4.5
42	6.8	2.8	6.8	−3.2	0.8	0.8	3.4	9.0	5.0	7.8	−1.0	3.0	1.8	3.6
43	10.0	2.0	4.4	0	0	0.4	0.4	8.5	3.2	5.1	−1.5	1.2	1.1	2.2
44	10.0	2.0	4.4	0	0	0.4	0.4	7.9	3.3	6.4	−2.1	1.3	2.4	3.5
45	9.2	4.0	8.0	−0.8	0	0	0.8	10.1	5.1	9.8	0.1	1.1	1.8	2.1
46	9.6	4.0	8.0	−0.4	0	0	0.4	10.1	5.4	9.8	0.1	1.4	1.8	2.3
47	9.2	4.0	4.4	−0.8	0	0.4	0.9	9.9	5.0	3.4	−0.1	1.0	−0.6	1.2
48	10.0	4.0	4.8	0	0	0.8	0.8	8.8	4.4	5.0	−1.2	0.4	1.0	1.6
49	8.8	6.4	8.0	−1.2	0.4	0	1.3	10.1	6.6	8.6	0.1	0.6	0.6	0.9
50	9.2	6.4	8.0	−0.8	0.4	0	0.9	10.1	8.3	9.2	0.1	2.3	1.2	2.6
51	10.0	6.4	6.8	0	0.4	0.8	0.9	10.1	9.2	8.1	0.1	3.2	2.1	3.8
52	10.0	6.0	6.4	0	0	0.4	0.4	10.1	6.7	8.0	0.1	0.7	2.0	2.1
53	7.6	6.0	3.6	−2.4	0	−0.4	2.4	8.1	6.2	3.4	−1.9	0.2	90−	2.0

(continued)

Table 7.16 (continued)

Number	VFH							TD						
	Location result (cm)			Location Error (cm)				Location result (cm)			Location Error (cm)			
	x	y	z	Δx	Δy	Δz	E	x	y	z	Δx	Δy	Δz	E
54	7.6	6.0	3.6	−2.4	0	−0.4	2.4	7.0	5.9	3.2	−3.0	−0.1	−0.8	3.1
55	8.4	10.0	9.6	0.4	0	1.6	1.6	9.3	10.0	10.0	1.3	0	2.0	2.4
56	8.4	10.0	9.6	0.4	0	1.6	1.6	8.7	10.1	10.0	0.7	0.1	2.0	2.1
57	8.0	10.0	7.2	0	0	1.2	1.2	7.6	9.3	7.1	−0.4	−0.7	1.1	1.3
58	8.0	10.0	7.6	0	0	1.6	1.6	7.8	9.5	7.2	−0.2	−0.5	1.2	1.3
59	8.0	10.0	4.8	0	0	0.8	0.8	7.4	8.0	5.5	−0.6	−2.0	1.5	2.6
60	7.2	8.8	4.0	−0.8	−1.2	0	1.4	7.4	8.0	5.5	−0.6	−2.0	1.5	2.6
61	5.6	8.0	8.8	−0.4	−0.2	0.8	2.2	6.9	10.1	9.7	0.9	0.1	1.7	1.9
62	6.4	10.0	9.2	0.4	0	1.2	1.3	6.9	10.1	9.7	0.9	0.1	1.7	1.9
63	5.6	9.6	4.8	−0.4	−0.4	0.8	1.0	5.9	9.5	5.7	−0.1	−0.5	1.7	1.8
64	5.6	9.6	4.8	−0.4	−0.4	0.8	1.0	5.9	9.5	5.7	−0.1	−0.5	1.7	1.8
65	4.4	8.0	8.8	0.4	−0.2	0.8	2.2	4.8	10.1	10.0	0.8	0.1	2.0	2.2
66	4.0	8.0	8.4	0	−0.2	0.4	2.0	3.3	7.7	8.0	−0.7	−2.3	0	2.5
67	3.2	2.8	8.4	0	0	0.4	0.4	3.6	2.6	8.8	0.4	−0.2	0.8	0.9
68	3.2	2.8	8.0	0	0	0	0	3.1	2.1	8.1	−0.1	−0.7	0.1	0.7
69	3.2	2.8	6.8	0	0	0.8	0.8	2.3	2.3	5.9	−0.9	−0.5	−0.1	1.0
70	3.2	2.8	6.8	0	0	0.8	0.8	1.0	1.3	5.2	−2.2	−1.5	−0.8	2.8

(continued)

Table 7.16 (continued)

Number	VFH							TD						
	Location result (cm)			Location Error (cm)				Location result (cm)			Location Error (cm)			
	x	*y*	*z*	Δ*x*	Δ*y*	Δ*z*	*E*	*x*	*y*	*z*	Δ*x*	Δ*y*	Δ*z*	*E*
71	3.2	2.8	4.8	0	0	0.8	0.8	0.5	−0.1	3.1	−2.7	−2.9	−0.9	4.1
72	3.2	2.8	5.2	0	0	1.2	1.2	3.7	3.7	4.3	0.5	0.9	0.3	1.1
73	7.2	6.8	8.0	0	0	0	0	10.1	8.9	9.3	2.9	2.1	1.3	3.8
74	7.2	6.8	8.4	0	0	0.4	0.4	10.0	8.7	10.0	2.8	1.9	2.0	3.9
75	7.2	6.8	6.4	0	0	0.4	0.4	8.0	7.3	6.4	0.8	0.5	0.4	1.0
76	7.2	6.8	6.8	0	0	0.8	0.8	8.0	7.3	7.1	0.8	0.5	1.1	1.4
77	7.2	6.8	4.0	0	0	0	0	7.5	6.3	3.7	0.3	−0.5	−0.3	0.7
78	7.2	6.8	5.2	0	0	1.2	1.2	7.8	7.4	4.9	0.6	0.6	0.9	1.2
Average error				−0.18	0.05	0.44	1.20				0.09	0.31	0.50	2.02

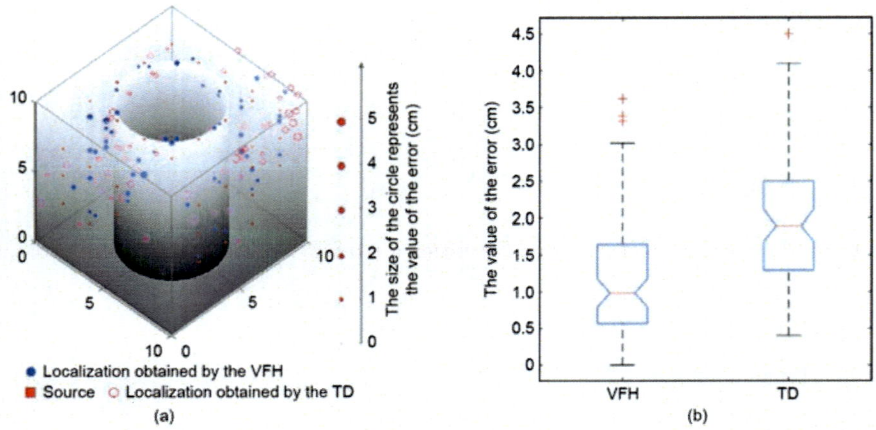

Fig. 7.26 Location result and error of AE sources obtained by two methods. **a** The visualization of the result and error; **b** the boxplot of the location error (from [17])

circle size represents the magnitude of the error of the source localization result. It can be clearly seen that the circle of the TD is much larger than that of the VFH. Figure 7.26b represents a boxplot of the location error E of the two methods. On each box, the central mark, the bottom edge of the box, and top edge of the box indicate the median, the 25th percentile, and the 75th percentile, respectively. The whiskers above and below the box show the most extreme data points, and the outliers are plotted individually using the "+" symbol. As shown in Fig. 7.26b, the median source location error obtained by using VFH is approximately 1.0 cm, while the median location error obtained by using TD is approximately 1.9 cm. Therefore, the location results of VFH are better. Other parameters of Fig. 7.26b can also validate this view. According to the location error of each lead-breaking test in the table, the average location error of the VFH is 1.20 cm, while the average location error of the TD is 2.02 cm. The average location accuracy of the VFH is nearly 40% greater than the TD. Therefore, in the complex 3-D structure, the location accuracy of the VFH is greatly improved comparing to the TD.

7.5 Conclusions

Three velocity-free methods which are designed for complex structures are introduced and based on the verification of the methods, some conclusions are drawn below.

(1) The ALM is applied to two dimensional complex structures with known geometric irregularities. The improved A* search algorithm was used to search the fastest travel path of the wave more effectively in the structure and make the path coincident with the propagation of wave. The ALM not only eliminates

the measurement of the wave velocity, but also avoids the localization error due to temporal and spatial errors of the pre-measured wave velocity. The virtual localization test and the lead-breaking test were carried out to verify the effectiveness of the ALM and the promising results were achieved. Comparing with FastWay, The ALM is not affected by the fluctuation error of the wave velocity and has a better location accuracy.

(2) When the empty area is unknown, the ALM becomes invalid. Fortunately, the SUEA overcomes the problem. It uses the improved A* search algorithm to search the shortest wave path, and matches with the potential empty region to estimate the size and position of the cavity. The lead break experiment was carried out and results show the average location accuracy of the SUEA increased by 78% compared to the results of the existing method TD.

(3) For the VFH, the A* search algorithm is improved and grid points are used to search for paths reflecting the actual propagation of acoustic waves. Therefore, it can achieve the source localization in 3-D complex structures with irregular spaces. Moreover, the wave velocity of the acoustic wave is considered as an unknown in the calculation to reduce the location error caused by the wave velocity change during the monitoring process. To evaluate the quasi-determination and effectiveness of the VFH, a lead-breaking test was carried out on hollow cubic concrete structures. The location results show that the average location error obtained by the VFH is 1.20 cm, and its average positioning accuracy is 40% higher than the TD. It can be concluded that the VFH can effectively adapt to geometrically irregular 3-D complex structures. The VFH overcomes the disadvantages of the existing localization method, which considers the impact of geometry, and is suitable for 3-D hole-containing structure.

References

1. Baxter MG, Pullin R, Holford KM, Evans SL (2007) Delta T source location for acoustic emission. Mech Syst Signal Process 21(3):1512–1520
2. Gollob S, Kocur GK, Schumacher T, Mhamdi L, Vogel T (2017) A novel multi-segment path analysis based on a heterogeneous velocity model for the localization of acoustic emission sources in complex propagation media. Ultrasonics 74:48–61
3. Dijkstra EW (1959) A note on two problems in connexion with graphs. Numer Math 1(1):269–271
4. Park B, Sohn H, Olson SE, DeSimio MP, Brown KS, Derriso MM (2012) Impact localization in complex structures using laser-based time reversal. Struct Health Monit 11(5):577–588
5. Eaton MJ, Pullin R, Holford KM (2012) Acoustic emission source location in composite materials using Delta T Mapping. Compos A Appl Sci Manuf 43(6):856–863
6. Hensman J, Mills R, Pierce S, Worden K, Eaton M (2010) Locating acoustic emission sources in complex structures using Gaussian processes. Mech Syst Signal Process 24(1):211–223
7. Liu G, Li T, Peng Y, Hou X (2005) The ant algorithm for solving robot path planning problem. In: Third international conference on information technology and applications (ICITA'05). IEEE, pp 25–27
8. Kothari M, Postlethwaite I (2013) A probabilistically robust path planning algorithm for UAVs using rapidly-exploring random trees. J Intell Rob Syst 71(2):231–253

9. Khatib O (1986) Real-time obstacle avoidance for manipulators and mobile robots. In: Autonomous robot vehicles. Springer, pp 396–404
10. Liu Y, Du J, Zhang Q (2017) Performance comparison between A~* algorithm and Dijkstra algorithm based on path optimization. Mod Electron Technol 13(40):181–183
11. Hu Q-C, Dong L-J (2020) Acoustic emission source location and experimental verification for two-dimensional irregular complex structure. IEEE Sens J 20(5):2679–2691. https://doi.org/10.1109/jsen.2019.2954200
12. Hart PE, Nilsson NJ, Raphael B (1968) A formal basis for the heuristic determination of minimum cost paths. IEEE Trans Syst Sci Cybernet 4(2):100–107
13. Hart PE, Nilsson NJ, Raphael B (1972) Correction to "a formal basis for the heuristic determination of minimum cost paths." ACM SIGART Bull 37:28–29
14. Djojo MA, Karyono K (2013) Computational load analysis of Dijkstra, A*, and Floyd-Warshall algorithms in mesh network. In: 2013 International conference on robotics, biomimetics, intelligent computational systems. IEEE, pp 104–108
15. Gaci S (2013) The use of wavelet-based denoising techniques to enhance the first-arrival picking on seismic traces. IEEE Trans Geosci Remote Sens 52(8):4558–4563
16. Dong L-J, Tao Q, Hu Q-C, Deng S-J, Chen Y-C, Luo Q-M, Zhang X-H (2022) Acoustic emission source location method and experimental verification for structures containing unknown empty areas. Int J Min Sci Technol 32(3):487–497. https://doi.org/10.1016/j.ijmst.2022.01.002
17. Dong L, Hu Q, Tong X, Liu Y (2020) Velocity-free MS/AE source location method for three-dimensional hole-containing structures. Engineering 6(7):827–834
18. Geiger L (1912) Probability method for the determination of earthquake epicenters from the arrival time only. Bull St Louis Univ 8(1):56–71
19. Li Q-Y, Dong L-J, Li X-B, Yin Z-Q, Liu X-L (2011) Effects of sonic speed on location accuracy of acoustic emission source in rocks. T Nonferr Metal Soc 21(12):2719–2726
20. Li X, Shang X, Morales-Esteban A, Wang Z (2017) Identifying P phase arrival of weak events: the Akaike Information Criterion picking application based on the Empirical Mode Decomposition. Comput Geosci 100:57–66
21. Liu M, Yang J, Cao Y, Fu W, Cao Y (2017) A new method for arrival time determination of impact signal based on HHT and AIC. Mech Syst Signal Process 86:177–187
22. Pearson MR, Eaton M, Featherston C, Pullin R, Holford K (2017) Improved acoustic emission source location during fatigue and impact events in metallic and composite structures. Struct Health Monit 16(4):382–399
23. Dong L, Tong X, Ma J (2021) Quantitative investigation of tomographic effects in abnormal regions of complex structures. Eng-Prc 7(7):1011–1022

Chapter 8
Application of Velocity-Free Localization Method in Hazard Analysis of slopes in Rare Earth Mine

Microseismic source localization is an important part of microseismic monitoring technology, which can provide an effective basis for identifying potential hazardous areas of rock instability in underground engineering, slope engineering, etc. At present, microseismic location technology is rarely used in the rare earth mine slope (REMS) due to the particularity of its structure and composition. In this chapter, we carried out source localization tests in REMS to determined the dangerous area in this chapter. The introduction of the method applied to analyze the slopes stability and hazard area is demonstrated in Sect. 8.1. The preparation and introduction of the field test is descripted in Sect. 8.2. The principle of risk analysis is demonstrated in Sect. 8.3. And peak ground acceleration (PGA) forward fitting is conducted by random forest method and gradient boosted decision tree (GBDT) method in Sect. 8.4. At last, the hazard analysis is descripted in Sect. 8.5.

8.1 Introduction

At present, there are many methods to analyze the stability and risk of REMS in insitu leaching production areas, mainly quantitative analysis methods including limit equilibrium method, numerical simulation method, and uncertainty analysis method [1, 2]. Some of these methods can only analyze the stability under static conditions, while others can analyze both the stability under static conditions and under external influence.

Limit equilibrium method is a traditional stability analysis method of earth rock slope, which is mature, quantifiable, simple and easy to operate. Based on the Mohr–Coulomb criterion, the limit equilibrium method takes the stress condition of the slope in the limit failure state as the research object. According to the static equilibrium condition, the ratio of the sliding moment to the resistance moment is characterized as the safety factor of the slope, and its stability is judged. This method is a deterministic method, which uses the determined mechanical parameters of soil and rock particles

L. Dong and X. Li, *Velocity-Free Localization Methodology for Acoustic and Microseismic Sources*, https://doi.org/10.1007/978-981-19-8610-9_8

of slope to conduct analysis. If the slope is divided into slices, the static equilibrium becomes a static indeterminate problem. To apply the analysis method in practice, several calculation methods with different assumptions have been produced. The commonly used stability analysis methods for slopes are shown below:

- Swedish method
  - Arc method
  - Fellenius method
- Janbu method
- Bishop method
- M-P method.

The limit equilibrium method ignores the internal deformation relationship of the structure in the process of failure. To accurately describe the stress–strain relationship of the slope, Clough and Chopra [3] began to use the finite element method to analyze the soil stability in 1966. The commonly used numerical simulation methods include finite element method, discrete element method, boundary element method, and finite difference method [1, 4].

The above methods are all deterministic analysis methods, but there are many factors affecting slope stability, such as the mechanical properties, properties, external environment of rock and soil particles, and deterministic methods cannot effectively analyze these uncertain factors. Therefore, uncertainty method is applied to slope engineering. The commonly used uncertainty methods include grey system evaluation method, reliability method, and fuzzy evaluation method.

Many scholars have proposed various methods to explore and study the stability and hazard area of the slope. Han et al. [5] constructed a numerical simulation model of landslide early warning based on failure index brittleness curve, and explored the influence of mine water level on the stability of the REMS. Li [6] used FLAC3D software to simulate the rock mass structure of the REMS in Tengchong, Yunnan. Wang et al. [7] studied the influence of liquid injection height on the slope stability of ionic rare earth mine by using Slide6.0 software. Yu [8] studied the landslide early warning method of rare earth mine slope in combination with laboratory tests and numerical simulation, and determined the early warning threshold of various indicators. Peng [9] studied the influence of leaching time on the physical and mechanical parameters of rare earth ore rock and soil mass, and analyzed the stability of REMS in combination with numerical simulation method and strength reduction method. Wu et al. [10] simulated the state of rare earth ore under long-term leaching with indoor soaking test, and analyzed the stability of REMS with Geo-Studio software. Wang et al. [11] built a grey prediction model to early warning the landslide accident of REMS with in-situ leaching.

It is apparent that for the stability and risk analysis of REMS, many scholars pay attention to the situation under the conditions of liquid injection and rainfall, and the stability and risk analysis of REMS under dynamic disturbance are rarely studied. In addition, these research methods cannot evaluate the real-time stability of REMS. Therefore, it is necessary to monitor and study the real-time dynamic stability of REMS. At present, many scholars have carried out research on the risk caused by vibration in the field of general slope or mine. To evaluate the ground vibration of underground mines caused by fault slip, the peak ground velocity and peak ground acceleration were inversed through seismic moment tensor, and based on these two indicators, an evaluation method was proposed to evaluate the severity of ground vibration of underground mines caused by fault slip [12]. Wang et al. [13] arranged vibration monitoring equipment in the earthquake stricken areas in the mountains. They studied the amplification effect of peak acceleration and arias intensity at different sites after the earthquake by monitoring the relevant dynamic parameters of aftershocks for a long time, providing reference for post disaster reconstruction on this basis. In order to analyze the stability of slope under dynamic disturbance, Chen et al. [14] studied the relationship between blasting vibration velocity, acceleration and stress of slope based on wave theory, and proposed a time history method based on equivalent acceleration and Salma method of limit equilibrium analysis to evaluate the dynamic stability of slope under the consideration of blasting dynamic load. Cong et al. [15] studied the impact of mining earthquakes on coal mine safety. They took the energy attenuation of mining earthquakes as the research object, obtained the vibration index by analyzing the attenuation law, and used tomography to evaluate the regional risk of mining earthquakes under dynamic disturbance. Xu et al. [16] proposed a rock mass damage evolution model based on microseismic monitoring. Combined with three-dimensional finite element method, they discussed the relationship between slope strength degradation and microseismic activities caused by rock mass damage in the process of slope dynamic instability, and quantitatively evaluated the slope stability. Lan et al. [17] classified a series of risk influencing factors, constructed a multi factor pattern recognition model using artificial intelligence technology, and predicted the risk probability of mine dynamic disasters through neural network and fuzzy mathematics. It can ensure the safety of mine operations to a certain extent. The above researches have promoted the development of stability analysis of rock mass structure under dynamic disturbance, but there is no research on the response of REMS to focal disturbance. As we all know, the wave which propagates from a vibration source to sensors contains the information about source parameters and propagation media. Source localization and tomograpy [18] are the tools to explore the information such as source coordinates and velocity distribution. Dong et al proposed series of velocity-free methods [19–24] which could avoid velocity error caused by pre-measured velocity. The accurate source localiztion would be helpful to determine hazardous areas.

8.2 Field Test in Huashan Rare EarthMine

The unexploited REMS selected in the test is located in the southwest of the South ore section of Huashan rare earth mine in Zhongshan Fuchuan County. The site photo and profile of the slope is shown in Fig. 8.1.

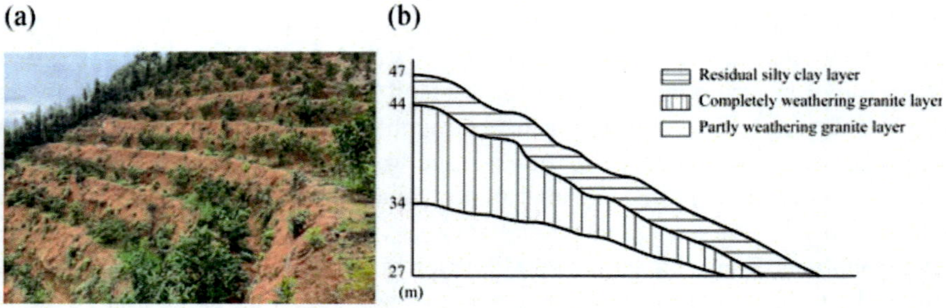

Fig. 8.1 **a** Site photo and **b** profile of the slope

The heights of the slope foot and slope top are 427 m and 447 m, respectively, with a height difference of 20 m. Judging from the actual situation, there are many classification methods of slopes, such as classification according to the formation mode, slope body composition, slope length, slope height, slope gradient, etc., while ionic REMS are generally classified according to the location of their underlying bedrock due to their special mineral formation and formation conditions. For bedrock exposed at the foot of the slope, it is a bare foot rare earth mine slope and for bedrock completely covered by upper soil, it is a fully covered slope of rare earth mine [25, 26]. It can be seen from Fig. 8.1b that the slope in the test belongs to full coverage slope. There are three strata from top to bottom, namely: residual silty clay layer, completely weathering granite layer, and partly weathering granite layer. The bedrock is unweathering or slightly weathering granite. The risk of bedrock is not considered due to its good stability. In the slope, rare earth minerals are mainly distributed in the completely weathering granite layer.

The slope surface is mainly covered by soil, and a small amount of vegetation and weeds. However, there is much rainfall in this area, especially in the rainy season (June and July every year). Therefore, it is necessary to do an assessment on potential dangerous areas and slope stability analysis before the in-situ leaching production process of rare earth mines in this area, so that the rare earth mines can be safely put into mining under the condition of ensuring the slope stability.

8.2.1 Preparation for Test

Before sensor installation, we laid the communication optical cable and sensor cable on the slope and set the serial port of dynamic acquisition instrument of each monitoring station. According to the specific situation of the slope, the test arranged four monitoring stations separately and a master station at a close distance from the four stations. Protective measures shall be taken for each station to prevent equipment damage caused by sudden rainfall. Considering the loose soil and large scale of the rare earth mine slope, 16 microseismic sensors are selected in this test, and each sensor is equipped with a 1 m long wave guide rod. The sensor is connected with

Fig. 8.2 Installation of wave guide rod in a sensor

one end of the wave guide rod in a spiral way. The other end of the wave guide rod is vertically inserted into the slope with exposure less than 10 cm, so that the sound wave propagating in the slope can be better received by the sensor, as shown in Fig. 8.2. The selected 16 microseismic sensors are scattered in the slope, which are shown in Fig. 8.3. The coordinates of 16 sensors (S_1, S_2, ..., S_{16}) are shown in Table 8.1. The 16 sensors are divided into four groups, which are respectively connected with the four stations to transmit the received signals. The sampling frequency is 6,000 Hz and the sampling resolution is 24 bits. Before the test, check the connection of microseismic equipment and the installation and coupling of each sensor, and test knock the rare earth mine slope to ensure that each sensor can effectively receive microseismic signals.

8.2.2 *Data Acquisition*

Microseismic monitoring uses sensors to monitor the surface and internal conditions of the structure. It is a dynamic monitoring technology, which can detect the geo-acoustic waves generated by the energy release caused by crack propagation, deformation and other reasons in the structure within a certain range. These geo-acoustic waves carry a lot of information closely related to the internal state of rock mass. Data acquisition is the process of converting the geo-acoustic wave generated by the rock mass under the change of external or internal environment into electrical signals, and storing the relevant data in the computer. Users can use data processing and analysis technology to explore the information stored in the electrical signal, and then infer the mechanism of generating the source signal. The data acquisition process of the multi-information intelligent monitoring for geotechnical engineering disaster and early warning system used in this test are shown in Fig. 8.4.

The hammering test was carried out on the slope, and the sound wave generated by heavy hammer percussion was used as the microseismic source of this test. The hit points were scattered on the slope, and the range covered the whole sensor layout

Fig. 8.3 Arrangement of sensors and stations

Table 8.1 Sensor coordinates

Sensor No	Coordinates (m)		
	x	y	z
S_1	28.424	15.236	36.602
S_2	18.608	7.994	36.822
S_3	26.015	26.569	29.908
S_4	16.337	19.782	30.211
S_5	40.505	32.598	36.996
S_6	33.105	20.544	36.797
S_7	34.437	41.996	30.289
S_8	28.667	30.702	29.939
S_9	41.073	7.597	45.783
S_{10}	34.437	-3.049	45.552
S_{11}	31.572	12.637	39.561
S_{12}	21.615	4.322	39.239
S_{13}	54.067	23.867	46.036
S_{14}	45.021	13.599	45.605
S_{15}	43.966	30.098	39.877
S_{16}	35.897	18.108	39.5

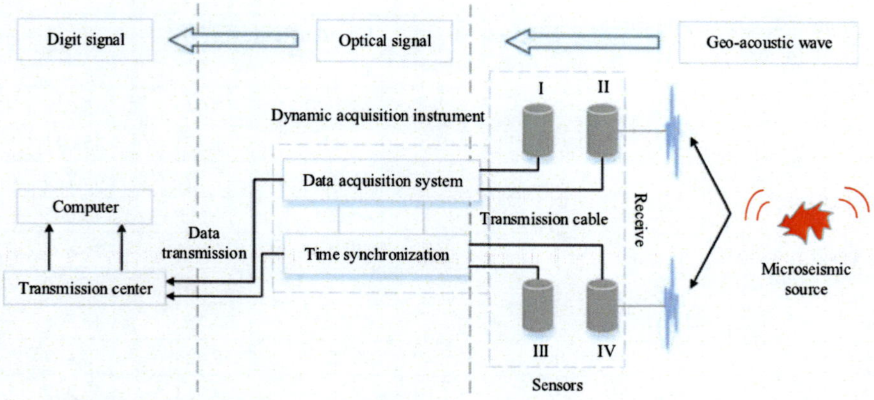

Fig. 8.4 Illustration of data acquisition process in the system

area. In the test, a total of 193 hits were made at 32 hit points, 136 effective hits were selected, and then one hit (a total of 32 hits) was selected at each hit point, which was recorded as No. 1, No. 2, ... and No. 32. These 32 hits were taken as the 32 source events, and their coordinates are shown in Table 8.2. Figure 8.5 is a three-dimensional diagram of the sensor and the hit point. During the knocking process, the microseismic signals are collected synchronously, and the characteristic parameters and waveforms of microseismic events are collected at the same time. Through the data collection, it is found that in the 32 hit points, at least four sensors are triggered each time, indicating that the data are effective.

The waveform of microseismic event at the No. 6 hit point obtained through data acquisition system is shown in Fig. 8.6.

8.2.3 *Hammering Test and Location Result*

In the mining field, microseismic source localization is mostly based on the method of measuring the wave velocity in advance, for instance, obtaining the average wave velocity of the measurement area by blasting. Traditional methods, such as STT and STD, substitute the obtained average wave velocity as the wave velocity of the whole monitoring area into the calculation formula. However, these methods have some defects. Because the rock mass in the monitoring area is not a uniform medium, the wave velocity of the wave propagating in such a medium is dynamic, and the average wave velocity will be different in different structural areas. In addition, the average wave velocity obtained by blasting method may deviate the average wave velocity of the area where the microseismic source is located in the rock mass. Therefore, the results of the traditional localization method do not have high accuracy.

In the test, the method used is the source localization method with the time difference of arrival as the dependent variable, which does not need to measure the velocity

Table 8.2 The coordinates of 32 hit points

Hit point No.	Coordinates (m)		
	x	y	z
1	34.437	−3.049	45.552
2	37.33	1.774	45.641
3	41.073	7.597	45.783
4	45.021	13.599	45.605
5	49.264	18.946	45.868
6	54.067	23.867	46.036
7	47.821	26.138	43.508
8	39.22	15.624	42.464
9	34.916	9.913	42.342
10	25.92	−0.772	42.956
11	21.615	4.322	39.239
12	26.648	7.933	39.334
13	31.572	12.637	39.561
14	35.897	18.108	39.5
15	40.016	23.92	39.665
16	43.966	30.098	39.877
17	40.505	32.598	36.996
18	36.687	26.175	36.839
19	33.105	20.544	36.797
20	28.424	15.236	36.602
21	24.669	11.291	36.661
22	18.608	7.994	36.822
23	18.578	13.866	33.515
24	29.071	22.999	33.35
25	32.581	27.795	33.58
26	36.97	34.758	33.819
27	34.437	41.996	30.289
28	32.031	37.183	30.198
29	28.667	30.702	29.939
30	26.015	26.569	29.908
31	21.753	22.438	29.951
32	16.337	19.782	30.211

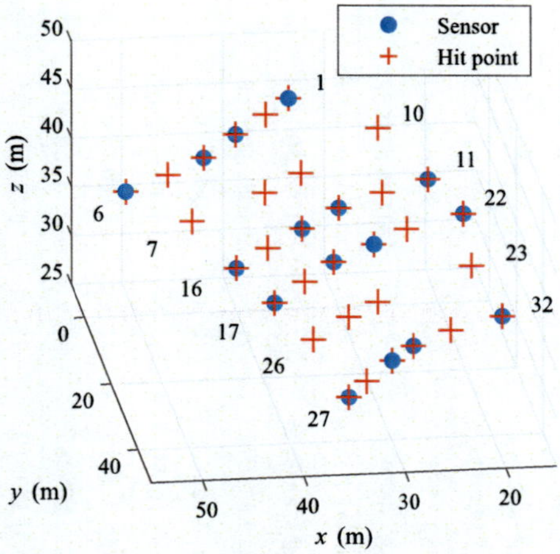

Fig. 8.5 Illustration of sensor and source position

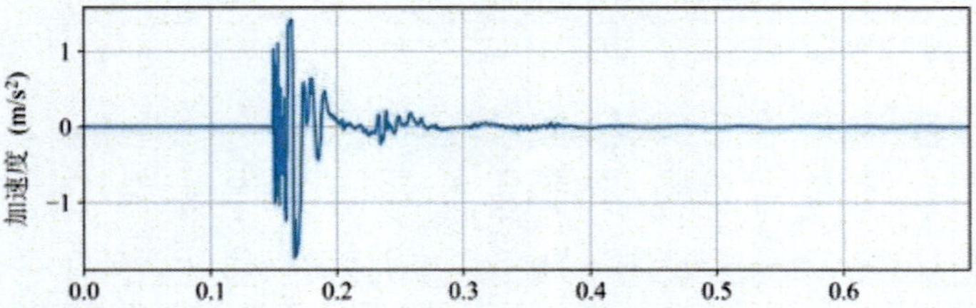

Fig. 8.6 The waveform generated by No. 6 hit

in advance [27]. This method no longer needs to measure the wave velocity in the monitoring area in advance, and regards the wave velocity as an unknown constant, and calculates the source coordinates and average wave velocity through algorithm iteration. This method overcomes the influence of the pre-measured wave velocity in the traditional positioning method on the location result, so it can effectively improve the location accuracy. We also improve the location accuracy by active source correction.

Based on the data collected by the microseismic monitoring technology, the arrival time of the microseismic signal generated by the percussion to its corresponding triggered microseismic sensor can be obtained, as shown in Table 8.3. The location result is shown in Table 8.4.

Table 8.3 Arrivals recorded by sensors

Arrivals (ms)		Arrivals (ms)		Arrivals (ms)		Arrivals (ms)		Arrivals (ms)	
t_1^{10}	613.0001083	t_8^{16}	435.6668534	t_{13}^{6}	845.66674	t_{18}^{16}	786.8335317	t_{25}^{6}	971.3336581
t_1^{1}	615.666775	t_8^{14}	441.3335201	t_{13}^{4}	872.0000733	t_{18}^{14}	792.6668651	t_{25}^{16}	982.8336581
t_1^{9}	640.8334417	t_8^{11}	444.8335201	t_{13}^{5}	853.5000734	t_{18}^{1}	792.3335317	t_{25}^{1}	987.5003248
t_1^{11}	649.8334417	t_8^{19}	446.3335201	t_{13}^{10}	217.0000433	t_{18}^{6}	797.8335317	t_{25}^{15}	987.8336581
t_1^{14}	661.666775	t_8^{15}	468.1668534	t_{14}^{16}	228.8333766	t_{19}^{6}	574.5002426	t_{25}^{5}	854.8336581
t_2^{2}	953.5000844	t_8^{6}	449.3335201	t_{14}^{6}	242.0000433	t_{19}^{16}	581.0002426	t_{25}^{7}	993.8336581
t_2^{11}	955.666751	t_8^{1}	456.1668534	t_{14}^{11}	237.6667099	t_{19}^{1}	583.3335759	t_{25}^{11}	995.6669914
t_2^{1}	960.166751	t_8^{2}	481.6668534	t_{14}^{14}	248.5000433	t_{19}^{11}	587.6669093	t_{25}^{14}	1026.833658
t_2^{10}	968.666751	t_8^{13}	487.0001867	t_{14}^{9}	257.5000433	t_{19}^{3}	605.0002426	t_{26}^{5}	926.0003018
t_2^{16}	981.0000844	t_8^{15}	574.6668534	t_{14}^{5}	143.0000433	t_{19}^{14}	627.5002426	t_{26}^{7}	1082.500302
t_2^{14}	1039.666751	t_8^{10}	471.8335201	t_{14}^{15}	255.0000433	t_{19}^{8}	610.3335759	t_{26}^{15}	1088.666968
t_3^{9}	869.0000589	t_8^{7}	508.0001868	t_{14}^{8}	265.0000433	t_{19}^{9}	629.1669093	t_{26}^{8}	1089.333635
t_3^{14}	885.8333922	t_9^{11}	181.5002231	t_{15}^{14}	1119.166681	t_{19}^{5}	535.1669093	t_{26}^{6}	1104.000302
t_3^{10}	900.6667256	t_9^{9}	187.6668898	t_{15}^{15}	1098.166681	t_{20}^{1}	935.3336192	t_{26}^{16}	1110.833635
t_3^{11}	901.3333922	t_9^{16}	191.5002231	t_{15}^{16}	1100.000015	t_{20}^{11}	946.5002858	t_{26}^{14}	1201.500302
t_3^{16}	903.1667256	t_9^{14}	198.3335565	t_{15}^{6}	1103.666681	t_{20}^{6}	956.3336192	t_{26}^{1}	1171.333635
t_3^{1}	915.5000589	t_9^{6}	202.0002231	t_{15}^{1}	1122.333348	t_{20}^{16}	959.0002858	t_{27}^{7}	290.8337166
t_3^{6}	922.6667256	t_9^{10}	195.3335565	t_{15}^{5}	1136.333348	t_{20}^{14}	989.3336192	t_{27}^{15}	338.8337166
t_3^{13}	927.1667256	t_9^{1}	196.8335565	t_{15}^{11}	1121.666681	t_{20}^{9}	986.0002859	t_{27}^{8}	322.5003833
t_4^{5}	848.1666903	t_9^{2}	215.8335565	t_{15}^{13}	1123.000015	t_{20}^{5}	1070.500286	t_{27}^{5}	206.16705

(continued)

Table 8.3 (continued)

Arrivals (ms)		Arrivals (ms)		Arrivals (ms)		Arrivals (ms)		Arrivals (ms)	
t_4^{16}	870.833357	t_9^5	178.3335565	t_{15}^8	1126.333348	t_{21}^1	833.000315	t_{28}^7	507.1670883
t_4^6	916.333357	t_{10}^{10}	159.5002543	t_{15}^2	1156.833348	t_{21}^{11}	839.500315	t_{28}^8	510.6670883
t_4^{11}	927.333357	t_{10}^2	176.0002543	t_{15}^9	132.5000145	t_{21}^{16}	856.000315	t_{28}^5	382.6670883
t_4^1	933.333357	t_{10}^9	188.8335877	t_{16}^{15}	231.1666606	t_{21}^6	853.8336483	t_{28}^{15}	539.0004216
t_4^{14}	941.333357	t_{10}^{11}	179.3335877	t_{16}^5	254.6666606	t_{21}^{10}	868.6669816	t_{28}^6	541.0004216
t_5^{15}	174.1666714	t_{10}^{14}	224.166921	t_{16}^{13}	269.3333273	t_{21}^{14}	885.000315	t_{29}^8	515.5004347
t_5^{16}	186.8333381	t_{11}^2	426.3334673	t_{16}^1	293.999994	t_{22}^1	381.3336704	t_{29}^6	546.0004347
t_5^6	191.1666714	t_{11}^1	450.6668006	t_{16}^7	281.6666606	t_{22}^{11}	387.6670037	t_{29}^5	426.333768
t_5^{14}	211.5000048	t_{11}^{16}	470.8334673	t_{16}^6	274.1666606	t_{22}^{16}	404.0003371	t_{29}^7	550.333768
t_5^1	212.8333381	t_{11}^{11}	451.500134	t_{16}^{16}	272.999994	t_{22}^6	401.3336704	t_{29}^1	557.1671014
t_5^{11}	218.0000048	t_{11}^{10}	456.500134	t_{16}^{14}	282.6666606	t_{22}^{14}	641.5003371	t_{30}^8	942.8337785
t_5^5	240.3333381	t_{11}^6	482.1668007	t_{16}^{11}	292.999994	t_{23}^1	200.8337398	t_{30}^6	960.6671118
t_6^{13}	841.8333072	t_{11}^{14}	623.500134	t_{16}^8	421.6666607	t_{23}^{11}	212.3337398	t_{30}^1	967.5004452
t_6^{15}	878.1666406	t_{12}^2	474.3334292	t_{17}^5	564.5001737	t_{23}^{16}	225.0004065	t_{30}^7	979.1671118
t_6^{14}	879.6666406	t_{12}^1	474.8334292	t_{17}^{15}	576.8335071	t_{23}^6	218.0004065	t_{31}^8	536.0004709
t_6^9	900.6666406	t_{12}^{11}	472.3334292	t_{17}^{16}	605.1668404	t_{23}^{14}	254.1670732	t_{31}^1	535.0004709
t_6^{16}	922.8333073	t_{12}^{10}	480.6667625	t_{17}^3	611.8335071	t_{24}^6	793.3337035	t_{31}^6	539.8338043
t_6^5	997.1666406	t_{12}^9	498.0000959	t_{17}^8	602.0001737	t_{24}^1	799.5003702	t_{31}^{16}	551.8338043
t_7^{15}	1140.000158	t_{12}^6	512.6667625	t_{17}^6	602.3335071	t_{24}^8	802.8337035	t_{31}^{14}	580.500471
t_7^{13}	1142.333491	t_{12}^{16}	493.0000959	t_{17}^7	601.0001737	t_{24}^{16}	806.6670368	t_{32}^1	874.0004826

(continued)

Table 8.3 (continued)

Arrivals (ms)		Arrivals (ms)		Arrivals (ms)		Arrivals (ms)		Arrivals (ms)	
t_7^5	1042.500158	t_{12}^{14}	511.3334292	t_{17}^{14}	624.6668404	t_{24}^{11}	811.1670368	t_{32}^{6}	884.833816
t_7^{14}	1157.000158	t_{13}^{10}	723.0000733	t_{17}^{1}	709.1668404	t_{24}^{14}	850.8337035	t_{32}^{11}	888.1671493
t_7^{16}	1162.000158	t_{13}^{9}	732.16674	t_{17}^{13}	750.3335071	t_{24}^{7}	836.8337035	t_{32}^{16}	895.333816
t_7^6	1168.500158	t_{13}^{11}	821.16674	t_{17}^{11}	860.8335071	t_{24}^{5}	740.1670368		
t_7^9	1198.000158	t_{13}^{1}	833.8334067	t_{18}^{11}	772.1668651	t_{24}^{9}	847.8337035		
t_7^{11}	1204.166825	t_{13}^{16}	841.5000733	t_{18}^{9}	781.3335317	t_{25}^{8}	968.0003248		

t_b^i indicates arrivals generated by bth hit point and recorded by ith sensor

Table 8.4 Location result and absolute error

Hit point No	Location result (m)			Absolute error (m)
	x	y	z	
1	35.75	−4.957	25.186	20.497
2	28.376	2.672	35.326	13.689
3	31.572	12.637	39.561	12.425
4	54.897	23.939	40.153	15.303
5	43.966	30.098	39.877	13.723
6	50.082	13.603	47.372	11.091
7	49.388	27.292	41.578	2.741
8	38.595	14.987	42.48	0.893
9	32.227	11.09	39.868	3.839
10	28.227	−4.996	45.46	5.425
11	15.179	0.169	36.822	8.032
12	22.879	1.009	42.463	8.482
13	34.437	−3.049	45.552	17.034
14	36.092	19.168	40.52	1.484
15	35.287	27.839	47.96	10.321
16	47.942	32.253	36.962	5.38
17	32.695	34.592	42.023	9.5
18	32.951	7	40.94	19.961
19	30.76	20.147	38.46	2.902
20	37.25	11.984	35.83	9.438
21	22.116	9.459	35.669	3.295
22	27.112	12.872	36.602	9.806
23	20.555	14.403	34.815	2.427
24	27.61	24.325	36.63	3.827
25	36.402	24.877	27.127	8.047
26	39.828	32.507	36.592	4.574
27	15.966	34.663	31.121	19.891
28	24.887	41.529	26.861	9.003
29	54.839	31.471	39.836	27.991
30	26.567	27.045	31.542	1.789
31	22.915	20.873	28.853	2.238
32	28.424	17.333	35.673	13.488
Average error				9.329

8.3 Basic Principle of Regional Risk Analysis of Rare Earth Mine Slope

8.3.1 Hazard Indicators of Slope Area

There are many ways to evaluate the regional risk of slope. The common one is to quantitatively calculate the risk of slope using the stability coefficient of slope, and the other is to calculate the probability of slope landslide by numerical simulation combined with mathematical methods. PGA (peak ground acceleration) is an important characteristic value to analyze and evaluate the impact of dynamic disturbance on the rock and soil structure of the slope. Therefore, PGA characteristic value is proposed to characterize the regional risk of rare earth mine slope.

In the field of microseismic, the dangerous area within the monitoring range is caused by the microseismic source generated by dynamic disturbance. By determining the spatial distribution of the PGA value of the wave emitted by the microseismic source within the monitoring range, we can predict the different damage degree of dynamic disturbance (i.e. microseismic) to the rock and soil structure, and then determine the regional danger.

The maximum acceleration on the path of a wave from a source to a sensor is the PGA of the source, which is usually used to measure the intensity of vibration caused by the source. Generally, the larger the PGA value of a source in the region, the greater the vibration intensity caused by the source, and the more serious the possible damage. At the same time, the wider the distribution range of large PGA value, the larger the influence range of the source on the monitoring area, and the greater the influence on the stability of rock and soil structures in the region. The relationship is shown in Fig. 8.7.

Fig. 8.7 Illustration of relationship between intensity of vibration and PGA

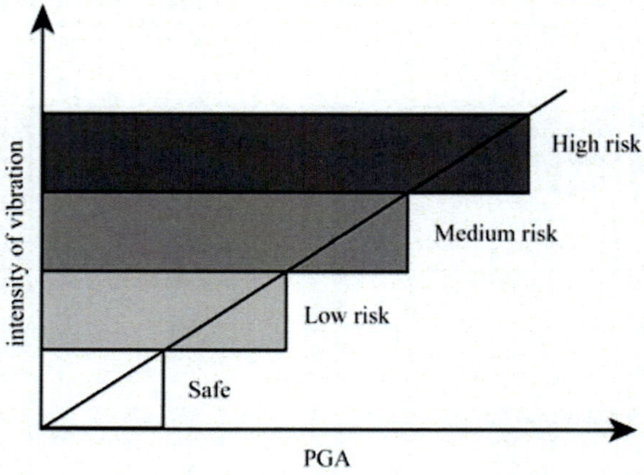

8.3.2 PGA Related Source Parameters

It is impossible to arrange a large number of sensors without separation in the actual project due to the limitation of conditions. Therefore, we converted the voltage time history curve into the acceleration time history curve based on the source data of the hammering disturbance collected by 16 microseismic sensors arranged on the slope, and obtained the actual PGA of each dynamic disturbance point (i.e. percussion point). In order to obtain the PGA in the whole monitoring area, the collected samples were trained and tested using multiple source parameters based on the methods of random forest and GBDT.

According to experience, PGA is affected by the propagation distance, energy, moment magnitude, and corner frequency of the source. Therefore, the source parameters required in the forward fitting process are as follows:

- The source coordinates at the dynamic disturbance point (x_0, y_0, z_0)
- Energy Q
- Moment magnitude M
- Corner frequency F.

The PGA values of sources 1–6 are shown in Table 8.5.

8.3.3 Classification of Dangerous Area of Rare Earth Mine Slope

Table 8.5 only lists the PGA values of some sources, and the actual effective percussion of the heavy hammer test is 136 times. Therefore, statistical analysis of the PGA from the 136 microseismic sources to the sensors is carried out, as shown in Fig. 8.8. The minimum value of the actual PGA in the whole monitoring area is 0.0064 m/s^2, the maximum value is 1.8175 m/s^2, and the average value is 0.1128 m/s^2. On the basis of statistical analysis, for the vibration caused by dynamic disturbance in the rare earth mine slope, taking PGA as the representative value of the risk level, four risk levels are divided, as shown in Table 8.6. It should be noted that the division of the risk area is relative.

8.4 PGA Forward Fitting

At any point i in the field, PGA meets formula:

$$PGA_i = f(x_i, y_i, z_i, x_0, y_0, z_0, L, Q, M, F), \tag{8.1}$$

where (x_i, y_i, z_i) represents the coordinates of any point in the field, and (x_0, y_0, z_0) the source coordinates. L, Q, M, and F indite the distance between the two points, energy, moment magnitude, and corner frequency, respectively.

Table 8.5 Parameters and PGA values of sources 1–6

Source No	Source coordinates (m)			Sensor coordinates (m)			Distance (m)	Acceleration (m/s^2)
	x	*y*	*z*	*x*	*y*	*z*		
1	34.437	−3.049	45.552	34.437	−3.049	45.552	0	1.758
				41.073	7.597	45.783	12.547	0.02
				45.021	13.599	45.605	19.728	0.035
				31.572	12.637	39.561	17.034	0.013
				28.424	15.236	36.602	21.227	0.01
2	37.33	1.774	45.641	18.608	7.994	36.822	21.61	0.012
				31.572	12.637	39.561	13.716	0.072
				28.424	15.236	36.602	18.5	0.031
				34.437	−3.049	45.552	5.625	0.01
				45.021	13.599	45.605	14.106	0.012
				35.897	18.108	39.5	17.509	0.014
3	41.073	7.597	45.783	41.073	7.597	45.783	0	1.361
				45.021	13.599	45.605	7.186	0.162
				34.437	−3.049	45.552	12.547	0.028
				31.572	12.637	39.561	12.425	0.034
				35.897	18.108	39.5	13.295	0.018
				28.424	15.236	36.602	17.397	0.012
				33.105	20.544	36.797	17.66	0.013
				54.067	23.867	46.036	20.824	0.013
4	45.021	13.599	45.605	35.897	18.108	39.5	11.868	0.265
				33.105	20.544	36.797	16.365	0.026
				31.572	12.637	39.561	14.776	0.058
				28.424	15.236	36.602	18.952	0.02
				45.021	13.599	45.605	0	0.013
5	49.264	18.946	45.868	45.021	13.599	45.605	6.831	0.149
				54.067	23.867	46.036	6.878	0.025
				43.966	30.098	39.877	13.723	0.014
				41.073	7.597	45.783	13.996	0.017
				35.897	18.108	39.5	14.83	0.011
				33.105	20.544	36.797	18.6	0.009
				31.572	12.637	39.561	19.814	0.009
				28.424	15.236	36.602	23.107	0.01
6	54.067	23.867	46.036	54.067	23.867	46.036	0	1.744
				45.021	13.599	45.605	13.691	0.045
				43.966	30.098	39.877	13.371	0.016

(continued)

Table 8.5 (continued)

Source No	Source coordinates (m)			Sensor coordinates (m)			Distance (m)	Acceleration (m/s^2)
	x	*y*	*z*	*x*	*y*	*z*		
				35.897	18.108	39.5	20.15	0.008
				41.073	7.597	45.783	20.824	0.01
				40.505	32.598	36.996	18.49	0.051

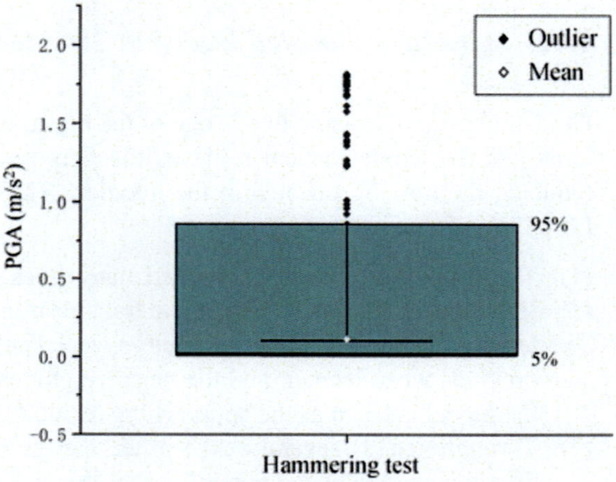

Fig. 8.8 Statistical analysis of the PGA from the 136 microseismic sources

Table 8.6 Hazard classification of REMS

Risk level	PGA percentile	PGA range	Description
Level 1	~70%	PGA ≤ a	Safe
Level 2	70% ~ 80%	a < PGA ≤ b	Low risk
Level 3	80% ~ 95%	b < PGA ≤ c	Medium risk
Level 4	95% ~	PGA > c	High risk

By substituting the source parameters collected in the slope monitoring area into Eq. (8.1) for forward fitting, the PGA at any point in the monitoring area can be calculated.

8.4.1 Random Forest Method

Random forest is an algorithm in the bagging ensemble learning algorithm series [28]. The basic principles of the algorithm are as follows [29]:

(1) The samples are repeatedly sampled by bootstrap for many times, and form multiple training sets which include sample data and attributes;

(2) The decision tree is constructed. Among several sample attributes, n attributes are selected as attribute subsets, and then the optimal attributes are selected as the nodes of the decision tree. The decision tree is not pruned to make it grow fully. After training, each decision tree outputs a weak learner result;
(3) The random forest is used for regression, and the final strong learner result is obtained through the average output of each decision tree;
(4) Put the test samples into the model for testing, and output the final results.

8.4.2 *Gradient Boosted Decision Tree Method*

Gradient boosted decision tree is one of the boosting algorithm series. This method is an additive model, which achieves the purpose of regression by continuously reducing the training residual in the iteration. The basic principles are as follows [30]:

(1) The samples are bootstrap sampled many times, and form multiple training sets;
(2) Initialize the first weak learner and train the training sets on the initialized weak learner. The next training generates a weak learner based on the residual of the previous weak learner, and the negative gradient value of the loss function in the model is taken as the approaching residual;
(3) Through step (2), several weak learners can be obtained, and these weak learners are obtained under the condition that the loss function is reduced as soon as possible, and the final strong learner is obtained by accumulating all the results of weak learners;
(4) Put the test samples into the strong learner for testing, and output the prediction results.

8.5 Hazard Analysis and Discussion of Slope Area

We use 907 microseismic signals to train the random forest algorithm for the model, and a source is randomly selected at four different heights as the test set. These four sources are regarded as the dynamic disturbance to the slope. The specific impact on the slope is shown in Figs. 8.9 and 8.10.

According to Fig. 8.9 and in combination with the risk levels classified in Table 8.6, the risk levals in the area of the red curve, yellow curve, and green curve are high, medium, and low, respectively. On the X–Y plane, it can be found that if the dynamic disturbance is located at the source 13, the overall impact on the rare earth mine slope is the largest. The maximum range of low-risk areas is about 24 m in the Y direction, about 20 m in the X direction, and the range of medium-risk areas is about 15 m in the Y direction and about 16 m in the X direction. When considering the slope stability protection, these areas need to be reinforced. When the dynamic disturbance is located at the position of source 27, the disturbance has the least impact on the overall slope. The maximum range of the low-risk area is about 21 m in the

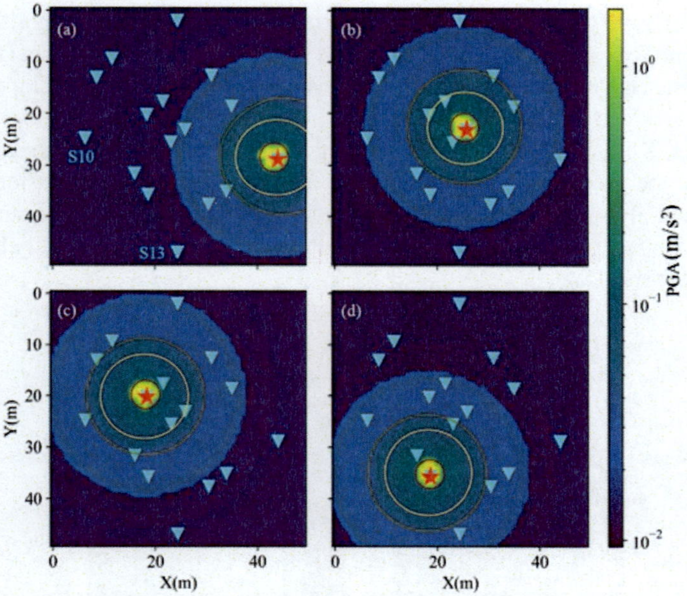

Fig. 8.9 Regional risk on X–Y plane with **a** source 27, **b** source 19, **c** source 13 and **d** source 4 disturbance using random forest algorithm

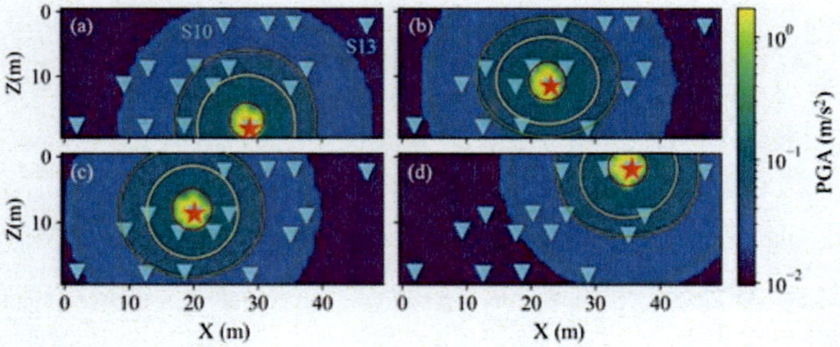

Fig. 8.10 Regional risk on X–Z plane with **a** source 27, **b** source 19, **c** source 13 and **d** source 4 disturbance using random forest algorithm

Y direction and about 10 m in the X direction. On the whole, the impact of dynamic disturbance located in the middle of the slope is greater than that of slope boundary disturbance.

Similarly, the risk level in the red curve is high, the yellow curve medium, and the green curve low in Fig. 8.10. It can be seen that on the X–Z plane, the dynamic disturbance at the same source 13 has the largest influence on the slope, and its influence basically covers the whole height of the slope, that is, once the dynamic disturbance occurs at this position, the whole slope will be in a dangerous state, including a dangerous area nearly 13 m high, which needs to be strengthened. In addition, the disturbance source 27 still has the least impact on the slope.

Similarly, 907 microseismic signals are used to train the GBDT algorithm for the model, and four sources that are the same as the random forest algorithm are selected as the test set. The impact of these four seismic disturbances on the slope is shown in Figs. 8.11 and 8.12.

In Fig. 8.11, the red curve shows high risk, the yellow curve shows medium risk, and the green curve shows low risk. Different from the conclusion obtained by random algorithm, the disturbance at source 19 under GBDT algorithm has the greatest impact on the slope, and almost the whole monitoring area belongs to at least a low-risk state.

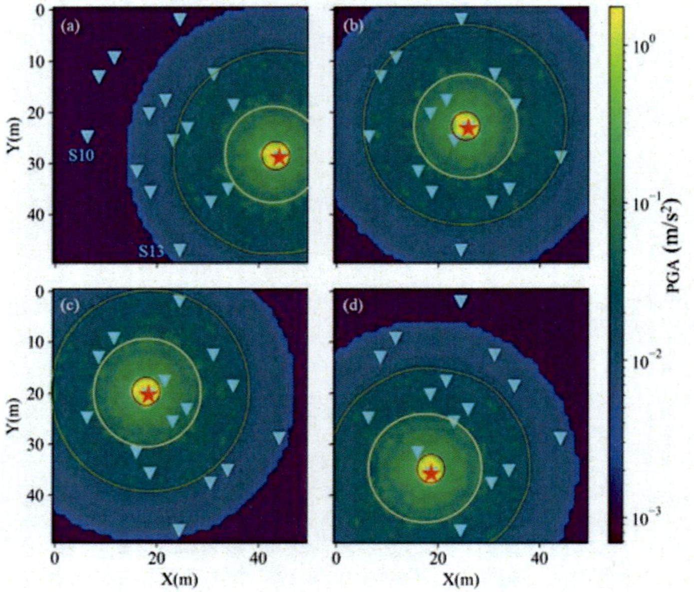

Fig. 8.11 Regional risk on X–Y plane with **a** source 27, **b** source 19, **c** source 13 and **d** source 4 disturbance using GBDT

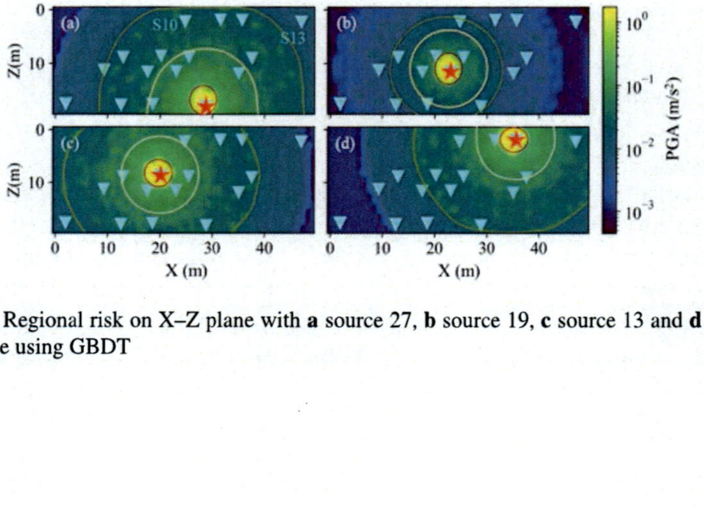

Fig. 8.12 Regional risk on X–Z plane with **a** source 27, **b** source 19, **c** source 13 and **d** source 4 disturbance using GBDT

As shown in Fig. 8.12, the areas within the red curve, yellow curve and green curve represent high risk, medium risk and low risk levels respectively. In the figure, the influence of the disturbance at source 19 is the smallest, the influence range of the disturbance of source 27 and source 13 on the rare earth mine slope is similar, and the low-risk area basically covers the whole slope.

Comparing Figs. 8.9, 8.10, 8.11, and 8.12, it is not difficult to find that the range of each dangerous area obtained by GBDT algorithm is larger than that obtained by random forest algorithm. The reason may be that GBDT algorithm is particularly sensitive to abnormal values in the sample, which leads to the overall larger results obtained by this algorithm, thereby improving the regional risk level of the slope. Therefore, the regional risk results in the test obtained by random forest algorithm are closer to the real situation.

8.6 Conclusions

The problem of low location accuracy in REMS can be effectively solved using the velocity-free method and active source correction. Using PGA to characterize the response of the slope to the disturbance source, the dynamic response and disaster development law of the REMS can be qualitatively evaluated. The PGA values in the monitoring area are obtained through forward model using random forest and GBDT algorithms. Then, the relative risk level of the slope is divided according to the statistical distribution of PGA. The hazard level of the slope under the dynamic disturbance at different positions is presented in the form of images. It is found that the slope is greatly affected by the disturbance in the center region, and the influence range is more than 20 m. Among the four randomly selected source disturbances, source 13 has the greatest disturbance impact on the slope, and source 27 has the least disturbance impact.

References

1. Raghuvanshi TK (2019) Plane failure in rock slopes–A review on stability analysis techniques. J King Saud Univ-Sci 31(1):101–109
2. Dong X-H, Liang G-L, Xu H-Z (2012) Summary and development trend of slope stability analysis methods. J Water Resour Architect Eng 10(5):100–102+106
3. Clough RW, Chopra AK (1966) Earthquake stress analysis in earth dams. J Eng Mech Div 92(2):197–211
4. Hu J, Liu H, Li L, Zhou S, Zhang Q, Sun S (2019) Stability analysis of dangerous rockmass considering rainfall and seismic activity with a case study in China's three gorges area. Polish J Environ Stud 28(2)
5. Han M, Rao Y-Z, Wang D, Zhang J-Y, Li Y (2020) Early warning model of water level landslide in rare earth mine based on fifc. ese Rare Earth 41(01):37–45. https://doi.org/10.16533/j.cnki.15-1099/tf.202001005
6. Li X-L (2021) Study on slope stability of complex rock mass structure in Diantan iron mine, Tengchong, Yunnan. Dissertation, Kunming University of Science and Technology

7. Wang X, Pan Y, Sun C-H, Deng Q-N, Hu W-J, Jiang H (2021) Study on slope stability and liquid injection height optimization of ionic rare earth mine. China Mining Mag 30(S1):311–315
8. Yu B (2016) Study on early warning technology of ionic rare earth mine landslide. Dissertation, Jiangxi University of Science and Technology
9. Peng H-D (2017) Study on the influence of leaching time on the slope stability of a rare earth mining area. Jiangxi University of Science and Technology
10. Wu C-F, Yao G-J, Chen L, Huang L, Tu Z-G (2016) Stability analysis of ionic rare earth mine slope based on geo studio. Chin J Geol Hazard Contr 27(2):72–77. https://doi.org/10.16031/j.cnki.issn.1003-8035.2016.02.11
11. Wang D, Yuan B-Y, Rao Y-Z, Rap R, Zhang Y-S (2015) Construction of slope monitoring scheme and grey prediction of landslide in rare earth mine. Nonferr Metals Sci Eng 6(3):116–120. https://doi.org/10.13264/j.cnki.ysjskx.2015.03.022
12. Ma J, Dong L-J, Zhao G-Y, Li X-B (2019) Qualitative method and case study for ground vibration of tunnels induced by fault-slip in underground mine. Rock Mech Rock Eng 52(6):1887–1901
13. Wang G-K, Wang Y-S, Liu J-W, Zhou Y, Su Y, Bi Y-Y, Xiang C (2022) Analysis of ground motion response characteristics of low mountain slopes in strong earthquake mountainous areas. J Geodesy Geodyn 42(2):187–192. https://doi.org/10.14075/j.jgg.2022.02.015
14. Chen M, Lu W, Yan P, Zhou C (2014) New method for dynamic stability analysis of rock slope under blasting vibration based on equivalent acceleration and Sarma method. Can Geotech J 51(4):441–448
15. Cong L, Cao A-Y, Zhou Y-H, Wang C-B, Chen F, Dong J-Y, Gu Y (2020) Comprehensive early warning method based on dynamic and static load rockburst hazard superposition. J Mining Saf Eng 37(4):767–776. https://doi.org/10.13545/j.cnki.jmse.2020.04.15
16. Xu N, Dai F, Liang Z, Zhou Z, Sha C, Tang C (2014) The dynamic evaluation of rock slope stability considering the effects of microseismic damage. Rock Mech Rock Eng 47(2):621–642
17. Lan T, Fan C, Li S, Zhang H, Batugin AS (2018) Probabilistic prediction of mine dynamic disaster risk based on multiple factor pattern recognition. Adv Civ Eng
18. Dong L, Pei Z, Xie X, Zhang Y, Yan X (2022) Early identification of abnormal regions in rock-mass using traveltime tomography. Engineering.
19. Dong L, Hu Q, Tong X, Liu Y (2020) Velocity-free MS/AE source location method for three-dimensional hole-containing structures. Engineering 6(7):827-834
20. Dong L, Zou W, Li X, Shu W, Wang Z (2019) Collaborative localization method using analytical and iterative solutions for microseismic/acoustic emission sources in the rockmass structure for underground mining. Eng Fract Mech 210(2019):95-112
21. Dong L, Tao Q, Hu Q, Deng S, Chen Y, Luo Q, Zhang X (2022) Acoustic emission source location method and experimental verification for structures containing unknown empty areas. Int J Min Sci Technol 32(3):487-497
22. Dong L, Sun D, Li X, Du K (2017) Theoretical and experimental studies of localization methodology for AE and microseismic sources without pre-measured wave velocity in mines. IEEE Access 5:16818-16828
23. Dong L, Shu W, Han G, Li X, Wang J (2017) A multi-step source localization method with narrowing velocity interval of cyber-physical systems in buildings. IEEE Access 5:20207-20219
24. Dong L-j, Li X-b, Zhou Z-l, Chen G-h, Ma J (2015) Three-dimensional analytical solution of acoustic emission source location for cuboid monitoring network without pre-measured wave velocity. Trans Nonferrous Met Soc China 25(1):293-302
25. Chi R-A, Tian J, Luo X-P, Xu Z-G, He Z-Y (2012) Basic study on rare earth ore of weathering crust leaching type. Nonferr Metals Sci Eng 3(4):1–13. https://doi.org/10.13264/j.cnki.ysjskx.2012.04.010
26. Wang G-S, Luo S-H, Hu S-L, Hong B-G, Deng X (2017) Seepage process and slope deformation of in-situ leaching in bare footed rare earth mines. Chin Rare Earth 38(03):35–46. https://doi.org/10.16533/j.cnki.15-1099/tf.201703005

27. Dong L-J, Li X-B, Tang L-Z, Gong F-Q (2011) Mathematical functions and parameters for microseismic source location without pre-measuring speed. Chin J Rock Mechan Eng 30(10):2057–2067
28. Breiman L (1999) 1 Random forests-random features. Mach Learn 45(1):5–32
29. Zhang K-A (2021) Multivalued random forest algorithm and its application in machine learning. Yunnan University of Finance and Economics
30. Friedman JH (2001) Greedy function approximation: a gradient boosting machine. Annals Stat:1189–1232

Chapter 9
Velocity-Free Localization of Trapped People

Many tunnels usually need to be excavated in underground engineering. In recent years, worker have been trapped due to geological hazards and engineering accidents from time to time. It is of great significance to accurately locate the trapped people and carry out rescue in time.

Microseismic monitoring system can realize disaster early warning based on microseismic technology. After the early warning, some workers may fail to escape the disaster area. When that happens, it is important to make full use of time to locate the position of trapped people and design rescue scheme. Therefore, using the sensors of microseismic monitoring system to locate the trapped people can not only improve the application value of the microseismic monitoring system, but also timely provide the position information of trapped people to rescuers, so as to win valuable time for the rescue and improve the rescue efficiency.

In this chapter, localization and rescue of trapped people after an accident or disaster is briefly introduced in Sect. 9.1. To verify the feasibility of applying the method to locate trapped people, simulation tests in site were conducted, which is introduced in Sect. 9.2. Then, Geiger method and velocity-free method (MSLM-MV) were used for the localization of trapped people. The location result is discussed in Sect. 9.3.

9.1 Introduction

Underground engineering mainly includes underground mining, tunneling, hydropower underground project and, urban underground engineering. During excavation or when the project is put into use, some geological hazards and engineering accidents often happen. It can be attribute to geotechnical uncertainty or human error [1]. Most geological hazards and engineering accidents, such as collapses, water and mud inrush, rockburst, object strike, toxic gas poisoning and explosion [2], cause casualties and economic losses.

L. Dong and X. Li, *Velocity-Free Localization Methodology for Acoustic and Microseismic Sources*, https://doi.org/10.1007/978-981-19-8610-9_9

Recently, microseismic monitoring technique has been widely used in underground engineering. It can provide support for risk assessment and warning of ground pressure hazards, such as collapse, caving, spalling, landslide, and rockburst [3]. However, it is difficult to give early warning for accidents caused by human errors and bad management. Furthermore, the warning time is finite. Therefore, some workers may fail to leave the dangerous region.

When there are workers trapped underground, the rescuers deliver oxygen, food, and water to the trapped people by surface drilling and then a large hole is drilled to lift the trapped people. There is no doubt that the position of trapped people is vital information during the rescue. Life detectors [4] and underground positioning system are applied to locate trapped people. Most of the systems are based on radio signal based systems (such as wireless local area networks and radio-frequency identification), non-radio signal based systems (such as acoustic systems and inertial navigation systems), and hybrid systems [5]. In recent years, Dong et al have developed some source localization methods without pre-measured velocity, including analytical solutions [6–8], iterative solutions [9–12], collaborative method [13], and methods considering complex structures [14–16]. The microseismic monitoring system using these methods makes the localization of trapped people feasible.

There are two advantages to apply microseismic technology in locating the trapped people, in view of the wide application of microseismic monitoring system. One is that it adds value to system and no additional investment. Another is that there is no requirement for the position of the trapped people, since the trapped people can be regarded as source in the system. Figure 9.1 shows the application of source localization technology in rescue process after a disaster.

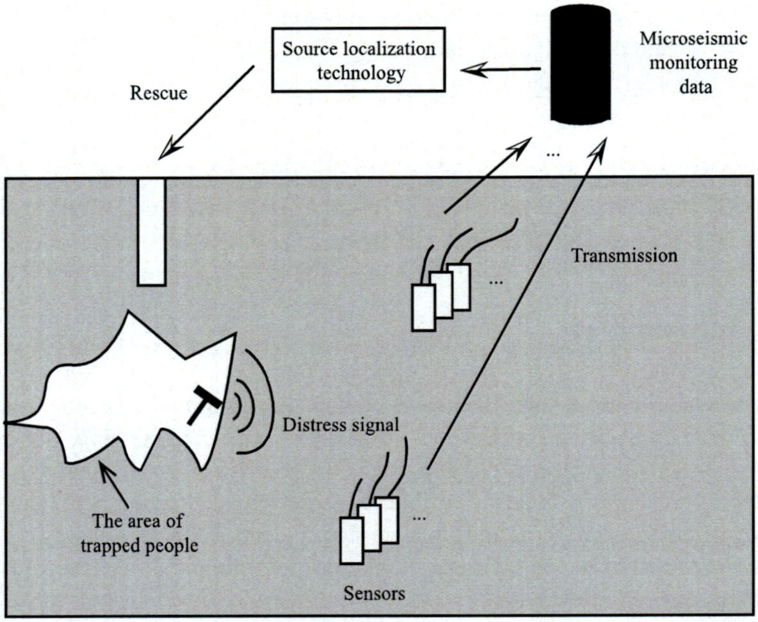

Fig. 9.1 Illustration of the rescue process with the assist of source localization technology

9.2 Simulation in Site

The blasting, drilling, and hitting tests are carried out in the Zhenao deposit, which is an underground lead–zinc mine in Shaanxi Province, China (Fig. 9.2). The mining district is mainly composed of phyllites and limestones. The limestones have a better quality than phyllites, which may be due to the constituents and weathering degree. The microseismic monitoring system, consisting of 26 single-component

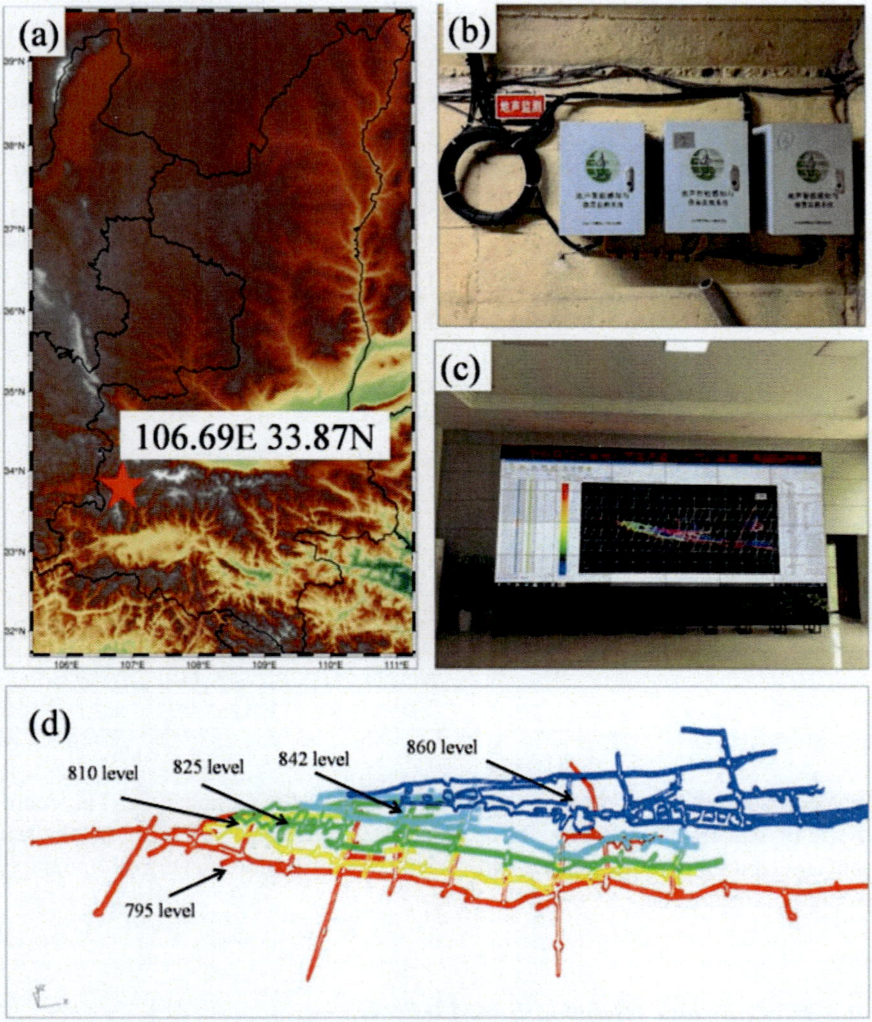

Fig. 9.2 **a** Location of Zhenao mine, **b** substation in 795 level, **c** the microseismic monitoring center in the Shaanxi Zhenao mine, and **d** layout of underground road way

Table 9.1 Coordinates of sensors

Sensor No	Coordinates (m)		
	x	*Y*	*z*
S_1	3,749,987.998	373,352.8885	800.0366068
S_2	3,749,962.803	373,321.5452	800.1807709
S_3	3,749,999.413	373,448.3944	799.5212198
S_4	3,749,924.78	373,402.8745	801.0689182
S_5	3,749,898.986	373,475.3981	803.33464
S_6	3,749,936.693	373,534.2606	802.5189061
S_7	3,749,906.373	373,641.4742	800.1596275
S_8	3,749,827.978	373,590.9545	800.5432843
S_9	3,749,810.609	373,710.7215	798.7421846
S_{10}	3,749,900.28	373,759.6184	798.0281053
S_{11}	3,749,839.501	373,810.9402	799.0030487
S_{12}	3,749,791.929	373,802.3092	799.0608521
S_{13}	3,749,940.84	373,530.7761	818.0744014
S_{14}	3,749,887.231	373,506.051	814.406404
S_{15}	3,749,953.311	373,596.9421	831.5875033
S_{16}	3,749,963.386	373,546.2261	829.2816903
S_{17}	3,749,957.88	373,545.2878	847.2466445
S_{18}	3,749,914.16	373,522.3726	846.5999844
S_{19}	3,749,947.61	373,593.0192	846.9040988
S_{20}	3,749,895.462	373,566.7504	845.9406428
S_{21}	3,749,819.209	373,698.7053	847.197921
S_{22}	3,749,827.522	373,811.7453	846.1384494
S_{23}	3,749,909.947	373,778.9286	862.8194675
S_{24}	3,749,811.937	373,809.6515	862.8542347
S_{25}	3,749,867.177	373,895.8101	863.8373512
S_{26}	3,749,817.505	373,863.4033	863.8193144

velocity sensors from 795 to 860 level, was deployed in March 2022. The coordinates of sensors are shown in Table 9.1.These facilities are mainly for long-term microseismic monitoring, calculating seismic event location, energy, tomography, multi-parameters, etc.

9.2.1 Simulate Distress Signal with Blasting

Night blasting tests have been carried out in site to simulate distress signal. The information about blasting tests is shown in Table 9.2. The positions and arrival

Table 9.2 The coordinates and date about blasting test

Blasting Sources No	Date	Coordinates (m)		
		x	y	z
P_1	15:26, April 1st, 2022	3,749,902.6	373,516.43	817.56
P_2	15:27, April 1st, 2022	3,749,971.46	373,550.71	796.62
P_3	15:31, April 1st, 2022	3,749,844.518	373,676.37	798.11
P_4	15:34, April 2st, 2022	3,749,828.73	373,688.61	798.54
P_5	23:50, April 2st, 2022	3,749,800.032	373,890.798	825.79
P_6	15:33, April 3st, 2022	3,749,832.8	373,688.278	798.23
P_7	23:50, April 3st, 2022	3,749,798.24	373,898.09	826.11
P_8	15:31, April 4st, 2022	3,749,838.57	373,687.57	798.63
P_9	23:37, April 4st, 2022	3,749,795.41	373,889.619	825.94

times of sensors recorded by microseismic monitoring system are shown in Table 9.3. Figure 9.3 shows the position of trapped people (simulated by blasting sources) and sensors. It can be found that P_1 has a better position than other sources, because it is located in the center area of sensor network and many sensors are around it. While P_2, P_5, P_7, and P_9 which are located in the edge of sensor network, have worse positions. Sources P_3, P_4, P_6, and P_8 are in the bottom of sensor network, but some sensors are around them (Fig. 9.3c).

Table 9.3 Arrival times recorded by sensors

Sensor No	Arrival Times of blasting test (s)								
	P_1	P_2	P_3	P_4	P_5	P_6	P_7	P_8	P_9
S_3	0.97417		1.03133	1.45833	0.78967			0.47617	
S_4		0.89750	1.48550	0.50117				0.47833	
S_5	0.96233		1.47400	0.80567				0.46750	
S_6	0.96150		1.46600	0.37617				0.45950	
S_7	0.97483	0.86883	0.99950	0.36633	0.75600		0.21120	0.44433	
S_8	0.97433	1.32417	1.00317	0.61783	0.76617		0.21980	0.44717	
S_9	0.99117	0.87200	0.99617		0.75000	0.14367	0.19850	0.34433	
S_{10}	0.99350		1.45600	0.35733	0.73983		0.19430	0.43750	
S_{11}	1.00250	0.84217	1.09050	0.37100	0.73600	0.19017	0.18350	0.43700	0.95100
S_{12}	1.00533	0.82450		0.48050	0.74183		0.18380		1.48317
S_{13}	0.95950	0.77300		0.46517			0.23500	0.46217	
S_{14}	0.96083			0.63650	0.77267			0.46033	
S_{15}	0.96983			0.80533	0.76300		0.22050	0.45483	
S_{16}	0.96533			0.63950	0.77183			0.46050	

(continued)

Table 9.3 (continued)

Sensor No	Arrival Times of blasting test (s)								
	P_1	P_2	P_3	P_4	P_5	P_6	P_7	P_8	P_9
S_{17}	0.96583			0.77450	0.77100		0.22880	0.46167	
S_{19}	0.96967			0.62183	0.76317		0.22120	0.45517	
S_{21}	0.98900	0.83267	1.45000	0.34983	0.73767	0.19367	0.19900	0.44000	
S_{22}	1.00350	0.84967	1.10783	0.37417	0.73250	0.22883	0.18280	0.44050	0.94950
S_{23}	0.99700	0.90583		0.39583			0.19420	0.45233	0.95700
S_{24}	1.00550	0.85567	1.05700	0.40083	0.72917	0.22817	0.18430	0.45300	0.95083
S_{25}	1.01467	0.89450	1.09417	0.39467	0.56250	0.32733	0.21420	0.46567	1.04967
S_{26}	1.01417	0.89783	1.08617	0.38083		0.23517	0.17720	0.46183	0.94383

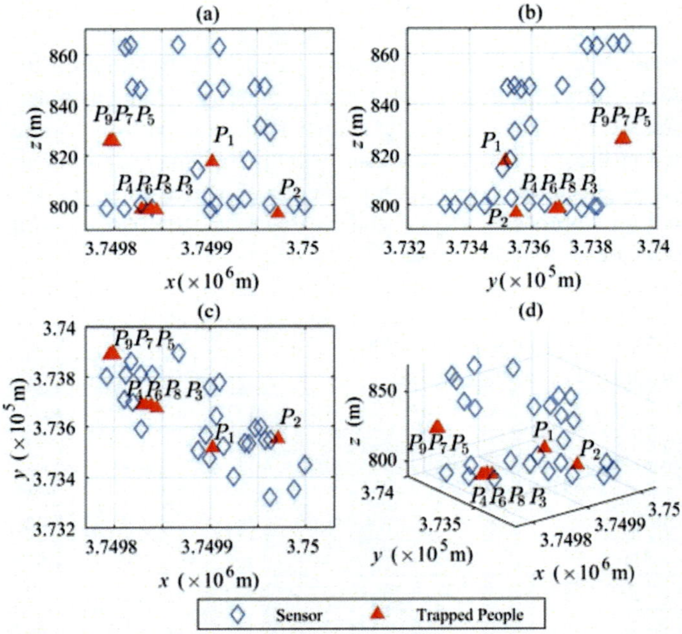

Fig. 9.3 **a** Front view, **b** side view, **c** top view, and **d** three-dimensional view of sensors and trapped people (simulated by blasting sources)

9.2.2 Simulate Distress Signal with Drilling

The monitoring data of drilling at three positions are used to simulate the distress signal sent by trapped people. Figure 9.4 shows three views and three-dimensional views of sensor position and simulated position of trapped people. According to the three views, P_2 is at the edge of the sensor array, with few sensors distributed around,

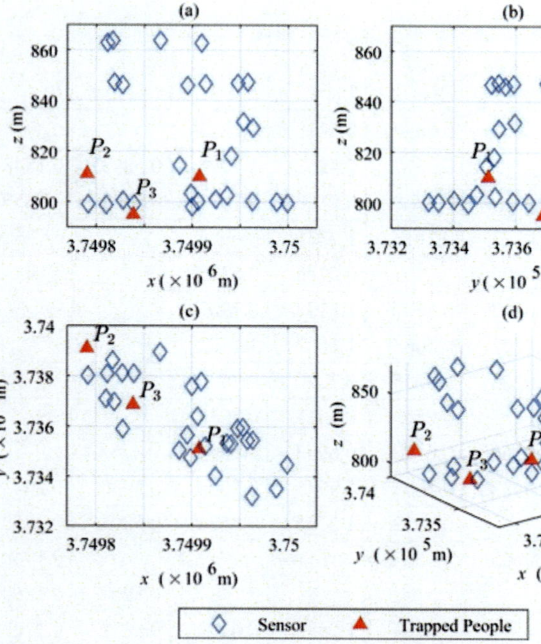

Fig. 9.4 **A** Front view, **b** side view, **c** top view, and **d** three-dimensional view of sensors and trapped people (simulated by drilling sources)

while P_1 and P_3 are relatively close to the middle of the sensor array and surrounded by more sensors. The arrival time obtained by microseismic monitoring system is shown in Table 9.4, in which P_{11}–P_{15}, P_{21}–P_{25}, and P_{31}–P_{35} represent the five distress signals sent by trapped people at P_1, P_2, and P_3, respectively. The coordinate of trapped people (unit: m) are P_1 (3,749,908, 373,512, 810), P_2 (3,749,791, 373,911, 811), and P_3 (3,749,839, 373,688, 795).

Table 9.4 Arrival time obtained by microseismic monitoring system

Sensor	Arrival time (s)							
	P_{11}	P_{12}	P_{13}	P_{14}	P_{15}	P_{21}	P_{22}	P_{23}
S_3		0.0268	0.1752	0.2953	1.5455	0.6510	0.2465	0.8288
S_4	0.6818	0.0262	0.1872	0.8198	1.2502			0.2123
S_5	0.6345	0.0147	0.0990	0.8277	0.5565			
S_6		0.0183	0.0468	1.2985	1.1410			
S_7	0.6347	0.0310	0.7265	0.6013	0.5495	0.6175	0.2128	0.1745
S_8	0.6812	0.0247	0.0948	0.3170	0.7098			
S_9	0.6965	0.0520	0.6575	1.1710	0.5835	0.6050	0.2003	0.1687

(continued)

Table 9.4 (continued)

Sensor	Arrival time (s)							
	P_{11}	P_{12}	P_{13}	P_{14}	P_{15}	P_{21}	P_{22}	P_{23}
S_{10}	0.6520	0.0563	0.6650	0.5118	0.7242			
S_{11}	0.6627		0.0577	0.4267	1.3098	0.5893	0.1848	0.1530
S_{12}	0.6523		0.0768	0.9120	1.1867	0.5902	0.1855	0.1538
S_{13}	0.6333	0.0113	0.2413	1.2335	0.6097			
S_{14}	0.5837	0.0177		0.2735	0.5322			
S_{15}	0.7693	0.0267	0.5507	1.1178	0.7903			
S_{16}		0.0222	0.6683	1.1882	1.1582			
S_{17}		0.0225	0.0655	0.3227	0.7500			
S_{18}			0.0608	0.4615	0.7878			
S_{19}	0.6002	0.0265	0.6480	0.6953	0.9093			
S_{21}	0.6522	0.0433	0.0605	0.4945	0.7068	0.5930	0.2023	0.1490
S_{22}	0.6642		0.2032	1.2725	0.5532	0.5895	0.1850	0.1467
S_{23}	0.6560	0.0535	0.6562				0.1948	
S_{24}						0.5907	0.1860	0.1482
S_{25}	0.6738		0.2755	1.0545	0.6292	0.5812	0.1767	0.1067
S_{26}	0.6562		0.2548			0.5830	0.1783	0.1403

Sensor	P_{24}	P_{25}	P_{31}	P_{32}	P_{33}	P_{34}	P_{35}
S_3	0.5738	0.8137		0.7613	1.5625	0.5522	0.3952
S_4	1.0077	0.9788	0.7298	0.7633	1.0378	0.7903	0.6795
S_5	0.9027	0.9142	0.4255	0.7523	1.0578	0.6852	0.6628
S_6	0.6067	1.5600	0.3988	0.7445	0.8807	0.6880	0.3833
S_7	0.3133	0.8038	0.3738	0.7298	0.8297	0.7440	0.3707
S_8	0.7287	1.0110	0.6962	0.7320	0.9393	0.7410	0.3562
S_9	0.2868	1.0248	0.2572	0.7310	0.7957	0.5497	0.3055
S_{10}	0.5997	0.8003	0.3745		0.8570	0.5932	0.4852
S_{11}	0.3042	1.1225	0.3557		1.6025	0.5585	0.4977
S_{12}	0.5648	0.9075	0.8757	0.7312	1.1918	0.7443	0.3710
S_{13}	0.4217	1.8307	0.7125	0.7472	0.9898	0.7207	0.3487
S_{14}	0.2943	0.8907	0.4253	0.7455	1.5347	0.7302	1.1065
S_{15}	0.5202	0.9048	0.3862	0.7403	1.3177	0.7070	0.3265
S_{16}	0.5690	0.8863	0.4185	0.7463	0.8412	0.5528	0.3797
S_{17}	0.3622	0.9060	0.4188	0.7470	0.8623	0.9887	0.5578
S_{18}	0.8922	0.8130			1.4002	1.0100	1.0120
S_{19}	0.3040	1.7235	0.7050	0.7407	1.1653	1.3535	0.3112
S_{21}	0.3223	0.7925	0.3275	0.7273	0.8353	0.5835	0.8257
S_{22}	0.2952	0.8047	0.4023	0.7282	0.8775	0.6172	1.1308

(continued)

Table 9.4 (continued)

	P_{24}	P_{25}	P_{31}	P_{32}	P_{33}	P_{34}	P_{35}
S_{23}	0.5563		0.3597				
S_{24}	1.1327		0.3715				
S_{25}	0.2963	0.7635	0.4018	0.7533	1.0362	1.6027	0.4537
S_{26}	0.2867		0.4113	0.7397			

9.3 Location Result and Discussion

The Geiger method and MSLM-MV are used for localization. The Geiger method has been introduced in Chap. 2, and some parameters should be initialized. The velocity is 4,900 m/s, the maximum iteration is 10,000 and the initial source position is the first triggered sensor position. The result is considered as the source position when $\|\Delta \mathbf{x}\|_2$ is less than 1×10^{-16}. For MSLM-MV which is introduced in Chap. 5, the initial source position is derived by Eqs. (3.3) and (3.4). And parameters μ and ε are 0.01 and 1×10^{-16}, respectively.

9.3.1 *Location Result of Simulation by Blasting*

Tables 9.5 shows the localization for trapped people simulated by blasting sources. The location results of Geiger method have large absolute error. And most of them are divergent. Comparing Fig. 9.5 with Fig. 9.3, it is found that sources in the center area of sensor network have low location error (P_1, P_6, and P_8) and sources in the edge of sensor network have high error (P_5, P_7, and P_9).The average location error of sources P_3, P_4, P_6, and P_8 is 41.6846 m and that of sources P_2, P_5, P_7, and P_9 is 58.0531 m. Therefore, when the people are trapped in the center area of sensor network, the location error is 16.9403 m. In the edge are of sensor network, the location error increases to 58.0531 m. If there are some sensors around them, it increases to 41.6846 m.

Table 9.5 Localization for trapped people simulated by blasting sources and location error of trapped people using Geiger method and MSLM-MV

Trapped People	x (m)	y (m)	z (m)	Absolute error (m)
Geiger method				
P_1	3,749,884	373,500.5	924.6691	109.8406
P_2	1.7E + 12	1.88E + 12	9.73E + 10	2.54E + 12
P_3	−1.6E + 11	8.04E + 12	2.7E + 12	8.49E + 12

(continued)

Table 9.5 (continued)

Trapped People	x (m)	y (m)	z (m)	Absolute error (m)
P_4	−7.7E + 12	9.25E + 11	−9.5E + 12	1.23E + 13
P_5	−1.5E + 12	3.05E + 12	1.26E + 12	3.63E + 12
P_6	2.04E + 12	-9.2E + 12	−1.9E + 12	9.66E + 12
P_7	3,749,822	373,792.3	832.9419	108.6693
P_8	−7.7E + 11	1.87E + 11	−7.5E + 11	1.1E + 12
P_9	4.79E + 12	2.84E + 12	4.87E + 11	5.59E + 12
MSLM-MV				
P_1	3,749,919.193	373,518.9882	815.3038982	16.94027
P_2	3,749,940.84	373,530.7761	818.0744	42.37023
P_3	3,749,810.609	373,710.7215	798.7422	48.27262
P_4	3,749,819.209	373,698.7053	847.1979	50.59798
P_5	3,749,867.177	373,895.8101	863.8374	77.33807
P_6	3,749,810.609	373,710.7215	798.7422	31.56602
P_7	3,749,795.371	373,844.8466	856.5811118	61.41318
P_8	3,749,810.609	373,710.7215	798.7422	36.30182
P_9	3,749,817.505	373,863.4033	863.8193	51.09103

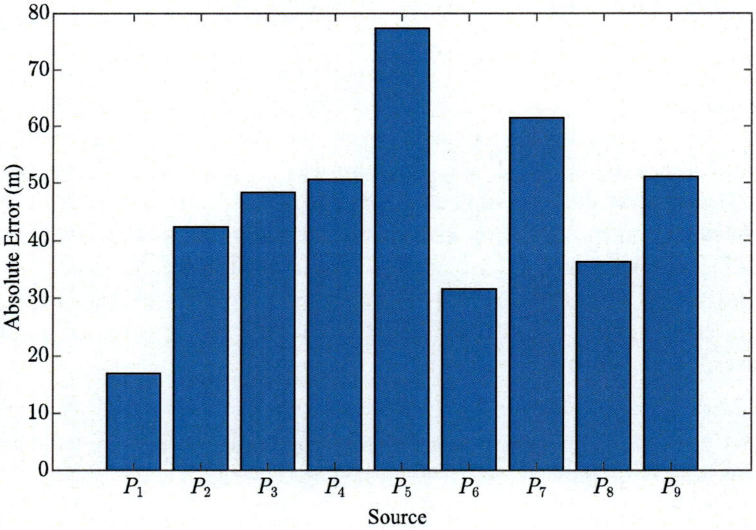

Fig. 9.5 Location error of each distress signal (simulated by blasting source) sent by trapped people using MSLM-MV

9.3.2 *Location Result of Simulation by Drilling*

The location of trapped people (simulated by drilling sources) are shown in Table 9.6. It can be found that most results of Geiger method greatly deviate from the position of trapped people. Three-dimensional views of location result using Geiger method and MSLM-MV are shown in Fig. 9.6. The location result of MSLM-MV is closer to the position of trapped people than the Geiger methods, and the located position basically surround the position of trapped people. Figure 9.7 shows the error of each distress signal using MSLM-MV. The location result of P_1 is the best (the minimum error can reach 15.70 m), that of P_3 is the second, and that of P_2 is the worst. The average of location error of P_1, P_2, and P_3 are 23.79 m, 84.80 m, and 35.80 m, respectively. Therefore, location error is significantly related to the number and spatial position of sensors around the trapped people.

Table 9.6 Localization for trapped people simulated by drilling sources and location error of trapped people using Geiger method and MSLM-MV

Trapped people	x (m)	y (m)	z (m)	Absolute error (m)
Geiger method				
P_{11}	3.7496E + 06	3.7339E + 05	1.4674E + 03	734.4236
P_{12}	3.7499E + 06	3.7352E + 05	7.6363E + 02	46.6429
P_{13}	−9.1668E + 11	−3.4780E + 12	2.1072E + 12	4.1686E + 12
P_{14}	6.0603E + 11	−1.2229E + 13	−4.6183E + 12	1.3086E + 13
P_{15}	−2.1206E + 12	4.0900E + 12	2.7151E + 11	4.6151E + 12
P_{21}	3.7499E + 06	3.7384E + 05	9.0234E + 02	151.7139
P_{22}	−3.6645E + 09	4.2240E + 10	−2.5899E + 10	4.9683E + 10
P_{23}	−7.0232E + 12	2.6247E + 13	−1.1978E + 11	2.7170E + 13
P_{24}	1.4322E + 12	4.8478E + 12	1.5576E + 11	5.0573E + 12
P_{25}	−4.5866E + 11	6.3870E + 12	−1.9762E + 12	6.7014E + 12
P_{31}	3.7499E + 06	3.7384E + 05	9.0234E + 02	191.7659
P_{32}	−3.6645E + 09	4.2240E + 10	−2.5899E + 10	4.9683E + 10
P_{33}	−7.0232E + 12	2.6247E + 13	−1.1978E + 11	2.7170E + 13
P_{34}	1.4322E + 12	4.8478E + 12	1.5576E + 11	5.0573E + 12
P_{35}	−4.5866E + 11	6.3870E + 12	−1.9762E + 12	6.7014E + 12

(continued)

Table 9.6 (continued)

Trapped people	x (m)	y (m)	z (m)	Absolute error (m)
MSLM-MV				
P_{11}	3,749,887.231	373,506.051	814.406404	22.04935427
P_{12}	3,749,922.628	373,514.3364	815.2082387	15.70210745
P_{13}	3,749,936.693	373,534.2606	802.5189061	37.07812408
P_{14}	3,749,887.231	373,506.051	814.406404	22.04935427
P_{15}	3,749,887.231	373,506.051	814.406404	22.04935428
P_{21}	3,749,855.256	373,866.6766	851.1382546	87.7752796
P_{22}	3,749,803.011	373,982.119	804.1589054	72.44984433
P_{23}	3,749,867.177	373,895.8101	863.8373512	93.94359446
P_{24}	3,749,817.505	373,863.4033	863.8193144	75.88049197
P_{25}	3,749,867.177	373,895.8101	863.8373512	93.94359446
P_{31}	3,749,810.609	373,710.7215	798.7421846	36.55602801
P_{32}	3,749,861.122	373,712.1438	794.1103836	32.75840232
P_{33}	3,749,810.609	373,710.7215	798.7421846	36.55602801
P_{34}	3,749,810.609	373,710.7215	798.7421846	36.55602801
P_{35}	3,749,810.609	373,710.7215	798.7421846	36.55602801

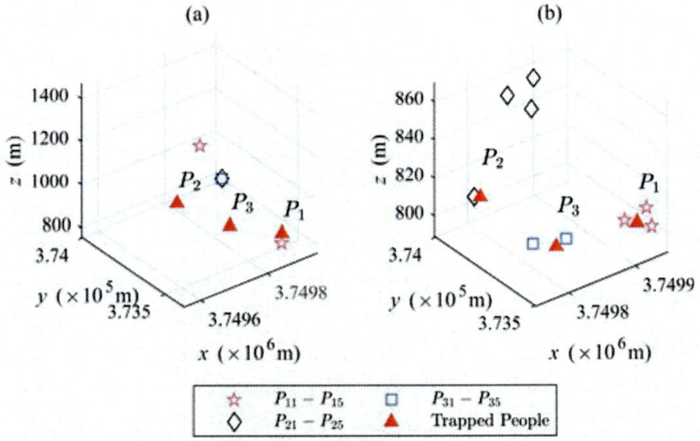

Fig. 9.6 Three-dimensional view of location result using **a** Geiger method (only four results with small absolute error are shown) and **b** MSLM-MV

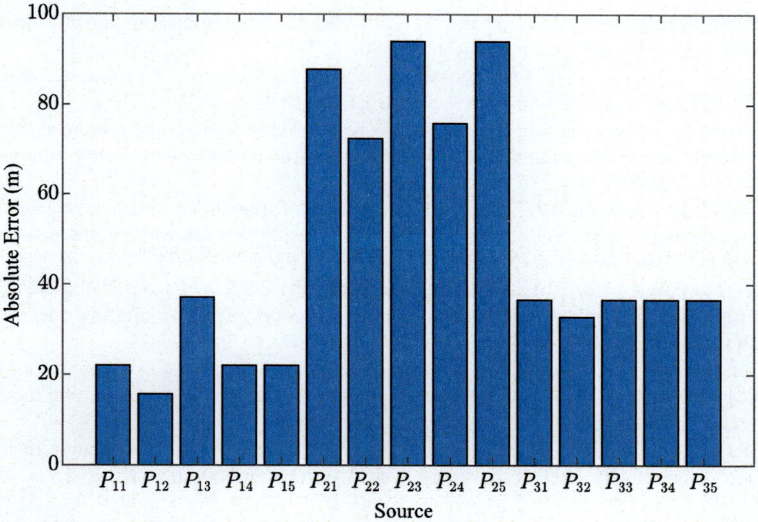

Fig. 9.7 Location error of each distress signal (simulated by drilling source) sent by trapped people using MSLM-MV

9.4 Conclusions

The monitoring data of blasting and drilling test at three positions (P_1–P_3) are used to simulate the trapped people knocking with tools to send out distress signals. The three-dimensional location methods are applied to locate the trapped people. The velocity-free method MSLM-MV has better location accuracy than traditional method (Geiger method). The average of location error of P_1, P_2, and P_3 are 23.79 m, 84.80 m, and 35.80 m, respectively. The different location error is mainly affected by the number and spatial distribution of surrounding sensors. Therefore, it is feasible to apply AE location method to the localization of trapped people. It can win more time for the rescue of the trapped people and make the microseismic monitoring system play a more important role.

References

1. Li S-C, Zhou Z-Q, Li L-P, Lin P, Xu Z-H, Shi S-S (2016) A new quantitative method for risk assessment of geological disasters in underground engineering: Attribute Interval Evaluation Theory (AIET). Tunn Undergr Space Technol 53:128–139
2. Qian Q-H, Lin P (2016) Safety risk management of underground engineering in China: progress, challenges and strategies. J Rock Mech Geotech Eng 8(4):423–442
3. Liu J-P, Si Y-T, Wei D-C, Shi H-X, Wang R (2021) Developments and prospects of microseismic monitoring technology in underground metal mines in China. J Central South Univ 28(10):3074–3098

4. Zheng X-Z, Wang H, Guo J, Zhang D (2020) Method for multi-information drilling detection after mining disasters. Comput Electr Eng 86:106726
5. Seguel F, Palacios-Játiva P, Azurdia-Meza CA, Krommenacker N, Charpentier P, Soto I (2022) Underground mine positioning: a review. IEEE Sens J 22(6):4755–4771
6. Dong L-j, Shu W-w, Li X-b, Han G-j, Zou W (2017) Three dimensional comprehensive analytical solutions for locating sources of sensor networks in unknown velocity mining system. IEEE Access 5:11337-11351
7. Dong L-j, Li X-b, Zhou Z-l, Chen G-h, Ma J (2015) Three-dimensional analytical solution of acoustic emission source location for cuboid monitoring network without pre-measured wave velocity. Trans Nonferrous Met Soc China 25(1):293-302
8. Dong L-j, Li X-b (2012) Three-dimensional analytical solution of acoustic emission or microseismic source location under cube monitoring network. Trans Nonferrous Met Soc China 22(12):3087-3094
9. Dong L, Sun D, Li X, Du K (2017) Theoretical and experimental studies of localization methodology for AE and microseismic sources without pre-measured wave velocity in mines. IEEE Access 5:16818-16828
10. Dong L, Shu W, Han G, Li X, Wang J (2017) A multi-step source localization method with narrowing velocity interval of cyber-physical systems in buildings. IEEE Access 5:20207-20219.
11. Dong L, Li X (2013) A Microseismic/Acoustic Emission Source Location Method Using Arrival Times of PS Waves for Unknown Velocity System. Int J Distrib Sens Netw 9(10):307489
12. Dong L, Li X, Tang L, Gong F (2011) Mathematical functions and parameters for microseismic source location without pre-measuring speed. Chin J Rock Mech Eng 30(10):2057-2067
13. Dong L, Zou W, Li X, Shu W, Wang Z (2019) Collaborative localization method using analytical and iterative solutions for microseismic/acoustic emission sources in the rockmass structure for underground mining. Eng Fract Mech 210(2019):95-112
14. Hu Q, Dong L (2020) Acoustic emission source location and experimental verification for two-dimensional irregular complex structure. IEEE Sens J 20(5):2679-2691
15. Dong L, Hu Q, Tong X, Liu Y (2020) Velocity-free MS/AE source location method for three-dimensional hole-containing structures. Engineering 6(7):827-834
16. Dong L, Tao Q, Hu Q, Deng S, Chen Y, Luo Q, Zhang X (2022) Acoustic emission source location method and experimental verification for structures containing unknown empty areas. Int J Min Sci Technol 32(3):487-497

Chapter 10
Velocity-Free Localization of Autonomous Driverless Vehicles

The rapid progress of science and technology has created favorable conditions for the development of autonomous driverless vehicles. However, the complex conditions in the cities greatly limit the application and promotion of autonomous driving technology. To solve the problem, the underground mine is recommended to apply as a pilot for the promotion of autonomous driverless vehicles. The autonomous vehicles are all connected to the internet to form a vehicle-to-vehicle (V2V) system. Because of the multi-levels and multi-stopes in underground mines, the key point is to obtain the precise locations of the vehicles in real-time. Therefore, the microseismic monitoring, image recognition technology, artificial intelligence training, and smart sensors are comprehensively utilized, based on the internet of things and cloud computing, to locate the autonomous driverless vehicles.

In this chapter, an introduction about autonomous driving technology is demonstrate in Sect. 10.1. In Sect. 10.2, virtual sources localization and pencil lead break tests are conducted to simulate and eliminate the effectiveness and accuracy of the velocity-free localization method. Then the location results are discussed in Sect. 10.4. The application and promotion of the autonomous driverless vehicles in underground mines can not only solve the problems of deep mining and reduce the frequent disasters caused by the harsh conditions but also can protect the life and property of the workers, as well as provide technical support for the safe and efficient recycling of deep resources.

10.1 Introduction

With the rapid development of the scientific and technological revolution, human society has entered a new time of intelligence. Autonomous driving technology has gradually become the focus of the industry under the rapid development of artificial intelligence and the automotive industry. Autonomous driving technology is a product of the high-performance computing chips, artificial intelligence, and the

L. Dong and X. Li, *Velocity-Free Localization Methodology for Acoustic and Microseismic Sources*, https://doi.org/10.1007/978-981-19-8610-9_10

Internet of Things. It is the main direction for the current development of intelligent and networked vehicles in the global automotive and transportation field.

The classifications of autonomous driving differ from different organizations. National Highway Traffic Safety Administration (NHTSA) classified the autonomous driving into five levels ($L_0 - L_4$) while Society of Automotive Engineers (SAE) classified it into six levels ($L_0 - L_5$) [1, 2]. The former four levels ($L_0 - L_3$) of the two classifications are the same. The only difference is that SAE distinguished the level L_4 of NHTSA into two levels (L_4 and L_5). The lowest and the highest levels indicate that the driver is in complete control of the vehicle at all times and the vehicle performs all functions in all conditions [3]. The details of the two classifications are listed in Table 10.1.

Although the autonomous driving technology has developed rapidly in recent years, a series of problems still exist in the process of application and promotion. Firstly, the lack of a high-level autonomous driving test theory and method limited the application of autonomous driving. A complete test system covering different levels of autonomous driving has not yet formed so far. Secondly, the city is densely populated and the traffic environment is complex. The road and other environments cannot meet the requirements of autonomous driving. Thirdly, the laws and regulations that support autonomous driving need to be improved urgently. Therefore, it would be very nice if a field can be used to test and promote autonomous driverless vehicles.

Traditional mining excavation and support work fully rely on manual drilling and blasting, with high labor intensity and low work efficiency. Various hazards threaten the safety of workers' lives and property during the whole work process [4–8]. With

Table 10.1 The classification of autonomous driving

NHTSA level	SAE level	SAE name	Function				Application	
			Driving subject	Sensing subject	Monitoring subject	function achieved	Road	Environment
Human monitoring		No automation	Human driver	Human driver	Human driver	Specific	Specific	Specific
L_0	L_0							
L_1	L_1	Driver assistance	Human & system	Human driver	Human driver	Specific	Specific	Specific
L_2	L_2	Partial automation	System	Human driver	Human driver	Specific	Specific	Specific
System monitoring		Conditional automation	System	System	Human driver	Specific	Specific	Specific
L_3	L_3							
L_4	L_4	High automation	System	System	System	Specific	Specific	Specific
	L_5	Full automation	System	System	System	All	All	All

the development of science and technology, people have developed rock drilling rigs that can be used in excavation and support work. The mining depth is getting deeper and deeper, and the conditions are becoming worse and worse. The harsh conditions are mainly characterized by high geo-stress, high temperature, and high osmotic pressure. During the excavation process, the dynamic disturbance may cause movement and suddenly release of the high geo-stress, resulting in disasters such as rockburst, and causing casualties and equipment damage. The real-time monitoring and early warning of potential instability disasters in mines, slopes, tunnels, and tailings dams have been widely used [9–12]. Besides, the toxic and harmful gases and dust generated during the operations seriously damage the health of the workers. Some innovative mining methods, such as non-explosive continuous mechanized mining, are urgently needed in such a situation [13]. Hence, if the autonomous driving can be applied to the mining and tunnel excavations, it may greatly improve the current conditions. Vehicle security was seriously regarded with the development of autonomous driving technology [14–16]. The mining area is a specific enclosed area with no legal barriers to autonomous driving, and the autonomous driverless vehicles do not even require a license plate number. Meanwhile, the autonomous driverless vehicles of the mining area are traveling along a regular route at a low speed which does not require high technologies. Moreover, the autonomous driverless vehicles will not feel tired while human drivers will feel after hours of works. The autonomous driverless vehicles can continue to work until the fuel or power is exhausted. The optimal charging scheduling policy based on deep reinforcement learning can be applied to extend the working time of autonomous driving vehicles [17]. Therefore, the mining area is a suitable environment for the accomplishment of autonomous driving. It is an important topic because the global demand for minerals is increasing and mining is becoming more challenging.

The autonomous driving has made big progress in some mining and machinery companies, especially some open-pit mines. Rio Tinto, the mining giant of Australia, is using huge self-driving trucks to transport iron ore and other materials in West Angelas, Australia [18, 19]. Rio Tinto isn't the only company employing autonomous vehicles. BHP Billiton and Fortescue have also been testing the new technology. BHP Billiton has deployed autonomous hauling at Jimblebar iron ore mine [20]. The productivity benefits of the autonomous haulage fleet have helped reduced costs by around 20 percent [21]. Caterpillar is also a pioneer of the advanced technology and is leading the industry with a portfolio of proven building block strategies including operator assist features, remote-controlled machines, semi-autonomous machines, fully autonomous machines, and a completely integrated worksite [22].

Benefiting from the advantages of GPS and Laser radar, the operating system can easily get the location of the autonomous driverless vehicles. Thus, the autonomous driving technology in the open-pit mines is more mature than that in the cities. However, GPS signals cannot spread in deep mining areas. Therefore, to apply the autonomous driverless vehicles into deep mining areas, precise locating of the autonomous driverless vehicles is a problem that must be solved. In this paper, not only the locating problem is considered, but also the pre-alarm of the disasters.

10.2 System Model Characteristics

Currently, the methods of locating surface autonomous driverless vehicles mainly include 3 types. The first one is to use high precision differential GPS and the inertial measurement unit (IMU). GPS can obtain a precise location, but the refresh rate is slow. The refresh rate of IMU is fast but the cumulative error affects the localization accuracy. The combination of the two technologies can quickly obtain accurate location information. The second method is to use Laser radar and high precision Map. The laser radar scanning point clouds of the surrounding environment are compared and matched with the high precision map to get the position information. The third method is to use the image data of the camera and a visual map. The position information can be obtained by comparing and matching the image data captured by the camera during driving with the visual map. However, because of the complex environment of multi-level, multi-stope and coexistence of ore and backfills, as shown in Fig. 10.1, it is impossible to get a high precision map of the deep mining area. Therefore, a new idea of locating the autonomous driverless vehicles must be proposed.

In recent years, data acquisition and transmission has made great progress, which can guarantee the timeliness, accuracy, and integrity of the data for locating the sources. Reasonable network configuration and communication technology can make up for the defects of sensors [23]. The progress achieved is not only reflected in the localization and control of ground vehicles, but also in underwater and underground. Han et al. reported a stratification-based data collection scheme for 3-D underwater acoustic sensor network to solve the problems of high power consumption, severe propagation delay, and so on [24]. To overcome the high and unbalanced energy consumption for underwater wireless sensor networks, an autonomous underwater vehicle location prediction based on data collection scheme was proposed [25]. Cao et al. presented a bio-inspired speed curve seeking and tracking control approach for subway trains [26]. A 3-D wideband geometry-based channel model was established to realize the function of multiple-input and multiple-output vehicle-to-vehicle communication in tunnel environments [27]. To tackle the issue of source location privacy, schemes based on the cloud and probabilistic analysis were proposed. It can protect the source location privacy by adding fake packets, phantom nodes, and fake sources into the wireless sensor networks [28, 29]. The probabilistic method is also used to deploy wireless mesh networks and evaluate its reliability. Nishiyama et al. established a hierarchical model of the moveable and deployable resource unit based on wireless mesh networks in which the probabilistic approach was applied to deploying a disaster response network [30]. Liu et al. developed a probabilistic region failure model to estimate the reliability of the wireless mesh networks [31]. The above mentioned positive results can provide beneficial information for the realization of autonomous driving in underground mines.

For the localization of the seismic sources in underground mines, Dong et al. proposed analytical and iterative velocity-free localization methods in complex and dynamic mining conditions [33–35]. The wave velocity of the proposed localization

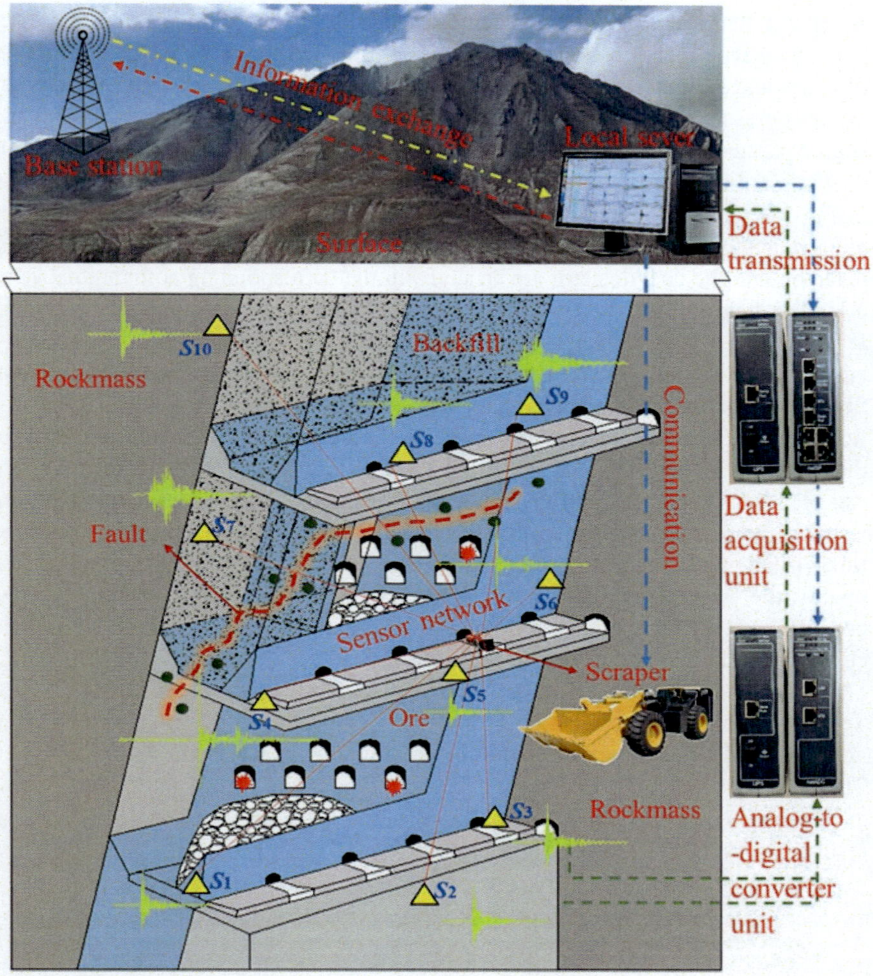

Fig. 10.1 The sketch map of an underground mine. The sensor network consists of ten sensors (S_1 to S_{10}), two data acquisition units, and three analog-to-digital converter units. The yellow triangles, the red stars, and the dark green points indicate the smart sensors, blasting sources, and microseismic sources, respectively. The scraper is now transporting the ore from the stope to the tunnels (from [32])

methods was set as an unknown parameter while the wave velocity was pre-measured and set as a constant in traditional localization methods. The arrival times of P wave received by the sensors networks were substituted into the nonlinear governing equations to solve the source parameters. As a result, the method successfully eliminated the location error and greatly improved the locating accuracy. Therefore, the velocity-free localization method can be applied to locate the autonomous driverless vehicles in underground mines. To obtain a continuous and regular source, the improved active source is installed at the bottom of the vehicles. The active source has an advantage that it can provide a specific wave with known parameters. In terms of the source type identification, Dong et al. investigated the different characteristics of blasts and seismic events and established the discriminators to identify the types of the sources [36, 37]. It can guarantee the obtaining of the accurate waveform and source parameters for picking the wave excited by the active sources.

The hierarchical structure of autonomous driving for underground mines is shown in Fig. 10.2, which contains three layers. For the sensor layer, the smart sensors are arranged around the tunnels. The digital camera, the laser radar, the ultrasonic detector, and the dust detector are arranged in the corresponding autonomous vehicles. The field information is collected by the units of the sensor layer and transferred to the decision layer through the optical cable. The network type used for autonomous driving vehicles in underground intelligent mine is single mode optical fiber communication. By analyzing the obtained information automatically, the artificial intelligence control system can locate the position of the vehicles and judge the surrounding environment. Based on the previous behavior, the control system can plan the next action and the route of the vehicles. The specific control signal is transmitted to the deep mining area through the optical cable and transmitted by distributed routers. To transmit multimedia data and guarantee the transmission quality, the wireless multimedia sensor networking is an efficient technology to be considered [38]. The vehicles in application layer receive the specific signal and conduct a certain job according to the information of the decision layer.

The application flowchart of the autonomous driving system is shown in Fig. 10.3. The initial monitoring data, including P wave velocity, digital images, ultrasonic data, and infrared data, are uploaded to the cloud computing platform through the internet. The received data are processed to locate the position of the autonomous vehicles, perceive the surrounding environment, and predict the behavior. The processed results are trained through artificial intelligence. The generalized prediction and

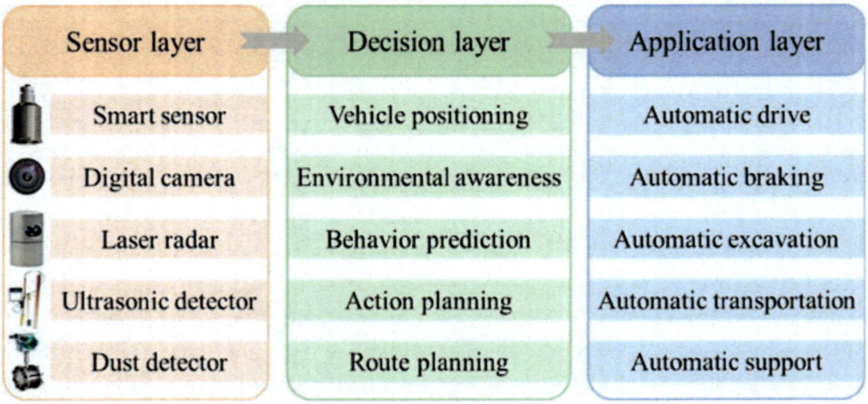

Fig. 10.2 The hierarchical structure of autonomous driving for the underground mines. The framework consists of three layers, which are the sensor layer, the decision layer, and the application layer (from [32])

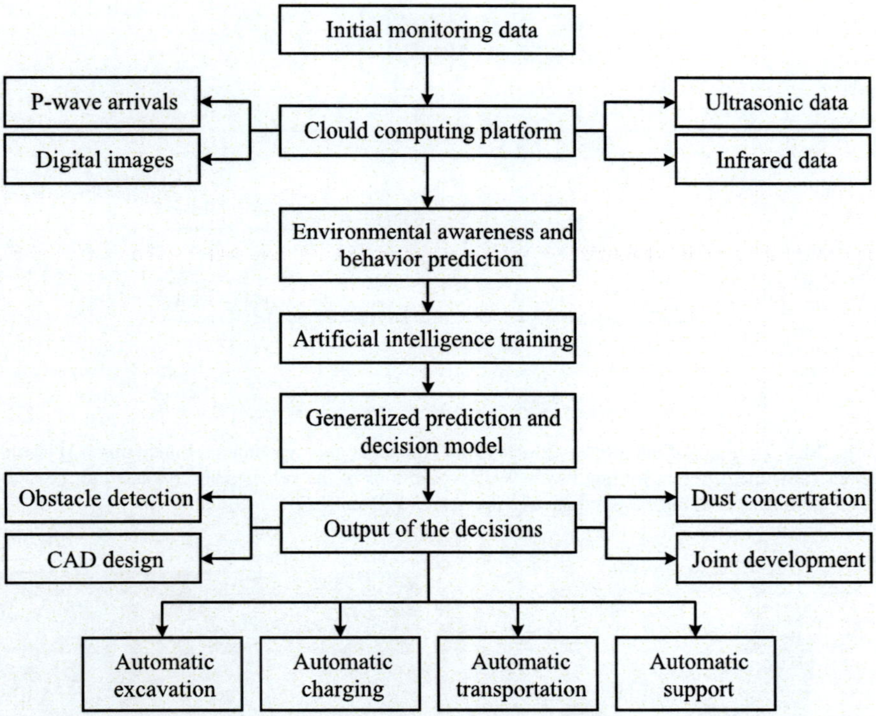

Fig. 10.3 The application flowchart of the autonomous driving system

decision model are applied to decide what to do in the next step for the autonomous vehicles, based on the initial design and the monitoring data of the deep mining area. The decisions are fed back to autonomous vehicles to perform specific functions.

For an underground mine, the excavation, blasting, transportation, and support work are of great importance to the safe and efficient mining. However, due to the huge demand for ore and materials, the mining depth increases rapidly. The increasingly harsh environment has flung down a challenge to the designers and the workers. The application of autonomous vehicles can improve this severe situation to a large extent. The autonomous driving vehicles are all connected to the internet to form a V2V system. Figure 10.4 shows the diagram of V2V model in underground intelligent mines. The blue lines indicate the different mining levels in underground mines. By forming a V2V system in underground intelligent mine, a certain autonomous driving vehicle can transmit the information of task accomplished and the work condition to the cloud computing platform. After analyzing the information received and optimizing task allocation, it can determine whether additional vehicles are needed to complete the work. The further decisions are fed back to the vehicles for different works. The autonomous driving system for an underground intelligent mine is shown in Fig. 10.5.

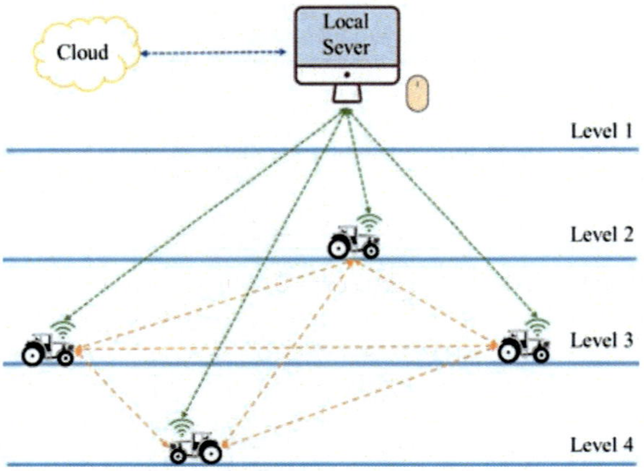

Fig. 10.4 Diagram of Vehicle-to-Vehicle (V2V) model in underground intelligent mines. The blue lines show the different mining levels. Work conditions of the vehicles are uploaded to the cloud and the further decisions are fed back to the vehicles (from [32])

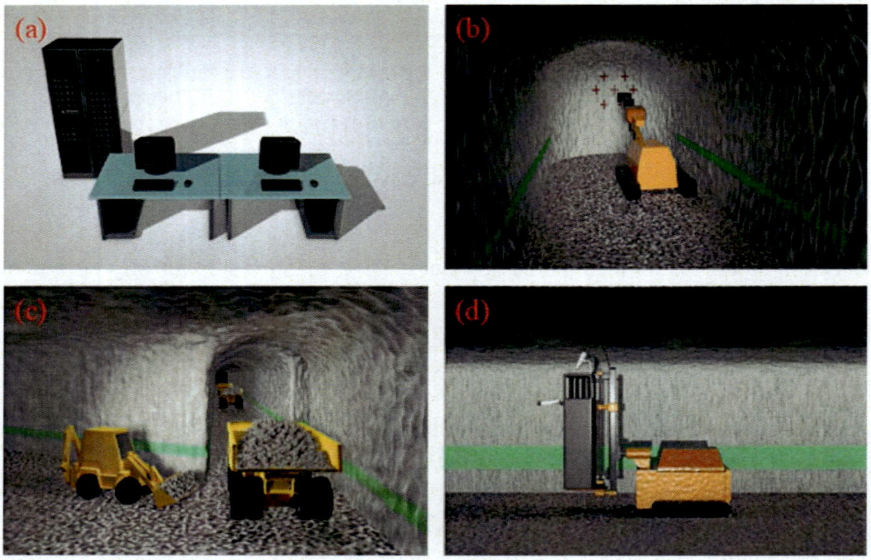

Fig. 10.5 The autonomous driving system for underground mine. The graphs **a**, **b**, **c**, and **d** indicate the cloud computing platform, the autonomous rock drilling jumbo and autonomous explosive charging vehicle, autonomous scraper and truck, and autonomous supporting vehicle (from [32])

10.2.1 The Cloud Computing Platform

The cloud computing platform, of which a simple concept figure is shown in Fig. 10.5a, is the most important part of the autonomous driving system. The function of the cloud computing platform is to receive and analyze the data acquired from the sensor layer and make decisions based on the analyzed results to control the autonomous driverless vehicles to perform specific works. The intelligent algorithms, such as the localization algorithm, the discrimination algorithm, the image recognition algorithm, and the boreholes search algorithm, are integrated in the cloud computing platform to locate the position of the autonomous vehicles, discriminate the specific waveform to pick the arrival times, confirm the driving route, and find the boreholes to install the bolts. The autonomous driverless vehicles are conducted certain jobs according to the instructions received from the cloud computing platform. Also, by analyzing the waveform parameters and source parameters of the acquired microseismic data, the seismic event with large magnitude can be predicted. The disaster pre-alarm system raises the alert with different level according to the predicted results to remind the managers and workers to aware of the potential disasters.

10.2.2 The Autonomous Rock Drilling Jumbo and Explosive Charging Vehicle

The rock drilling jumbo is used to drill a hole in the tunnel face for the arrangement of the explosives. Unlike the traditional rock drilling jumbo, the autonomous rock drilling jumbo has to locate its position before it works. Figure 10.5b shows a simple model of the autonomous rock drilling jumbo. Three improved pneumatic active sources, in which a spindle-shaped striker is added to obtain a half-sine wave [39], are arranged at the bottom of the autonomous rock drilling jumbo to locate the position of the jumbo. Two laser rangefinders are arranged on both sides of the autonomous rock drilling jumbo to measure the distance between the vehicle and the two sides of the tunnels. A proximity sensor is arranged in the drill to measure the depth of the boreholes. During the working process of autonomous rock drilling jumbo, the smart sensors arranged in the drill can test the size of loose circle and the fracture inside rocks. The measured results of the laser rangefinders and the proximity sensor are transmitted to the cloud computing platform in real-time. When the autonomous rock drilling jumbo is operating, the three active sources are activated synchronously at regular intervals. The produced half-sine waves are received by the smart sensors network and are transmitted to the cloud computing platform. The arrivals of the sine waves are picked by the workstation and are used to locate the source of the active sources. The locations of the sources can reflect the position of the autonomous rock drilling jumbo. Then the three sources are fitted as a straight line to indicate the direction of the vehicles. The fitted results are compared with the initial CAD design to calculate the deviation. Thereafter a corrected direction for the vehicle is decided. The corrected direction and the information of the boreholes are transmitted to the autonomous rock drilling jumbo to conduct the next works.

The autonomous explosive charging vehicle is similar to the autonomous rock drilling jumbo. The only difference is that an automatic extensible mechanical arm, in which some smart sensors with different function were configured, is installed in the front of the drill arm to search the boreholes and fill the explosives and the detonating cord into the boreholes automatically. Also, the autonomous explosive charging vehicle is controlled by the cloud computing platform. The field data collected by the sensors of the mechanical arm are uploaded to the cloud computing platform for behavior prediction, and the next instruction is transmitted to the autonomous explosive charging vehicle for further operation. After filling the boreholes with explosive, a dedicated worker should connect explosives, detonating cords, detonators, and leads together. Only after the workers and vehicles are evacuated to a safe area, could the dedicated worker detonate the explosives.

10.2.3 The Autonomous Scraper and Autonomous Truck

The scraper is used to transport the blasted ore from the stope to the trucks and tramcars in the main haulage roadway. Three skilled drivers are engaged in the regular work of a manually controlled scraper for the traditional three shifts work mode while no drivers are needed for the autonomous scraper. For the first month, the autonomous scraper travels through the tunnels at a low speed. The active sources are also applied to locate the position of the scraper. The position information is uploaded to the cloud computing platform and the marching route is recorded and optimized by the artificial intelligence system. After a lot of practice and training, the autonomous scraper can work in the optimum path with a faster speed. The obstacle detector installed around the scraper can ensure the scraper from crashing into other objects. When the autonomous scraper enters into the stope, the scraper can easily approach the ore dump by comparing the positions between the scraper and blasting sources. When the autonomous scraper is close to the trucks, as shown in Fig. 10.5c, the ultrasonic detector starts to confirm the precise position of the carriages and load the ore into it. The processes proceed in a quick and orderly way.

The autonomous truck is designed to transport the ore to the concentrator and the ore passes without the need for manual labor. Because the loading and unloading places are fixed positions, the high-quality images of the loading place, unloading place, and all the corners are collected and uploaded to the cloud computing platform as the initial information before the autonomous truck works. Four intelligent force sensors are arranged in the bottom of the autonomous truck to determine whether it is empty or filled with ore. The high definition cameras are installed in the autonomous truck to obtain the surrounding environment while the truck is moving. The position information of the autonomous truck is uploaded to the cloud computing platform and compared with the initial information in real-time to decide the next move.

The marching route of the autonomous truck is also recorded and optimized by the artificial intelligence system.

10.2.4 The Autonomous Supporting Vehicle

Rational and effective support works are the basic guarantee for the safe production of the underground mines. The support works are always lagging behind the excavation due to the ventilation after blasting. The delay of support works has little influence in intact rockmass while it may cause severe disasters in fractured rockmass. The autonomous supporting vehicle, whose positioning pattern is the same as the autonomous rock drilling jumbo, can perform normal support works without the need for ventilation once the blasting is finished. The workbench consists of three parts, of which the first one is to drill boreholes for the bolts, the second one is to install the bolt into the boreholes, and the third one is to inject the coupling agent such as resin and cement paste into the boreholes. As shown in Fig. 10.5d, a bolt reserve box, two high definition cameras, a dust detector, and an ultrasonic detector are designed in front of the autonomous supporting vehicle. The rotatable drill arm ensures the drilling of boreholes with different orientations to support the roof and both sides of the tunnels. A mechanical hand is installed beside the drill arm to deliver the smart bolt from the bolt reserve box to the workbench. The sensors arranged in the smart bolt could monitor the distribution of internal stress of rocks. These monitoring data are transmitted to cloud computing platform to evaluate the possibility of rock instability disasters and provide information of disaster pre-alarming. The ultrasonic detector is installed to detect the crack propagation and the range of the loose zone. The obtained data are uploaded to the cloud computing platform to calculate the length and space of the bolts. The high definition cameras are designed to take photos for checking whether there are isolated loose stones and to confirm the support effect of the bolts. To ensure the clarity of the photo, a dust detector is installed to measure the dust concentration. The spray dust reducer will start work if the dust concentration is larger than the threshold.

10.3 Simulation and Performance Evaluation

As we all know, the wave velocity should be premeasured before calculating the source coordinates in traditional localization methods. However, the wave velocity is not a constant in anisotropic materials [40, 41]. The locating results are not accurate if the constant wave velocity is used to calculate the source coordinates. Therefore, the wave velocity in the governing equation should be an unknown parameter to improve the locating accuracy when calculating the source coordinates. Hence, we proposed a velocity-free localization method to locate the sources in anisotropic materials. In the velocity-free localization method, the governing equation of the difference of arrival times for sensor i and sensor j is expressed as:

$$\Delta t_{ij} = t_i - t_j = \frac{d_i - d_j}{v} \tag{10.1}$$

where the t_i and t_j are the arrival times of sensor i and sensor j, d_i and d_j are the distances of the propagation paths from the source to sensor i and sensor j, and v is an unknown parameter indicating the wave velocity along different propagation paths. The propagation paths are calculated using the A* search algorithm [42].

In Eq. (10.1), d_i and d_j are functions of the known independent variables (x_i, y_i, z_i), (x_j, y_j, z_j) and unknown source coordinates (x, y, z), respectively. Therefore, Eq. (10.1) can be expressed as

$$\Delta t_{ij} = f\left(x_i, y_i, z_i, x_j, y_j, z_j, x, y, z, v\right) \tag{10.2}$$

As a nonlinear fitting problem, a fitting value can be obtained using all of the monitored data. The obtained fitting value is a function of the monitored data and is noted as $\Delta\widehat{t_{ij}}$.

The degree of deviation between the fitting value and the monitored value, which reflects the effect of the fitting results, can be obtained by taking the square of the difference for Δt_{ij} and $\Delta\widehat{t_{ij}}$. A smaller deviation corresponds to a better fitting result. Considering that we have 24 sensors, the sum of the deviations between all of the fitting values and the monitored values can be expressed as

$$g(x, y, z, v) = \sum_{i,j=1}^{n} \left(\Delta t_{ij} - \Delta\hat{t}_{ij}\right)^2. \tag{10.3}$$

The sum of the deviations is a function of the source coordinates (x, y, z) and the wave velocity. While the fitting values and the monitored values get the best fitting result, the sum of the deviations reaches a minimum. Hence, the solution of the minimum for Eq. (10.3) can be considered as the source coordinates.

10.3.1 Localization for Virtual Sources

An example of locating the virtual sources was conducted to analyze the accuracy and reliability of the velocity-free localization method, which will be applied to locate the autonomous driverless vehicles in underground intelligent mines. The virtual locating system consists of 16 virtual sensors, whose coordinates are randomly distributed in a plane. The coordinates of the virtual sensors are listed in Table 10.2. The virtual sources include 6 sources (A, B, C, D, E, and F) located inside the sensor network and four sources (M, N, P, and Q) located outside the sensor network. The average wave velocity of the virtual locating system was assumed as 4600 m/s. The arrival times of

Table 10.2 Coordinates of the virtual sensors

Sensor No.	X (cm)	Y (cm)	Sensor No.	X (cm)	Y (cm)
1	0	0	9	158	178
2	200	0	10	130	100
3	0	200	11	18	130
4	38	55	12	53	173
5	60	115	13	83	87
6	96	28	14	180	62
7	105	163	15	170	128
8	135	27	16	200	200

the sensors corresponding to each virtual source were obtained by means of dividing the distance between the virtual source and each sensor by the average wave velocity. Considering the error in wave velocity measurement, the average wave velocity was reduced or increased by 5% and 10%. As a result, 5 wave velocity gradients (V1: 4,140 m/s, V2: 4,370 m/s, V: 4,600 m/s, V3: 4,830 m/s, and V4: 5,060 m/s) were obtained to locate the virtual sources. The wave velocity is inversely proportional to the arrival times when the distance between the virtual source and the sensor is fixed. Therefore, the errors of the arrival times are also taken into account by calculating the errors of the wave velocity. The real coordinates, locating results, and locating errors are listed in Table 10.3. The coordinates of the virtual sources are solved using the above-mentioned velocity-free localization method.

From Table 10.3, it clearly shows that the locating results are in good agreement with the real coordinates of the virtual sources. The locating errors of the virtual sources are all less than 5 cm, among which the maximum error and the minimum error are 4.613 cm and 0.052 cm, respectively. For the coverage of the sensor network, the maximum error only accounts for 2.31% of the monitoring scale. The average locating errors under the 5 wave velocity gradients are 1.516, 1.519, 1.513, 1.514, and 1.517 separately, which account for about 0.76% of the dimensional of the sensor network. It proves that the localization method is accurate for locating the virtual sources.

The locating results of the virtual sources under different wave velocity gradients are shown in Fig. 10.6. The blue circles, red stars, yellow triangles, orange triangles, pink triangles, light blue triangles, and blue squares in the figure indicate the virtual sensors, virtual sources, locating results under V1, V2, V, V3, and V4, respectively. The gray dashed lines determined the monitoring scope. Because the locating results of the virtual sources under different wave velocity gradients are all approximately equal to the real coordinates, the markers indicating different values are overlapped. Therefore, the locating results of virtual source *B* are enlarged to display it clearly. Figure 10.6 clearly shows that not only the virtual sources inside the sensor network but also the sources outside the sensor network are all located accurately.

Table 10.3 Locating results and errors of the virtual sources under different velocity gradients (V1, V2, V, V3, and V4)

Virtual sources	Real coordinates (cm)		Locating results under V1 (cm)			Locating results under V2 (cm)			Locating results under V (cm)		
	X	Y	X	Y	Error	X	Y	Error	X	Y	Error
A	63	142	63.035	143.118	1.119	63.039	143.112	1.113	63.038	143.128	1.128
B	148	118	148.067	114.222	3.779	148.074	114.219	3.782	148.07	114.222	3.779
C	102	83	103.065	83.231	1.090	103.079	83.229	1.103	103.064	83.231	1.089
D	83	100	87.047	102.192	4.603	87.056	102.197	4.613	87.043	102.183	4.595
E	52	64	51.946	64.119	0.130	51.946	64.114	0.127	51.943	64.117	0.130
F	97	135	94.056	135.159	2.949	94.061	135.17	2.944	94.058	135.172	2.947
M	45	241	44.936	241.004	0.064	44.933	241.019	0.069	44.953	240.978	0.052
N	205	108	205.004	107.181	0.819	205.021	107.195	0.805	205.008	107.185	0.815
P	157	208	157.051	208.422	0.425	157.043	208.436	0.438	157.032	208.419	0.420
Q	246	246	246.070	246.169	0.183	246.1	246.168	0.196	246.081	246.162	0.181
Average locating errors (cm)			1.516			1.519			1.513		

(continued)

Table 10.3 (continued)

Virtual sources	Real coordinates (cm)		Locating results under V3 (cm)			Locating results under V4 (cm)		
	X	Y	X	Y	Error	X	Y	Error
A	63	142	63.046	143.117	1.118	63.049	143.116	1.117
B	148	118	148.068	114.214	3.786	148.078	114.225	3.775
C	102	83	103.068	83.221	1.091	103.07	83.226	1.093
D	83	100	87.052	102.193	4.608	87.048	102.187	4.601
E	52	64	51.952	64.098	0.109	51.945	64.118	0.130
F	97	135	94.053	135.175	2.952	94.066	135.156	2.938
M	45	241	44.945	241.001	0.055	44.927	241.013	0.074
N	205	108	205.02	107.19	0.810	205.011	107.182	0.818
P	157	208	157.044	208.422	0.424	157.049	208.421	0.424
Q	246	246	246.078	246.172	0.189	246.108	246.170	0.201
Average locating errors (cm)			1.514			1.517		

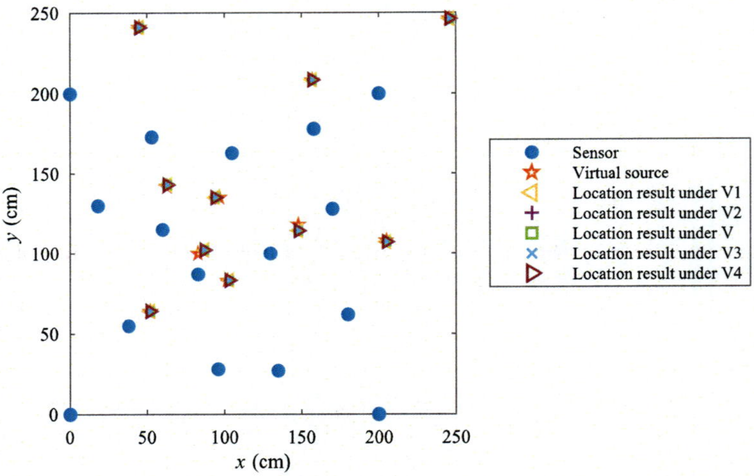

Fig. 10.6 Locating results of the virtual sources under different wave velocity gradients

Locating errors of the virtual sources under different wave velocity gradients are shown in Fig. 10.7. The colors red, light green, light blue, pink, and gold are the locating errors corresponding to the wave velocity gradients V1, V2, V, V3, and V4, respectively. From Fig. 10.7, it can be found that the maximum errors and minimum errors under different wave velocity gradients occur at source *D* and source *M*, respectively. For a certain virtual source, the locating errors under different wave velocity gradients are almost equal to each other. It demonstrates that the velocity-free localization method is stable and reliable for locating the virtual sources.

Fig. 10.7 Locating errors of the virtual sources under different wave velocity gradients

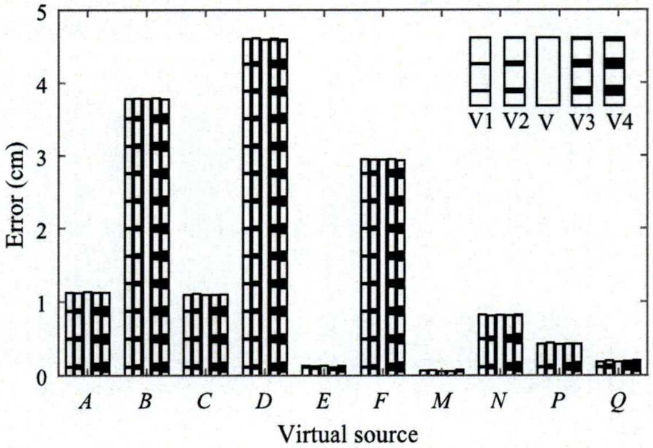

10.3.2 Pencil Lead Break Tests (PLB)

To validate the effectiveness and accuracy of the method being used for locating the autonomous driverless vehicles in underground mines, pencil lead break tests were conducted to simulate the pneumatic active sources assembled in the autonomous driverless vehicles. The test materials and acquisition system are shown in Fig. 10.8. A 60 cm × 60 cm × 1.2 cm granite slate with three excavated areas of different shapes was prepared to show the complex and irregular underground goafs. The surface of the granite slate was meshed into grids with the dimension of 2 cm × 2 cm to obtain the accurate coordinates of the sensors and the pencil lead break sources. The Germany Vallen AMSY-6 AE system with 24 channels, as shown in Fig. 10.8b, was used to acquire the signals of the pencil lead break tests. The natural frequency and the response frequency of the sensors are 150 kHz and 100 to 450 kHz, respectively. The sample rate of the acquisition system was set to 20 MHz. Seventeen sensors were arranged in the observe side to represent the sensors in the work face of the tunnels while seven sensors were arranged in the reverse side to indicate sensors installed in the inside of the rocks. The actual layout of the sensors in the observer side is shown in Fig. 10.8a. The boundary of the tunnels was obtained by connecting the centers of the sensor locations. The route of the pencil lead break tests simulated the process of a V-turn for an autonomous driverless mine car with several carriages.

The autonomous driverless vehicles are invariably traveling on the bottom of the tunnels. The work level of the vehicle is preset before it starts to work. Therefore, we only need to solve the *X* and *Y* coordinates to improve computing efficiency. The ray paths and source coordinates are solved using the A* search algorithm introduced in [42] and the velocity-free localization method [34, 40], respectively.

The locating results and errors of pencil lead break sources are listed in Table 10.4 and shown in Fig. 10.9. The errors in Table 10.4 are obtained by calculating

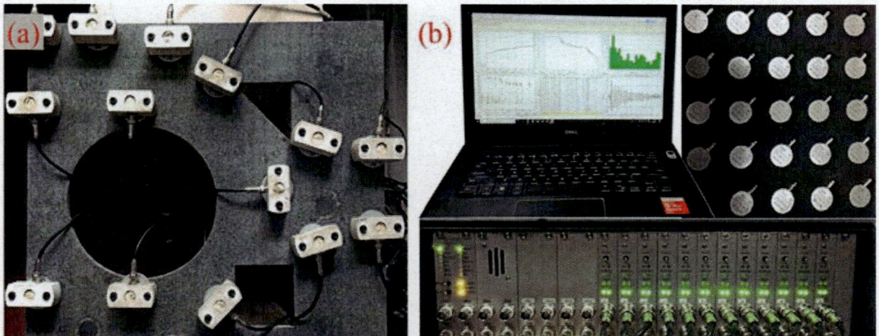

Fig. 10.8 The test materials and acquisition system. **a** The granite slate and the layout of the AE sensors in the observer side, and **b** The Germany Vallen AMSY-6 AE system with 24 channels (from [32])

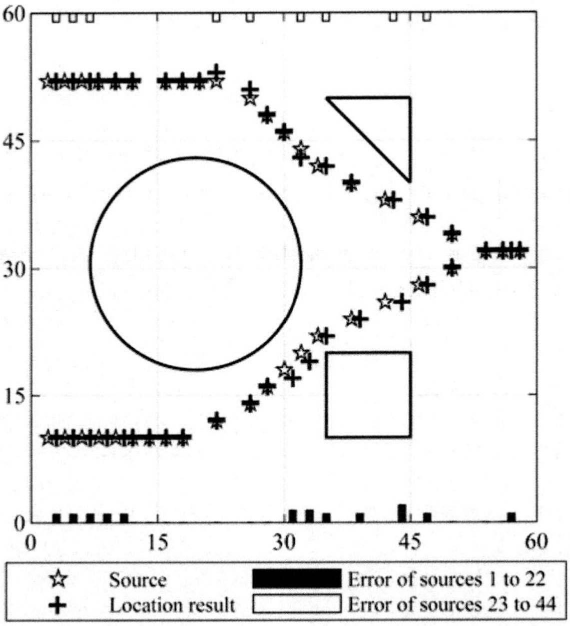

Fig. 10.9 Locating results and errors of pencil lead break sources where the white circle, triangle, and square indicate the areas excavated. The stars and plus signs indicate the real pencil lead break sources and location results, respectively. The bars draw on bottom and top indicate location error of sources 1 to 22 and sources 23 to 44 respectively

Table 10.4 Locating results and errors of the PLB sources

Test No.	Real coordinates		Locating results		Errors		
	X (cm)	*Y* (cm)	*X* (cm)	*Y* (cm)	X_{error} (cm)	Y_{error} (cm)	D_{error} (cm)
1	2	10	3	10	1	0	1
2	4	10	5	10	1	0	1
3	6	10	7	10	1	0	1
4	8	10	9	10	1	0	1
5	10	10	11	10	1	0	1
6	12	10	12	10	0	0	0
7	14	10	14	10	0	0	0
8	16	10	16	10	0	0	0
9	18	10	18	10	0	0	0
10	22	12	22	12	0	0	0
11	26	14	26	14	0	0	0
12	28	16	28	16	0	0	0
13	30	18	31	17	1	1	1.414
14	32	20	33	19	1	1	1.414

(continued)

Table 10.4 (continued)

Test No.	Real coordinates		Locating results		Errors		
	X (cm)	Y (cm)	X (cm)	Y (cm)	X_{error} (cm)	Y_{error} (cm)	D_{error} (cm)
15	34	22	35	22	1	0	1
16	38	24	39	24	1	0	1
17	42	26	44	26	2	0	2
18	46	28	47	28	1	0	1
19	50	30	50	30	0	0	0
20	54	32	54	32	0	0	0
21	56	32	56	32	0	0	0
22	58	32	57	32	1	0	1
23	58	32	58	32	0	0	0
24	56	32	56	32	0	0	0
25	54	32	54	32	0	0	0
26	50	34	50	34	0	0	0
27	46	36	47	36	1	0	1
28	42	38	43	38	1	0	1
29	38	40	38	40	0	0	0
30	34	42	35	42	1	0	1
31	32	44	32	43	0	1	1
32	30	46	30	46	0	0	0
33	28	48	28	48	0	0	0
34	26	50	26	51	0	1	1
35	22	52	22	53	0	1	1
36	20	52	20	52	0	0	0
37	18	52	18	52	0	0	0
38	16	52	16	52	0	0	0
39	12	52	12	52	0	0	0
40	10	52	10	52	0	0	0
41	8	52	8	52	0	0	0
42	6	52	7	52	1	0	1
43	4	52	5	52	1	0	1
44	2	52	3	52	1	0	1
Average locating errors					0.4318	0.1136	0.5188

the absolute value between the real coordinate and the located result of each source. D_{error} indicates the ADE for the located result of a source. From the data in Table 10.2, we can count that 23 out of 44 sources are precisely located whose D_{error} is 0 cm. Besides, the locating errors of 18 sources are merely 1 cm. Only three locating errors of the sources are bigger than 1 cm which are perfectly acceptable for an underground mine. The average locating errors of X coordinates, Y coordinates, and absolute distance are 0.4318 cm, 0.1136 cm, and 0.5188 cm, respectively. These results were obtained using a single source. For an actual autonomous driverless vehicle, the location was obtained by fitting the results of three pneumatic active sources which can provide a better localization result.

Figure 10.9 clearly shows the locating results and errors of the pencil lead break sources. The white circle, triangle, and square indicate the excavated areas in an underground mine. The green triangles and squares are the AE sensors arranged in observe side and reverse side to indicate the microseismic sensors used in underground mines. The blue stars give out the real coordinates of the pencil lead break sources which were used to simulate the active sources during vehicles operation. The colorful dots show the locating results of the pencil lead break sources with the colorbar indicating the different errors between the real coordinates and the located coordinates. From Fig. 10.9, it can be seen that the locating results are all within the envelope of the sensors. Majority of the locating results are overlapped with the real pencil lead break sources which indicate that accuracy of the locating results is enough for locating the autonomous driverless vehicles in underground mines.

The wave velocity along different propagation paths is calculated and shown in Fig. 10.10. In Fig. 10.10, Δl indicates the differences between the distances of the source to the first triggered sensor and the distances of the source to other sensors. Δt indicates the differences between the arrival of the first triggered sensor and the arrivals of other sensors corresponding to Δl. The different colors of the dots show the wave velocities of the corresponding paths. From Fig. 10.10, it shows that the wave velocity varies along different propagation paths but all fall within the area of the blue dashed line and the coordinate axis. The majority of the wave velocities are concentrated in the interval of [2,000, 5,000] m/s. This phenomenon demonstrates that the traditional localization methods using pre-measured wave velocity are unreasonable which may cause unacceptable errors. The velocity-free localization method eliminated the errors caused by the premeasured wave velocity. Therefore, the localization method used in this chapter is more rational for the localization of autonomous driverless vehicles.

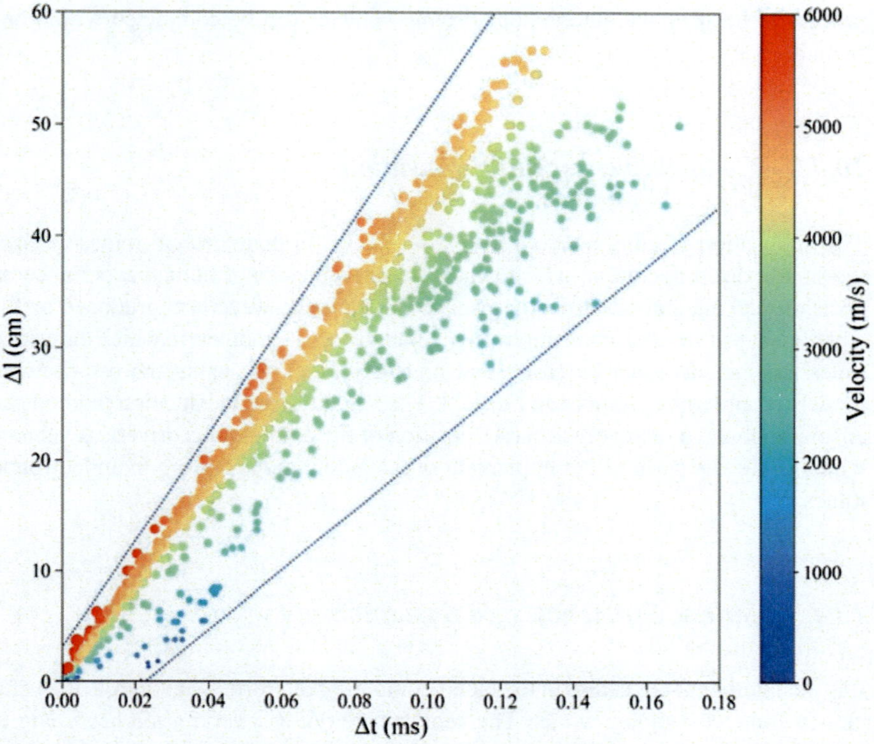

Fig. 10.10 The wave velocity along different propagation paths (from [32])

10.4 Discussions

10.4.1 Timeliness of the Proposed Localization Method

Considering the attenuation of the wave signal in the cable, the distance between a sensor and the corresponding data acquisition unit is less than 500 m. The propagation speed of the electric signal in the cable is about 2 × 108 m/s. The travel time from a sensor to the data acquisition unit can be ignored. The analog signal is converted into digital signal in analog-to-digital converter unit and then transmitted to the local server. Traditionally, the distance between the data acquisition unit and the local server is about 1,000 m. Considering the transmission of digital signal in optical cable, the travel time from the data acquisition to the local server can be ignored. Therefore, the timeliness of the localization method for autonomous driving vehicles mainly depends on the travel time of the signal from the autonomous driving vehicle to the sensor. Suppose that the distance between an autonomous driving vehicle to the sensor is 600 m and the propagation velocity of P-wave in rock is 4,000 m/s, the travel time is about 0.15 s. Take the time required for location algorithm, the whole process of locating the autonomous driving vehicles can be completed within one second.

10.4.2 Competitiveness and Innovation

The application of autonomous driverless vehicles in underground mines realizes the standardized operation, which can avoid the influence of human experience on decision-making and construction results. The half-sine waveform produced by the spindle-shaped strikers of the improved pneumatic active sources, in which the arrival times are included, can be easily and precisely identified using the discriminant models proposed by Dong et al. [36, 37]. The velocity-free localization method can ensure to obtain a relatively accurate position of the autonomous driverless vehicle, which is the key point of the implement of the autonomous driving in underground mines.

10.4.3 Safety, Efficiency, and Sustainability

It is the intelligent algorithm in the cloud computing platform that controls the vehicles to conduct complex works. The vehicles are doing a certain job according to the instructions of the intelligent algorithm. By locating the position, calculating the velocity of the vehicle, and perceiving the surrounding environment, the intelligent algorithm can plan the next actions for different vehicles. The artificial intelligence training system and the image recognition technology can protect the safe and efficient operations of the autonomous scraper and autonomous truck. Immediately after the blasting, the ore and other materials can be transported to the designated location.

Traditional support design is based on the investigated results of the joints development and the rockmass classification, which have cost a lot of manpower and resources and miss the best time for support work. Application of the ultrasonic detector in autonomous support vehicle could obtain the fracture results of the rockmass to give a scientific basis for the support design, which commendably avoids the delay of the best occasion. The work efficiency is greatly improved due to the 24 h' full-time work of the autonomous driverless vehicles.

10.4.4 Harmonization and Coordination

A worker can control dozens of autonomous driverless vehicles in the surface monitoring room where the surrounding condition is conformable to live. Also, workers can control the vehicles to drive to a specific site to conduct a job if necessary. The priority of the vehicles is set in advance to avoid the crash, for example, the empty vehicle should dodge the truck full of ore. The operating conditions of the autonomous driverless vehicles are displayed on the monitoring screen in real-time. If a vehicle is in trouble, the emergency lights will flash to remind the workers and the manual driving vehicles to keep away. The alarm system in the cloud computing system will be activated and pop up a fault alarm to alert the monitoring personnel to arrange professional workers to eliminate the trouble.

10.5 Conclusions

The utilization of the improved active sources, laser rangefinders, proximity sensor, and automatic extensible mechanical arm, in conjunction with the application of the velocity-free localization method, make the accurate localization of autonomous rock drilling jumbo and explosive charging vehicle come true. It can not only overcome the obstacles that GPS signals cannot spread in deep underground mines but also solve the technical problem that cannot provide high-precision maps as the underground conditions becoming increasingly complicated with the mining operations.

The automatic operations of the autonomous scraper and autonomous truck are realized through the use of image recognition technology and artificial intelligence training. The ultrasonic detector installed in the autonomous supporting vehicle can provide data of rockmass fracture for support design. The application of the autonomous supporting vehicle can avoid missing the best support occasion and wasting a lot of manpower and resources for drilling investigation and rockmass classification.

Virtual sources localization and laboratory pencil lead break tests were conducted to assess the reliability, effectiveness, and accuracy of the velocity-free localization method used to locate the autonomous driverless vehicles. Results of the virtual source localization proved that the average locating errors under different wave velocity gradients are 1.516 cm, 1.519 cm, 1.513 cm, 1.514 cm, and 1.517 cm. It demonstrates that the velocity-free localization method is stable and reliable. Results of the pencil lead break tests show that more than half of the sources are precisely located. Only 3 sources have a positioning error greater than 1 cm which is fully acceptable for an underground mine. The average locating errors of X coordinates, Y coordinates, and absolute distance are 0.4318 cm, 0.1136 cm, and 0.5188 cm, respectively.

The technical problems of underground deep mining (such as the high temperature, high geo-stress, and high osmotic pressure) are solved and the frequent disasters are controlled through the application and promotion of the autonomous driverless vehicles. It not only protects the safety of workers and equipment but also provides technical support for the safe and efficient recycling of deep resources.

References

1. National Highway Traffic Safety Administration (2013) U.S. department of transportation releases policy on automated vehicle development. http://carsandracingstuff.com/library/articles/22707.php. Accessed 30 May 2013
2. SAE standards (2018) Taxonomy and definitions for terms related to driving automation systems for on-road motor vehicles. https://www.sae.org/standards/content/j3016_201806/. Accessed Jun 2018
3. Schreurs MA, Steuwer SD (2015) Autonomous driving-political, legal, social, and sustainability dimensions. In: Autonomes fahren. Springer, pp 151–173
4. Dong L-J, Sun D-Y, Li X-B, Zhou Z-L (2017) Interval non-probabilistic reliability of a surrounding jointed rockmass in underground engineering: a case study. IEEE Access 5:18804–18817
5. Dong L-J, Sun D-Y, Li X-B, Ma J, Zhang L-Y, Tong X-J (2018) Interval non-probabilistic reliability of surrounding jointed rockmass considering microseismic loads in mining tunnels. Tunn Undergr Space Technol 81:326–335
6. Dong L-J, Shu W-W, Li X-B, Zhou Z-L, Gong F-Q, Liu X-L (2017) Quantitative evaluation and case study of risk degree for underground goafs with multiple indexes considering uncertain factors in mines. Geofluids
7. Dong L-J, Shu W-W, Li X-B, Zhang J-M (2018) Quantitative evaluation and case studies of cleaner mining with multiple indexes considering uncertainty factors for phosphorus mines. J Clean Prod 183:319–334
8. Feng G-L, Feng X-T, Chen B-R, Xiao Y-X, Zhao Z-N (2019) Effects of structural planes on the microseismicity associated with rockburst development processes in deep tunnels of the Jinping-II Hydropower Station, China. Tunn Undergr Space Technol 84:273–280
9. Dong L-J, Shu W-W, Sun D-Y, Li X-B, Zhang L-Y (2017) Pre-alarm system based on real-time monitoring and numerical simulation using internet of things and cloud computing for tailings dam in mines. IEEE Access 5:21080–21089
10. Wang B, Wu C, Kang L, Reniers G, Huang L (2018) Work safety in China's thirteenth five-year plan period (2016–2020): current status, new challenges and future tasks. Saf Sci 104:164–178
11. Wang B, Wu C, Huang L, Kang L, Lei Y (2020) Safety science as a new discipline in China. Saf Sci 121:201–214
12. Feng G-L, Feng X-T, Chen B-r, Xiao Y-X, Yu Y (2015) A microseismic method for dynamic warning of rockburst development processes in tunnels. Rock Mech Rock Eng 48(5):2061–2076
13. Wang S, Li X, Yao J, Gong F, Li X, Du K, Tao M, Huang L, Du S (2019) Experimental investigation of rock breakage by a conical pick and its application to non-explosive mechanized mining in deep hard rock. Int J Rock Mech Min Sci 122:104063
14. Xun Y, Liu J, Kato N, Fang Y, Zhang Y (2019) Automobile driver fingerprinting: a new machine learning based authentication scheme. IEEE Trans Industr Inf 16(2):1417–1426
15. Xun Y, Liu J, Zhang Y (2019) Side-channel analysis for intelligent and connected vehicle security: a new perspective. IEEE Network 34(2):150–157
16. Luo Q, Liu J (2018) Wireless telematics systems in emerging intelligent and connected vehicles: threats and solutions. IEEE Wirel Commun 25(6):113–119
17. Liu J, Guo H, Xiong J, Kato N, Zhang J, Zhang Y (2019) Smart and resilient EV charging in SDN-enhanced vehicular edge computing networks. IEEE J Sel Areas Commun 38(1):217–228
18. Chalfie DSCaM (2012) Rio Tinto invests $518M in autonomous trains for Pilbara iron ore rail network in Western Australia. Mining World 9(2):31
19. Gershgorn D (2016) Self-Driving, 416-ton Trucks are Hauling Raw Materials around Australia. https://qz.com/874589/rio-tinto-is-using-self-driving-416-ton-trucks-to-haul-raw-materials-around-australia/. Accessed Dec 2016
20. Dyson N (2017) BHP to double autonomous trucks at Jimblebar. Mining Magazine. https://www.miningmagazine.com/innovation/news/1331400/bhp-to-double-autonomous-trucks-at-jim Accessed Jul 2017

21. Jurgens D (2017) Creating the future of mining–integration and automation. IMARC keynote speech in Melbourne, Australia.https://www.bhp.com/media-and-insights/reports-and-presentations/2017/10/creating-the-future-of-mining-integration-and-automation. Accessed Nov 2017
22. Glover M (2016) Caterpillar's autonomous journey-the argument for autonomy. SAE Technical Paper. https://doi.org/10.4271/2016-01-8005.Accessed:Sep.2016
23. Wang J, Liu J, Kato N (2018) Networking and communications in autonomous driving: a survey. IEEE Commun Surv Tutor 21(2):1243–1274
24. Han G, Shen S, Song H, Yang T, Zhang W (2018) A stratification-based data collection scheme in underwater acoustic sensor networks. IEEE Trans Veh Technol 67(11):10671–10682
25. Han G, Long X, Zhu C, Guizani M, Bi Y, Zhang W (2019) An AUV location prediction-based data collection scheme for underwater wireless sensor networks. IEEE Trans Veh Technol 68(6):6037–6049
26. Cao Y, Wang Z-C, Liu F, Li P, Xie G (2019) Bio-inspired speed curve optimization and sliding mode tracking control for subway trains. IEEE Trans Veh Technol 68(7):6331–6342
27. Jiang H, Zhang Z, Wu L, Dang J, Gui G (2019) A 3-D non-stationary wideband geometry-based channel model for MIMO vehicle-to-vehicle communications in tunnel environments. IEEE Trans Veh Technol 68(7):6257–6271
28. Han G, Miao X, Wang H, Guizani M, Zhang W (2019) CPSLP: A cloud-based scheme for protecting source location privacy in wireless sensor networks using multi-sinks. IEEE Trans Veh Technol 68(3):2739–2750
29. Wang H, Han G, Zhang W, Guizani M, Chan S (2019) A probabilistic source location privacy protection scheme in wireless sensor networks. IEEE Trans Veh Technol 68(6):5917–5927
30. Nishiyama H, Rodrigues TG, Liu J (2018) A Probabilistic approach to deploying disaster response network. IEEE Trans Veh Technol 67(12):12086–12094
31. Liu J, Jiang X, Nishiyama H, Kato N (2011) Reliability assessment for wireless mesh networks under probabilistic region failure model. IEEE Trans Veh Technol 60(5):2253–2264
32. Dong L, Sun D, Han G, Li X, Hu Q, Shu L (2020) Velocity-free localization of autonomous driverless vehicles in underground intelligent mines. IEEE Trans Veh Technol 69(9):9292–9303. https://doi.org/10.1109/tvt.2020.2970842
33. Dong L-J, Shu W-W, Li X-B, Han G-J, Zou W (2017) Three dimensional comprehensive analytical solutions for locating sources of sensor networks in unknown velocity mining system. IEEE Access 5:11337–11351
34. Dong L-J, Sun D-Y, Li X-B, Du K (2017) Theoretical and experimental studies of localization methodology for AE and microseismic sources without pre-measured wave velocity in mines. IEEE Access 5:16818–16828
35. Dong L-J, Zou W, Li X-B, Shu WW, Wang Z-W (2019) Collaborative localization method using analytical and iterative solutions for microseismic/acoustic emission sources in the rockmass structure for underground mining. Eng Fract Mech 210:95–112
36. Dong L-J, Wesseloo J, Potvin Y, Li X-B (2016) Discriminant models of blasts and seismic events in mine seismology. Int J Rock Mech Min Sci 86:282–291
37. Dong L-J, Wesseloo J, Potvin Y, Li X-B (2016) Discrimination of mine seismic events and blasts using the fisher classifier, naive bayesian classifier and logistic regression. Rock Mech Rock Eng 49(1):183–211
38. Han G-F, Jiang J-F, Guizani M, Rodrigues JJC (2016) Green routing protocols for wireless multimedia sensor networks. IEEE Wirel Commun 23(6):140–146
39. Li X-B (2014) Foundation and application of rock dynamics. Science Press, Beijing
40. Dong L, Hu Q, Tong X, Liu Y (2020) Velocity-free MS/AE source location method for three-dimensional hole-containing structures. Engineering 6(7):827-834
41. Hu Q-C, Dong L-J (2019) Acoustic emission source location and experimental verification for two-dimensional irregular complex structure. IEEE Sens J 20(5):2679–2691
42. Hart PE, Nilsson NJ, Raphael B (1968) A formal basis for the heuristic determination of minimum cost paths. IEEE Trans Syst Sci Cybernet 4(2):100–107

Chapter 11
Application of Velocity-Free Methods in Micro-Crack Mechanism and Instability Precursors

To explore the potential precursors of rock instability, it is necessary to clarify the mechanism of micro-crack from fracturing to failure, which involves the evolution of fracture size, orientation, source model, and their relationships to the loading. The waveforms of AE recorded by the sensor network attached rock sample during laboratory tests provide a data basis for solving these problems, since these observations are directly related to the characteristics of the fracturing sources. In this chapter, the source position was obtained by velocity-free localization method, and then moment tensor for each event was calculated considering the uncertainties of P-wave polarity, azimuth, and the takeoff angles of the rays. In Sect. 11.1, an introduction about failure mechanism of rocks is demonstrated. A uniaxial compression test of a granite sample was conducted and descripted in Sect. 11.2. The characteristic parameters obtained by sensors are demonstrated in Sect. 11.3 and the entire process of uniaxial compression of granite was divided into five stages. Then the evolution of different types of cracks in granite using moment tensor inversion is analyzed in Sect. 11.4.

11.1 Introduction

The macroscopic failure of the rock is a result of a series of complex and diverse micro-cracks accumulation, nucleation, and interconnection. The models of rock micro-cracking under compressive stress include grain crushing, sliding along pre-existing cracks, stress concentrations around pores, elastic mismatch between grains, dislocation movement, and combinations of these mechanisms [1]. These micro-cracking models not only respectively dominate the rock fracturing process at different stages of loading, but also demonstrate the fracturing scale and type. From the perspective of fracture mechanics, rock cracks are usually characterized as a combination of basic types of tension and shear [2]. Exploring the rock fracture mechanism is helpful to understand the causal mechanism of rock engineering disasters, which plays an important role in predicting, controlling, and reducing emergencies.

L. Dong and X. Li, *Velocity-Free Localization Methodology for Acoustic and Microseismic Sources*, https://doi.org/10.1007/978-981-19-8610-9_11

Acoustic emission analysis provides spatial–temporal information about the evolution process of rock fracture [3–6], where the AE waveforms emitted under different fracture modes have obvious differences [7]. Combining the time domain and frequency domain characteristics of AE signals is a significant means to characterize rock fracture types. Generally, tensile crack forms AE waveforms with shorter rise time and higher average frequency, whereas, the shearing process always leads to the AE waveforms with longer duration, lower average frequency, and more delayed maximum amplitude [8, 9].

In the time domain, the gradients of an ascending part of the AE waveform can evaluate the evolution process of different rock cracks in shear and bending test, which relate to three types of rock fracture: tension, shear, and mixing [7]. The gradient of the waveform rise refers to the ratio of the rise time to the amplitude (RA). In frequency domain, the change of AE frequency not only reveals the size of the crack scale [10], but also the fracture type of AE source. The statistical analysis of the AE dominant frequency is meaningful for understanding the microscopic failure mechanism of rocks. Surrounding the phenomenon of dual dominant frequency bands in the process of rock fracture, the AE signals of high frequency and low frequency are considered to be generated by shear and tension fracture, respectively [11]. Based on the above-mentioned basic viewpoints, some researchers have discussed the evolution process of shear and tension cracks in white marble under direct tensile test [11], uniaxial compression [12], and Brazilian test [13]. The dominant frequency identification method is referential, but the complexity of the rock fracture mechanism should still be considered. The main energy of a micro-fracture is not necessarily all concentrated in the dominant frequency [14], and the entire fracture process exists in more than two dominant frequency bands [15]. Combining the perspectives of time domain and frequency domain, the classification of concrete crack types by the rise angle and the average frequency (RA-AF) characteristic parameters has been extensively studied and discussed, and the analysis method of RA-AF characteristic parameters has been verified by SiGMA analysis [16, 17]. This method has gradually developed to the definition of crack types in different rocks under various mechanical environments [18–20]. However, there is still a lack of quantitative characterization methods to determine the transition line between the two types of fracture. The existing empirical methods include: (1) the test division method to carry out a single fracture type-led test, such as three-point bending tests, direct shear tests, and determine the transition line of the fracture type is determined by the RA-AF values distribution of the AE signals in each test [19, 21]; (2) the dominant frequency feature division method is based on the AE dominant frequency to define the type of shear and tension fracture [22].

In addition to the methods mentioned above, there are many studies using moment tensor to identify the type of rock failure [23–27]. Moment tensor is a matrix of nine force couples describing the state of stress at the source. Cautious analysis of the moment tensor matrix makes it possible to infer the rock mass fracture process. Zhang et al. [23] classified the micro-fractures into five types (compression, shear-compression, shear, shear-tensile, tensile), based on the tensile angle derived from moment tensor interpretation. Yamamoto et al. [24] classified hydraulic fracturing into shear-dominant events and tensile/compression-dominant events, by the consistency coefficient, which is a function of the classic moment tensor decomposition (isotropic (ISO), double-couple (DC), and compensated linear vector dipole (CLVD)). Hampton et al. [28] categorized AE into shear, tensile, and mixed mode, through manipulating the eigenvalues of the moment tensor matrix. This method has been widely used to identify rock failure under different stress conditions [29–31]. Actually, this method was first proposed for an in situ hydro-fracturing test [32]. The author developed this method [33] and compared it to the parameter-based method [16].

Theoretically, the mining explosions are associated with abrupt volume changes. In terms of the moment tensor, it can be represented by three orthogonal linear force dipoles, which constitute a pure ISO source [34]. A cavity collapse reproduces the effects of rockfalls caused by gravity or more energetic rockbursts, where the source mode produces a moment tensor consisting of a negative ISO component and a CLVD with vertical symmetry [34]. For a tensile failure of the roof rock, horizontal dipoles could be used to describe the source radiation patterns. A fault slip can be represented by a pure DC. In fact, the proposed mechanism may occur simultaneously, which leads to a superposition of the ISO, DC, and CLVD components. The elemental failure models in the mining environment, their diagonalized moment tensors, beachballs, and their corresponding positions in the source type diagram are shown in Fig. 11.1. And the acquisition and processing of the multiple monitoring data is shown in Fig. 11.2.

11.2 Experiment

11.2.1 Instruments and Rock Samples

The experimental instruments include two parts: stress loading equipment and AE monitoring system (Fig. 11.3). The loading equipment is the true tri-axial electro-hydraulic servo mutagenesis test system with model TRW-3000, which can perform dynamic and static rock mechanics experiments on samples of various specifications. The test system can achieve independent loading in both vertical and horizontal directions and the uniaxial compression test is carried out using the vertical loading frame during the experiment. It records load, stress, displacement, and strain values in real time, and simultaneously draws load–displacement and stress–strain curves. An AE monitoring system equipped with the highest standard international acquisition card is AMSY-6 multi-channel equipment from Vallen, Germany (consisting of parallel measuring channels) and achieves synchronous AE characteristic parameters and waveform acquisition during the experiment. Each block has 2 independent channels. The channel ADC, the accuracy, and the wideband operating frequency are 40 MHz, 18 bits, 18 kHz–2.4 MHz, respectively. The sensor in the experiment was the VS45-H sensor, with a response frequency of 20 kHz–450 kHz.

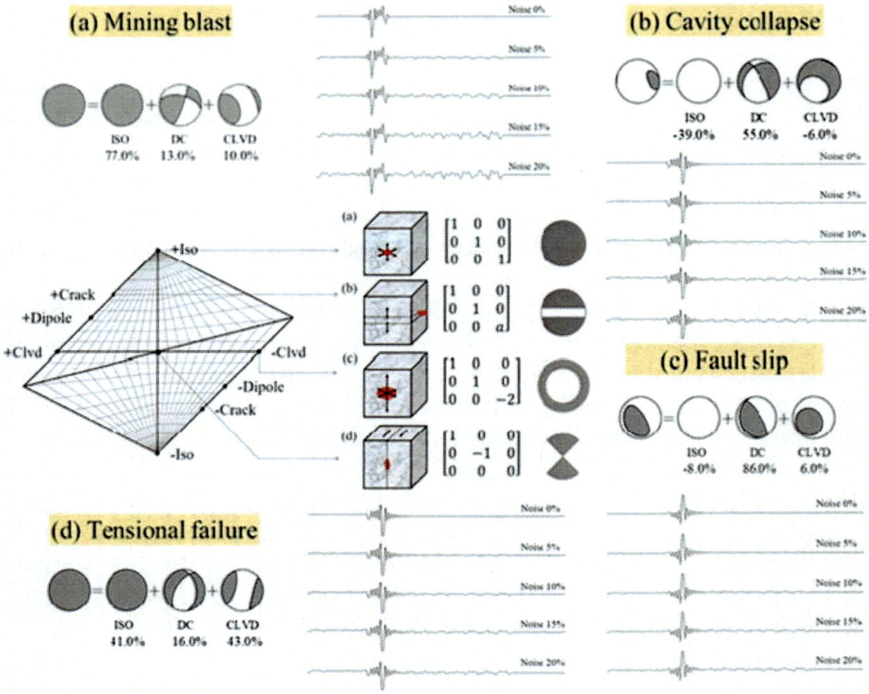

Fig. 11.1 Schematic diagram of possible failure models, their moment tensors, beachballs, and corresponding positions in the Hudson plot (from [35])

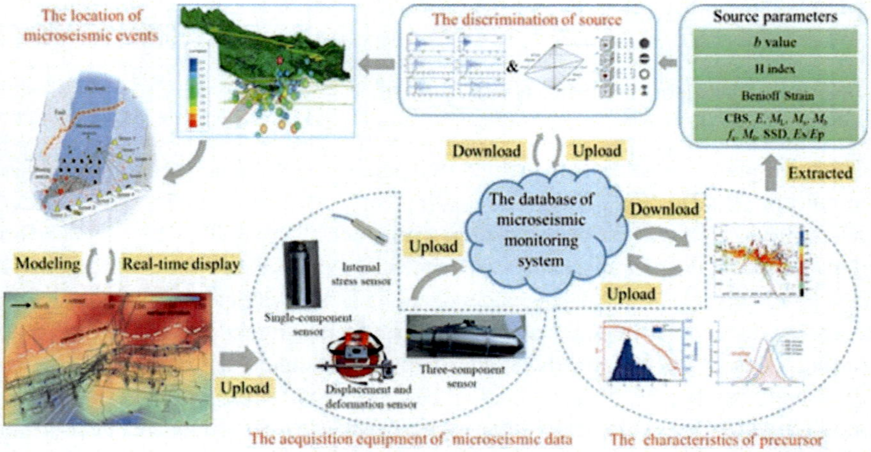

Fig. 11.2 Diagram of real-time acquisition and processing of the multiple monitoring data (from [36])

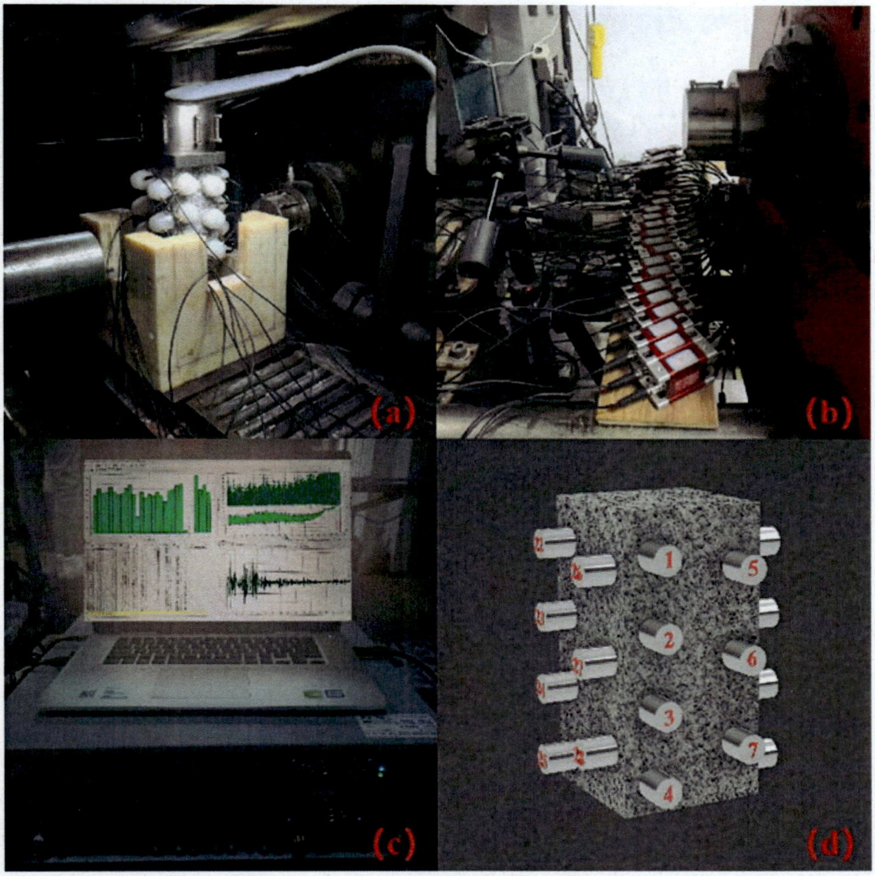

Fig. 11.3 Testing system and sensor arrangement: **a** Sample, AE sensor, and loading device; **b** Preamplifier; **c** AE collection interface and host machine; **d** AE sensor layout (from [37])

Granite, taken from the depth of 100 m underground in the Linglong gold mine, was selected, which was made into a standard cuboid rock sample of 100 mm × 100 mm × 200 mm. The surface of the rock sample meets the parallelism, flatness, and perpendicularity requirements of ISRM recommendations. Water was added by 1/4 gradient in the process of soaking granite to make saturated samples. In order to minimize the differences among samples, all samples were taken from one granite rock.

11.2.2 Sensor Arrangement and Loading Procedure

In the experiment, 28 AE sensors were arranged on the surface of the rock sample, utilizing alternating dislocation. The sensor was fixed by a fixing device, to avoid falling off during the loading process. A couplant was used in the contact area between the sample and the sensor for the sake of the coupling effect. According to the site environment of the experiment, the pre-amplifier, pre-filter, threshold, and sampling frequency of the AE experimental analysis system are set to 40 dB, 95–450 kHz, 55 dB, 10 MHz, respectively.

Before the experiment, the hidden problems in the experiment were eliminated by checking the installation of the sample, the coupling degree of AE sensor and debugging the equipment. The stress loading rate set in the uniaxial compression experiment is 500 N/s. A preload of 1–2 KN is applied to the rock sample to ensure that the rock sample is in full contact with the upper compression plate, to eliminate the noise generated when the rock sample contacts the plate. The loading test system and AE monitoring system are synchronized, and the actual data are recorded in real time during the experiment. The experiment was carried out until the granite lost bearing capacity completely.

11.3 Results

A single AE event triggers multiple channels to record AE hits at the same time, the AE waveforms and parameter characteristics of which are often highly consistent. We define the characteristic parameter value of AE event as the average value of its corresponding characteristic parameter value of AE hits. AE event rate and amplitude respectively represent the degree and scale of micro-fracture in the process of granite damage accumulation. The dominant frequency of AE events reflects the type and mechanism of granite damage and fracture. The above characteristic parameters can effectively characterize the progressive state of crack under uniaxial compression.

In Fig. 11.4, the accumulation trend of AE events indicates that the initiation, nucleation, and expansion of granite fracture are positively correlated with the change of uniaxial stress loading, and the AE amplitude is distributed to the high amplitude interval during the process of uniaxial compression of granite damage evolution. The dominant frequency of AE events appears in three continuous dense bands: high frequency band (280 kHz–320 kHz), low frequency band (85 kHz–125 kHz), and intermediate frequency band (145 kHz–195 kHz).

According to the evolution laws of stress and AE, the entire process of uniaxial compression of granite can be divided into five stages:

- Stage A (the compaction stage during 0–795 s): the characteristics of AE event are low amplitude, low frequency, and low event rate. The main sources of AE signals originate from the primary cracks closed and the friction dislocation between grains.

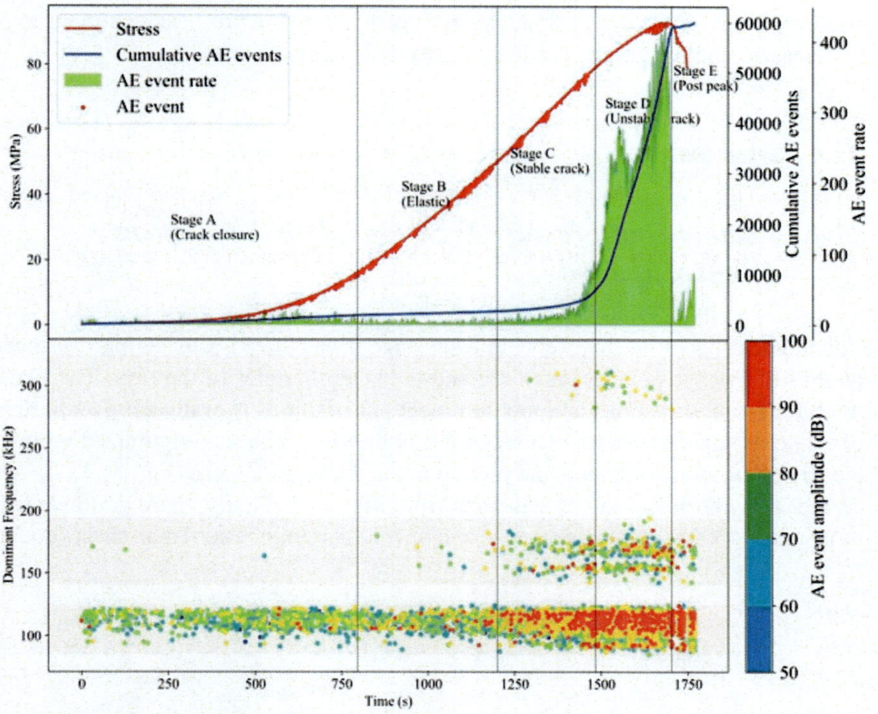

Fig. 11.4 The AE timing and spectrum characteristics of granite in uniaxial compression (from [37])

- Stage B (the elastic stage during 796–1200 s): the AE activity is almost quiet, and the characteristics are similar to the previous stage.
- Stage C (the stable crack propagation stage during 1201–1484 s): the increase of AE event rate is accompanied by a large number of AE events with high amplitude and intermediate and high frequency. The primary cracks and secondary cracks with different scales and multiple fracture modes propagate stably in granite.
- Stage D (the unstable crack propagation stage during 1485–1702s): the AE events of high energy level increase sharply, and the low frequency band becomes wider, and the low, intermediate and high frequency bands coexist, which reveal that the crack propagation of granite is accelerated, and micro-crack initiation and large crack penetration are synchronous.
- Stage E (the post-peak stage during 1703–1770s): AE events with high-frequency basically disappear, and AE events with low-frequency and high-amplitude dominate the rock damage. The crack penetration leads to the formation of discontinuity surfaces inside the granite, and the weakened coupling between the sensor and the rock surface, which causes the attenuation of the acquisition wave frequency, especially the influence on the AE signal with high-frequency. With the further

penetration of macroscopic cracks, the energy stored in pre-peak continues to be released, and the granite gradually loses its bearing capacity.

11.4 Discussion

11.4.1 Analysis of Fracture Types Based on the Moment Tensor Method

The moment tensor of the events in each stage is calculated considering the uncertainties of P-wave polarity, azimuth, and the takeoff angles of the rays. The likelihood for the observations and measurement uncertainties is evaluated over a range of random moment tensors in a Bayesian framework. The final optimized moment tensor solutions are obtained and plotted in the Hudson diagram (Fig. 11.5).

We draw the following conclusions from Fig. 11.5. Firstly, there are obviously more crack events (+crack for tensile and -crack for compression) than shear (double

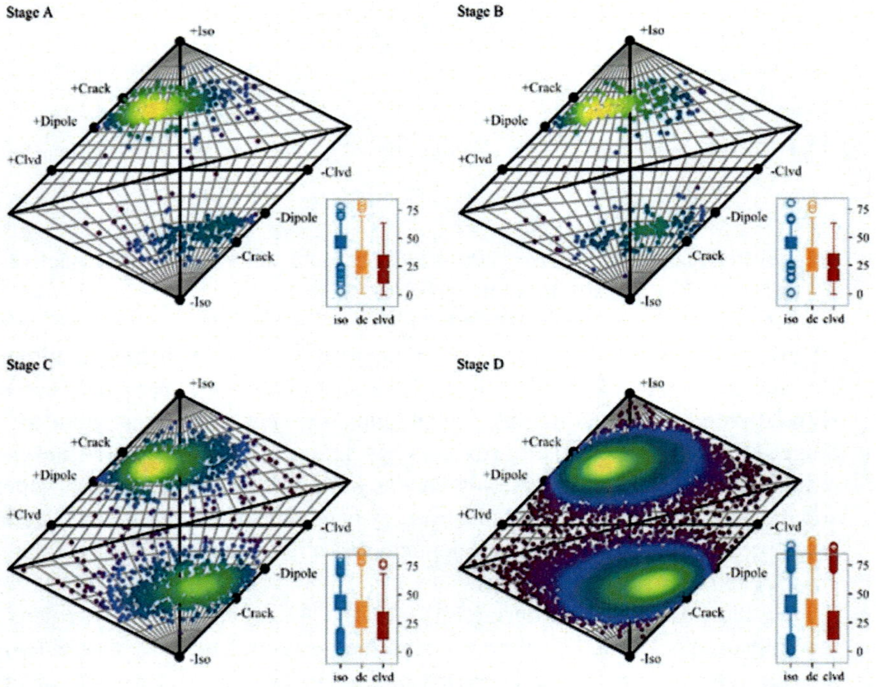

Fig. 11.5 Hudson diagram of the final optimized moment tensor solutions. The horizontal axis describes the type of constant volume component in the source (DC vs. CLVD), while the vertical axis presents the proportion of the volumetric variations. The box plot at right bottom describes the statistics of the specific ratio of each component in the moment tensor decomposition (from [37])

Table 11.1 The proportion of events with DC component proportion larger than a certain value (using all triggered sensors)

	Stage A [%]	Stage B [%]	Stage C [%]	Stage D [%]
DC Proportion > 60%	2.81	2.30	5.05	6.94
DC Proportion > 50%	6.19	6.62	14.36	17.09
DC Proportion > 40%	23.63	26.51	32.57	36.64

couple, at the center of the Hudson diagram) events in these four stages. Secondly, there are more tensile events than compression events in the first three stages. Thirdly, the shear events began to increase in the stage D, but the corresponding proportion was still very small.

By comparing the statistics of the moment tensor decompositions of these four stages, it can be seen that the proportion of the ISO component decreases slightly as the loading increases (with mediums of 46.88, 46.06, 44.10, and 42.47, respectively). The shear composition gradually increases from 28.89% (medium) for stage A to 34.23% for stage D. There is no significant change for the CLVD component. We can draw a rough hypothesis here: the decreasing of ISO and the increment of DC can be used as a sign of rock crack development, and it may be a precursor of rock instability.

To verify the above hypothesis, we counted the number of events with shear components greater than a certain percentage at each stage (Table 11.1). When the DC component is greater than 40% in the moment tensor, the destructive effect of shear can no longer be ignored. It can be seen that in the first two loading stages, this type of events is about one-fifth, and it increases to more than one-third in the latter two stages. When the DC component is greater than 50%, the shear begins to play a leading role in the failure. It can be seen that, in the first two stages, this type of events counts about 6%, and it increases to 14.36% in the Stage C and to 17.09% in the Stage D. When the DC component is greater than 60%, the fracturing is dominated by shearing. This type of events varies from 2.81% in Stage A to 6.94% in Stage D. This shows that the increased shearing has the potential for predict instability.

Ma et al. [35] proposed a probabilistic model for classifying events, based on the event position in the Hudson diagram. According to the probabilistic model and the moment tensor inversion results of this study (Fig. 11.5), the types of events are

Table 11.2 The source types classified based on the event position in the Hudson diagram (using all triggered sensors)

	Stage A [%]	Stage B [%]	Stage C [%]	Stage D [%]
Explosive dominated	0.37	0	0.39	0.25
Tensile dominated	64.35	61.09	47.20	44.07
Shear dominated	3.75	6.05	10.50	14.55
Compression dominated	31.14	32.56	41.62	40.88
Implosive dominated	0.37	0.28	0.26	0.21

Table 11.3 The proportion of events with DC component proportion larger than a certain value (using only the first 6 triggered sensors)

	Stage A [%]	Stage B [%]	Stage C [%]	Stage D [%]
DC Proportion > 60%	8.44	14.98	20.07	21.58
DC Proportion > 50%	21.20	28.81	37.96	37.51
DC Proportion > 40%	38.46	51.58	56.58	56.18

obtained (Table 11.2). The rock fracture evolution trends obtained by the 2 analysis methods are consistent, and the shear cracks or DC components are gradually increasing. However, the proportions of crack types in each stage are different in the results. This reflects two issues: firstly, the classification criteria for specific types are affected by experience; secondly, the 2 methods respectively classify cracks based on the initial movement direction of the P-wave and time-frequency characteristics, which results in different perspectives for extracting information from the AE waveform.

11.4.2 Uncertainty of the Moment Tensor Method

There is inevitable uncertainty using P-wave first motion to analyze the full moment tensor. The first aspect is the inaccuracy P-wave picking, and the spatial distribution of the sensors. The second aspect is that this method is essentially derived from earthquake fault analysis. The corresponding fault slip forms the tension and compression zones, and there is a difference in polarity. When it is generalized to the full moment tensor inversion, the uncertainty will increase, although the probability framework is adopted. In order to clarify the uncertainty caused by the inaccuracy of the polarity, we used the first six sensors to do another analysis, and the results are shown in Fig. 11.6.

It can be seen through comparing Figs. 11.5 and 11.6 that when the constraint (the number of sensors used) is less, the shear component in the moment tensor inversion result will increase. The medium of shear component increases to 35.57, 41.65, 43.32, and 43.24 when only considering the first 6 triggered sensors. The proportion of events with DC component proportion larger than a certain value also increases (Table 11.3). It can also be seen from Fig. 11.5 that when there are fewer constraints, the inversion results will be more scattered (Fig. 11.6). When only the first six sensors are used, the same trends are observed: the effect of shearing on failure becomes greater as the loading increases.

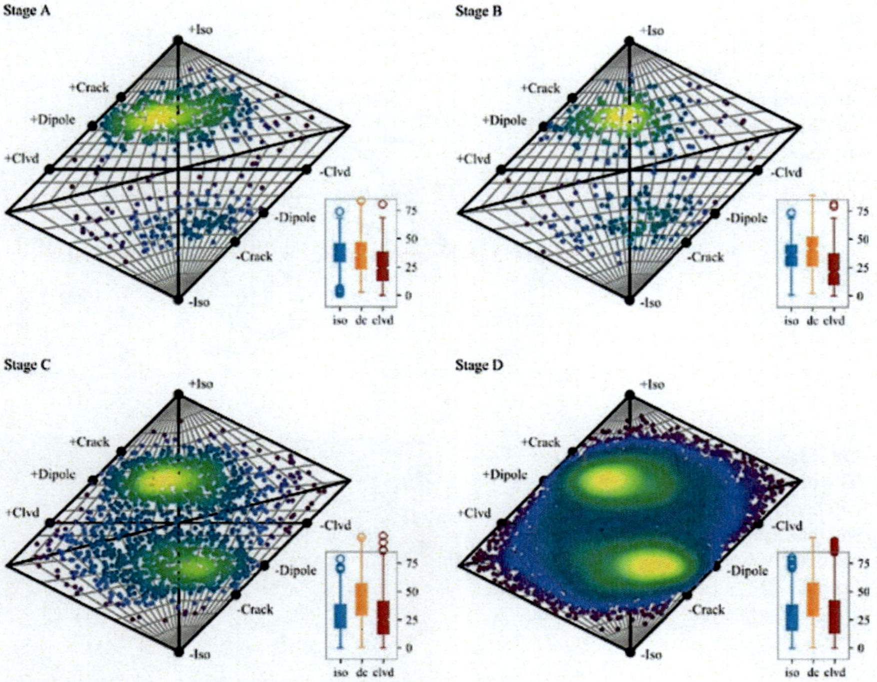

Fig. 11.6 Hudson diagram of the final optimized moment tensor solutions consider only the first six triggered sensors. The box plot at right bottom describes the statistics of the specific ratio of each component in the moment tensor decomposition (from [37])

11.4.3 Fracture Types of Granite in Post-Peak

As indicated by the first four stages, shear-dominated events are gradually increasing with the instability intensifies. This phenomenon continues to Stage E (Fig. 11.7). At this stage, the proportion of shear crack reaches 21.50%, and the medium of the DC component is 37.93%, which continues to grow on the basis of the previous stage. Instead of absolutely classifying the events into a certain source type, we should be concerned more about the proportions of individual components of the moment tensor and their evolutions over time.

In the last stage of loading (Stage E), the rock sample has shown visible macroscopic cracks. The positioning results of AE events are obtained by the localization method without using premeasured velocity based on the A* search algorithm [38]. We projected the events with DC components greater than 60% onto the rock sample (Fig. 11.8) to analyze the effect of such events on crack development. In terms of spatial distribution, the shear-dominated events are all located near the visible cracks. We concluded that the failure is a process of event accumulation. In this process, shearing develops, from being unimportant to playing a leading role. Finally, under the action of shearing, the rock eventually fails.

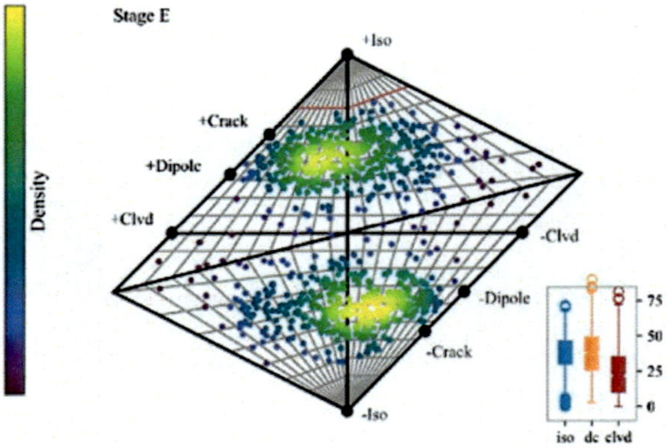

Fig. 11.7 Classification of crack mode based on Hudson diagram of the final optimized moment tensor solutions for the Stage E (from [37])

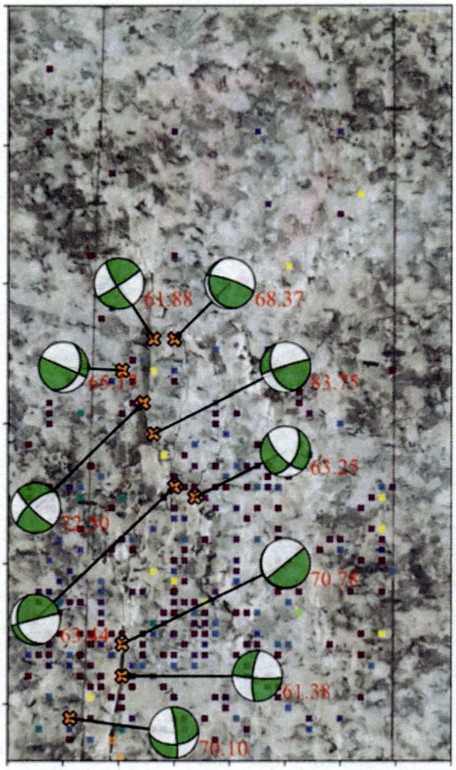

Fig. 11.8 Events (in Stage E) with double couple component larger than 60% and their spatial position along the visible cracks (from [37])

11.5 Conclusions

Moment tensor analysis method was applied to investigate the micro-crack mechanism after using velocity-free method for source localization. The conclusions are drawn below.

The proportion of shear cracks gradually increases during the whole loading process of uniaxial compression. The results demonstrate that the increasing of shear cracks is a significant sign of unstable crack propagation and coalescence. Water not only promotes the propagation of rock cracks, but also affects the fracture pattern of the rock.

Visible macroscopic cracks are mainly caused by shear fracturing. Instead of absolutely classifying the events into a certain source type, we need to pay more attention to the proportions of individual components of the moment tensor and their evolutions over time.

References

1. Kemeny JM, Cook NG (1991) Micromechanics of deformation in rocks. In: Shah SP (ed) Toughening mechanisms in quasi-brittle materials. Springer, pp 155–188
2. Zhao Y, Wang Y, Tang L (2019) The compressive-shear fracture strength of rock containing water based on Druker-Prager failure criterion. Arab J Geosci 12(15):1–8
3. Chmel A (2013) A comparative acoustic emission study of compression and impact fracture in granite. Int J Rock Mech Min Sci 64:56–59
4. Hu Q-C, Dong L-J (2019) Acoustic emission source location and experimental verification for two-dimensional irregular complex structure. IEEE Sensor Journal 20(5):2679–2691
5. Dong L-J, Sun D-Y, Han G-J, Li X-B, Hu Q-C, Shu L (2020) Velocity-free localization of autonomous driverless vehicles in underground intelligent mines. IEEE Trans Veh Technol 69(9):9292–9303
6. Dong L-J, Zou W, Li X-B, Shu W-W, Wang Z-W (2019) Collaborative localization method using analytical and iterative solutions for microseismic/acoustic emission sources in the rockmass structure for underground mining. Eng Fract Mech 210:95–112
7. Shiotani T, Ohtsu M, Ikeda K (2001) Detection and evaluation of AE waves due to rock deformation. Constr Build Mater 15(5–6):235–246
8. Aggelis D, Matikas T, Shiotani T (2010) Advanced acoustic techniques for health monitoring of concrete structures. In: Kim SH, Ann KY (eds) The Song's handbook of concrete durability. Middleton Publishing Inc:331–378
9. Aggelis DG (2011) Classification of cracking mode in concrete by acoustic emission parameters. Mech Res Commun 38(3):153–157
10. He M-C, Fei Z, Yu Z, Shuai D, Lei G (2015) Feature evolution of dominant frequency components in acoustic emissions of instantaneous strain-type granitic rockburst simulation tests. Rock and Soil Mechanics 36(1):1–8
11. Li L, Deng J, Zheng L, Liu J (2017) Dominant frequency characteristics of acoustic emissions in white marble during direct tensile tests. Rock Mech Rock Eng 50(5):1337–1346
12. Zhang Z-H, Deng J-H, Zhu J-B, Li L-R (2018) An experimental investigation of the failure mechanisms of jointed and intact marble under compression based on quantitative analysis of acoustic emission waveforms. Rock Mech Rock Eng 51(7):2299–2307
13. Wang Y-S, Deng J-H, Li L-R, Zhang Z-H (2019) Micro-failure analysis of direct and flat loading Brazilian tensile tests. Rock Mech Rock Eng 52(11):4175–4187
14. Zhang Y-B, Liang P, Sun L, Tian B-Z, Yao X-L, Liu X-X (2019) Spectral characteristics of acoustic emission in the process of saturated granite fracture under uniaxial compress. Rock Soil Mech. 40(7):10

15. Zhang Y, Zhang X, Liang P, Cheng S, Sun L, Yao X, Liu X, Liang J (2019) Experimental research on time-frequency characteristics of AE P-wave and S-wave of granite under failure process. Chin J Rock Mech Eng Geotech 38:3554–3564
16. Ohno K, Ohtsu M (2010) Crack classification in concrete based on acoustic emission. Constr Build Mater 24(12):2339–2346
17. Shahidan S, Pulin R, Bunnori NM, Holford KM (2013) Damage classification in reinforced concrete beam by acoustic emission signal analysis. Constr Build Mater 45:78–86
18. Wang M, Tan C, Meng J, Yang B, Li Y (2017) Crack classification and evolution in anisotropic shale during cyclic loading tests by acoustic emission. J Geophys Eng 14(4):930–938
19. Du K, Li X, Tao M, Wang S (2020) Experimental study on acoustic emission (AE) characteristics and crack classification during rock fracture in several basic lab tests. Int J Rock Mech Min Sci 133:104411
20. Liu X-L, Liu Z, Li X-B, Han M-S, Yang L-Q (2019) Acoustic emission and micro-fracture characteristics of rocks under Brazilian splittingload. Chin J Eng 041(011):1422–1432
21. Hu X, Su G, Chen G, Mei S, Feng X, Mei G, Huang X (2019) Experiment on rockburst process of borehole and its acoustic emission characteristics. Rock Mech Rock Eng 52(3):783–802
22. Zhang Z-H, Deng J-H (2020) A new method for determining the crack classification criterion in acoustic emission parameter analysis. Int J Rock Mech Min Sci 130:104323
23. Zhang P, Yu Q, Li L, Xu T, Yang T, Zhu W (2020) The radiation energy of AE sources with different tensile angles and implication for the rock failure process. Pure Appl Geophys 177(7):3407–3419
24. Yamamoto K, Naoi M, Chen Y, Nishihara K, Yano S, Kawakata H, Akai T, Kurosawa I, Ishida T (2019) Moment tensor analysis of acoustic emissions induced by laboratory-based hydraulic fracturing in granite. Geophys J Int 216(3):1507–1516
25. McLaskey GC, Lockner DA (2018) Shear failure of a granite pin traversing a sawcut fault. Int J Rock Mech Min Sci 110:97–110
26. Zhang X-P, Zhang Q (2017) Distinction of crack nature in brittle rock-like materials: a numerical study based on moment tensors. Rock Mech Rock Eng 50(10):2837–2845
27. Stierle E, Vavryčuk V, Kwiatek G, Charalampidou E-M, Bohnhoff M (2016) Seismic moment tensors of acoustic emissions recorded during laboratory rock deformation experiments: sensitivity to attenuation and anisotropy. Geophys Suppl Month Not Royal Astron Soc 205(1):38–50
28. Hampton J, Gutierrez M, Matzar L, Hu D, Frash L (2018) Acoustic emission characterization of microcracking in laboratory-scale hydraulic fracturing tests. J Rock Mech Geotech Eng 10(5):805–817
29. Liu Q, Liu Q, Pan Y, Liu X, Kong X, Deng P (2018) Microcracking Mechanism Analysis of Rock Failure in Diametral Compression Tests. J Mater Civ Eng 30(6):04018082
30. Wong LNY, Xiong Q-Q (2018) A method for multiscale interpretation of fracture processes in Carrara marble specimen containing a single flaw under uniaxial compression. J Geophys Res: Solid Earth 123(8):6459–6490
31. Liu J-P, Li Y-H, Xu S, Jin C-Y, Liu Z-S (2015) Moment tensor analysis of acoustic emission for cracking mechanisms in rock with a pre-cut circular hole under uniaxial compression. Eng Fract Mech 135:206–218
32. Ohtsu M (1991) Simplified moment tensor analysis and unified decomposition of acoustic emission source: application to in situ hydrofracturing test. J Geophys Res: Solid Earth 96(B4):6211–6221
33. Ohtsu M, Okamoto T, Yuyama S (1998) Moment tensor analysis of acoustic emission for cracking mechanisms in concrete. Struct J 95(2):87–95
34. Cesca S, Rohr A, Dahm T (2012) Discrimination of induced seismicity by full moment tensor inversion and decomposition. J Seismolog 17:147–163
35. Ma J, Dong L-J, Zhao G-Y, Li X-B (2018) Discrimination of seismic sources in an underground mine using full waveform inversion. Int J Rock Mech Min Sci 106:213–222

36. Dong L-J (2021) Localization and discrimination of microseismic/AE sources in mining: from data to information. In: Advances in acoustic emission technology. Springer, pp 3–16
37. Dong L-J, Zhang Y-H, Ma J (2020) Micro-crack mechanism in the fracture evolution of saturated granite and enlightenment to the precursors of instability. Sensors-Basel 20(16):4595
38. Dong L-J, Hu Q-C, Tong X-J, Liu Y-F (2020) Velocity-free MS/AE source location method for three-dimensional hole-containing structures. Engineering 6:827–834

Chapter 12
The Case of the Velocity Field Imaging in Mine—The Prediction of Rock Instability Risk

Rockburst induced by mining activities may cause strong vibrations and severe rock damage. Combining the microseismic (MS) muti-parameter and tomography may be a better approach for seismic hazard recognition in mining. The MS events used in the chapter are recorded by MS monitoring system in Zhen'Ao Dingsheng Mining Co., Ltd. In Sect. 12.1, the tectonics of the mining and arrangement of sensors are introduced. Then velocity-free localization method is used for MS source location in Sect. 12.2. Based on the location result, tomography analysis is conducted for several layers in Sect. 12.3. Considering the indication of the tomography results, multi-parameters, in-site surveys, mining parameters, and records of daily works as a warning identification of hazards, we provide a new method for pre-warning in underground mining.

12.1 The Sensors Network Distribution of the Microseismic Monitoring System

Zhen'Ao Dingsheng Mining Co., Ltd is located in Liufengguan Town, Fengxian County, Baoji City, Shanxi Province and belongs to the Dongtangzi lead–zinc deposit. The tectonics in the area is simple, with a northwest-west strike and an overall south dip direction. North–south gullies develop and the strata are transected. The overall terrain is high in the north and the east, and low in the south and the west. The highest elevation in the north is 2,051 m, while that in the east is 1,300 m. The difference of the general elevation is 300–600 m and the terrain slope is 20°–40°, which are shown in the Fig. 12.1.

In the lead–zinc mine, No. 1 ore body is located in the north wing of the Beizhi secondary anticline, and No. 2 ore body in the saddle and two wings of the Nanzhi secondary anticline. The ore body is bedded, which is thick and large in the saddle, and gradually becomes thin to sharp along the east and depth. The main ore body of the mine is No. 2 ore body whose resources account for 93% of the total ore resources,

L. Dong and X. Li, *Velocity-Free Localization Methodology for Acoustic and Microseismic Sources*, https://doi.org/10.1007/978-981-19-8610-9_12

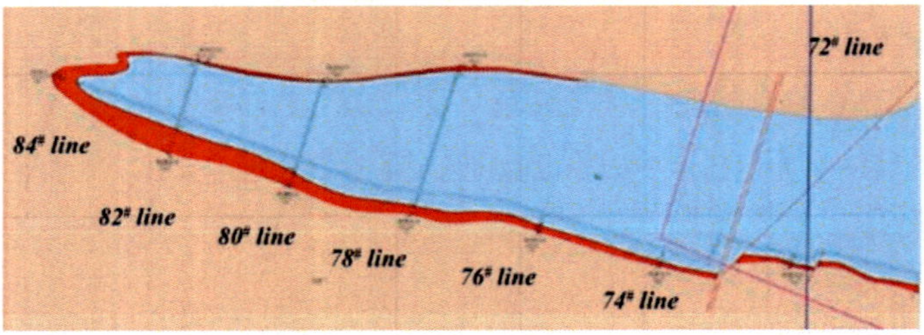

Fig. 12.1 The plan figure of the deposit

while No. 1 ore body accounts for 7% of that. The hydrogeological conditions of the whole mining area are simple, mainly karst fissure filled with water. The engineering geological type is a layered deposit dominated by hard-semihard rock strata, with medium engineering geological conditions, good rock integrity and stability of top and bottom slabs. The current underground mining method is a combination of a main flat cave and a blind inclined shaft. The middle sections of 1,060 m and 1,010 m have been mined.

The mining area is facing greater pressure on ground pressure management as it enters the stage of deep mining. Stress changes in stopes and a large number of remaining goafs caused by mining disturbance have brought significant hidden dangers to production. A multi-band geoacoustic intelligent perception system covering the middle section of 795 to 860 and line 80 to 90 has been built, which is helpful to comprehensively capture effective information before rock mass disasters occur. It has important engineering application value for the prevention and control of ground pressure zoning, the exploration of disaster development regulation, and the early warning of stope disaster.

The geoacoustic intelligent sensing and microseismic monitoring system consists of 26 single-component geoacoustic intelligent sensors. A 26-channel high-precision geoacoustic sensing and monitoring system was built with the six and seven middle sections as the core monitoring areas, of which four sensors were installed in 860 middle section, six in 842 middle section, two in 825 middle section, two in 810 middle section, and 12 in 795 middle sections. The sensor monitoring network of the geoacoustic intelligent sensing system is shown in the Fig. 12.2.

12.2 The Application of Passive Source Localization Without Pre-velocity Method

To verify the location accuracy of the monitoring network, a blasting test was carried out. The arrival times and source position were shown in Table 12.1. According to the arrival times and the coordinates of sensors, the source position of the monitored blasting signals is calculated by traditional method with premeasured velocity,

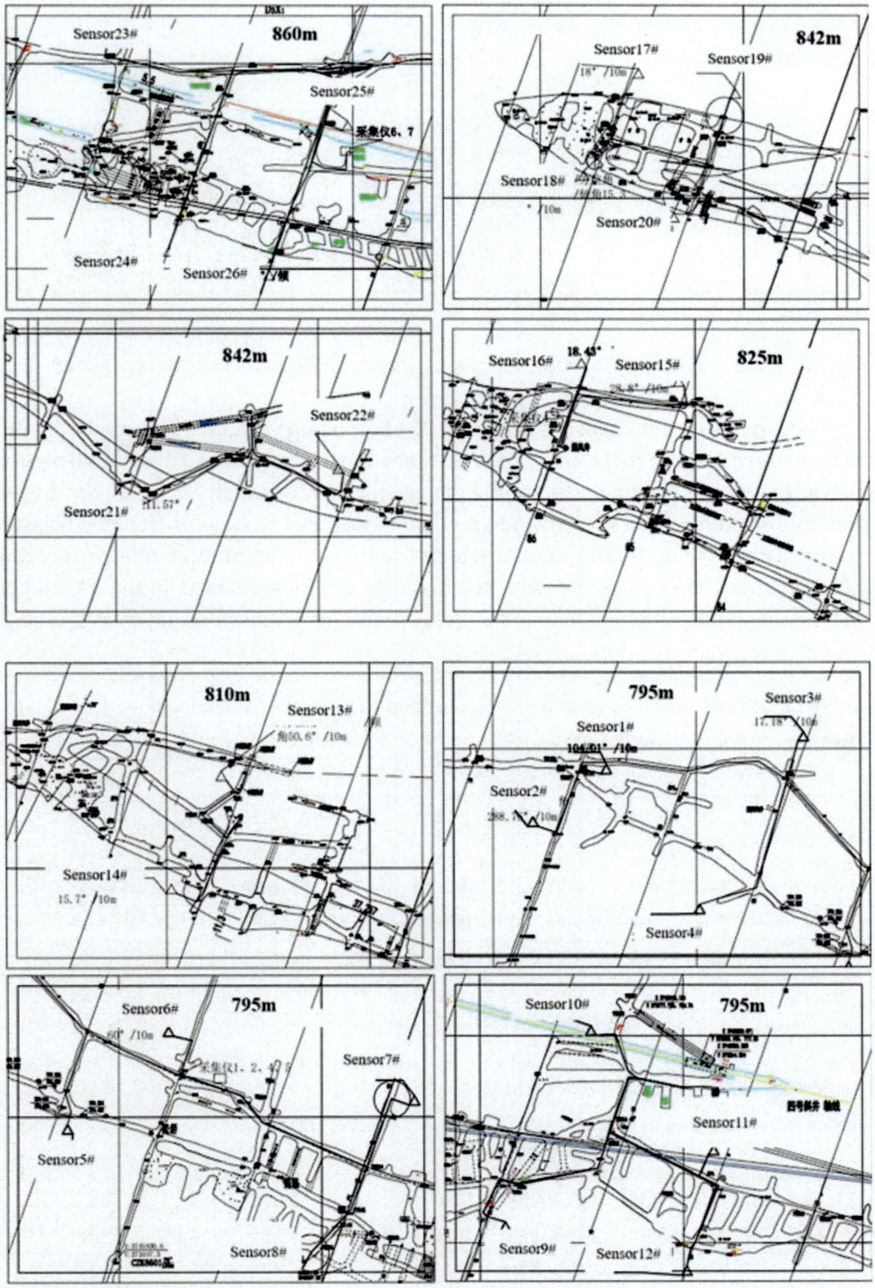

Fig. 12.2 The sensor monitoring network of the geoacoustic intelligent sensing system

Table 12.1 The time and the location of the blasts

Event No	Time	Source position (m)		
		x	*y*	*z*
1	2022.4.1 15:27	3,749,971.46	373,550.71	796.62
2	2022.4.2 23:32	3,749,800.03	373,890.79	825.79
3	2022.4.3 23:50	3,749,798.24	373,898.09	826.11
4	2022.4.4 23:37	3,749,795.41	373,889.61	825.94
5	2022.8.9 01:23	3,749,901.76	373,687.54	860.99

CLMAI, and LM. The location results of three methods are shown in Tables 12.2, 12.3, and 12.4.

The location accuracy of No. 4 event is the best using velcocity-free method with a location error of 8.15 m. However, the best location accuracy using traditional method is No. 1 event. The main reason is that the pre-measured velocity is fit for No. 1 event but not for others. The velocity-free methods avoid the effect of difference between pre-measured velocity and actual velocity and show the similar relative location accuracy between events. Because the position of No. 4 event is in the area where the sensor monitoring network is relatively dense, its location accuracy is the best.

Table 12.2 The location result of traditional method with premeasured velocity

Event No	Location result (m)			Error (m)			
	x	*y*	*z*	*x*	*y*	*z*	Absolute error
1	3,749,939.18	373,575.05	842.34	32.28	24.34	45.72	61.04
2	3,749,899.22	373,836.79	948.07	99.19	54.01	122.28	166.46
3	3,749,924.00	374,014.25	451.18	125.76	116.16	374.93	412.17
4	3,750,222.08	373,851.41	1006.07	426.67	38.21	180.13	464.71
5	3,749,933.98	373,724.53	492.11	32.22	36.99	368.88	372.13
Average							295.29

Table 12.3 The location result of CLMAI

Event No	Location result (m)			Error (m)			
	x	*y*	*z*	*x*	*y*	*z*	Absolute error
1	3,749,929.36	373,494.92	803.85	42.10	55.79	7.23	70.27
2	3,749,868.95	373,840.28	1080.72	68.92	50.52	254.93	268.87
3	3,749,867.18	373,895.81	863.84	68.94	2.28	37.73	78.62
4	3,749,775.11	373,951.51	817.87	20.30	61.89	8.07	65.63
5	3,749,984.64	373,648.84	836.52	82.88	38.71	24.47	94.69
Average							115.62

Table 12.4 The location result of LM

Event No	Location result (m)			Error (m)			
	x	*y*	*z*	*x*	*y*	*z*	Absolute error
1	3,749,960.24	373,562.77	783.69	11.22	12.06	12.93	20.94
2	3,749,785.71	373,893.21	810.79	14.33	2.41	15.00	20.88
3	3,749,786.73	373,893.21	815.92	11.51	4.88	10.19	16.13
4	3,749,790.86	373,893.21	820.21	4.55	3.59	5.73	8.15
5	3,749,907.38	373,682.88	881.28	5.61	4.66	20.29	21.56
Average							17.53

The monitoring system is composed of geoacoustic intelligent sensors, data set acquisition instruments, signal processor, underground data center, communication cables, optical fibers, and surface dispatching monitoring center. The visual monitoring software can record and analyze the source parameters as well as calculate the source location in real time, which can provide important information for mine safety analysis and accident warning.

The three weeks of geoacoustic safety monitoring were taken as examples. The source localization algorithm without pre-measured velocity was used to invert the position of source events in the monitoring area. The three-dimensional view of the roadway structure and the source location in the mining area are shown in Figs. 12.3, 12.4, and 12.5.

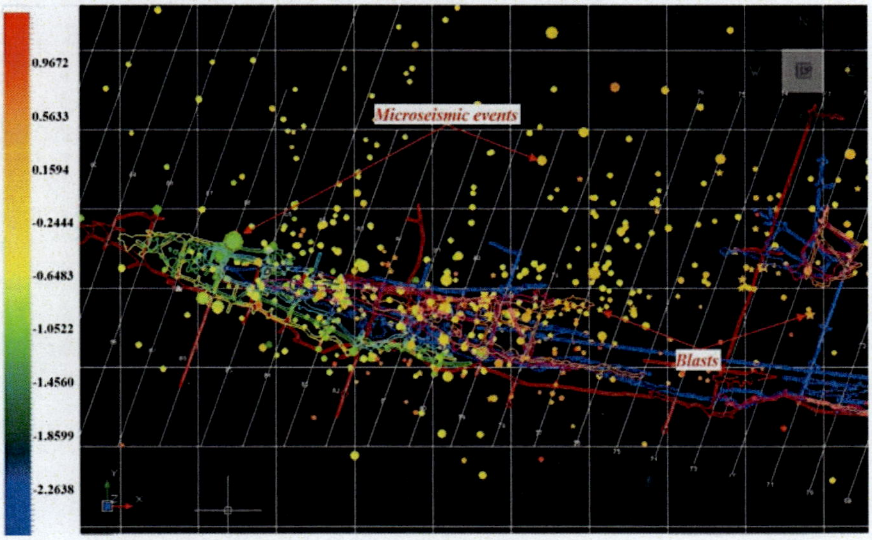

Fig. 12.3 The three-dimensional view of the roadway structure and the source location in the mining area (the first week)

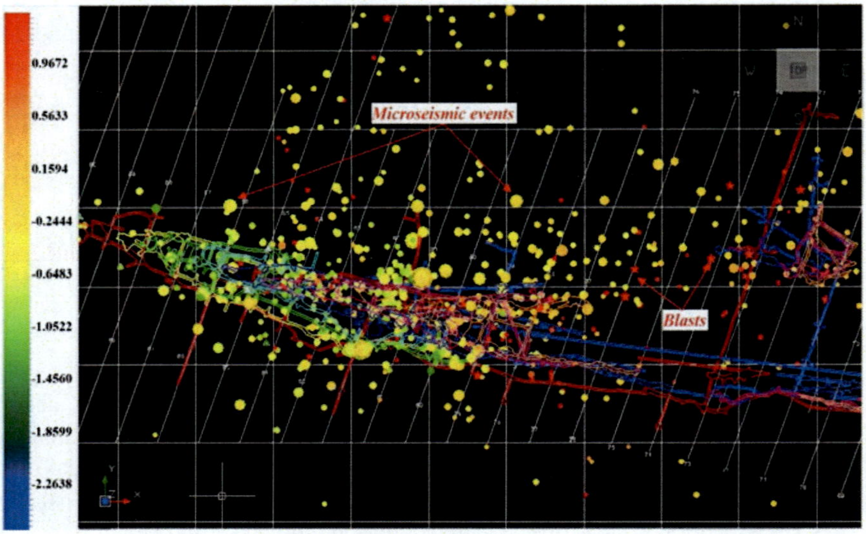

Fig. 12.4 The three-dimensional view of the roadway structure and the source location in the mining area (the second week)

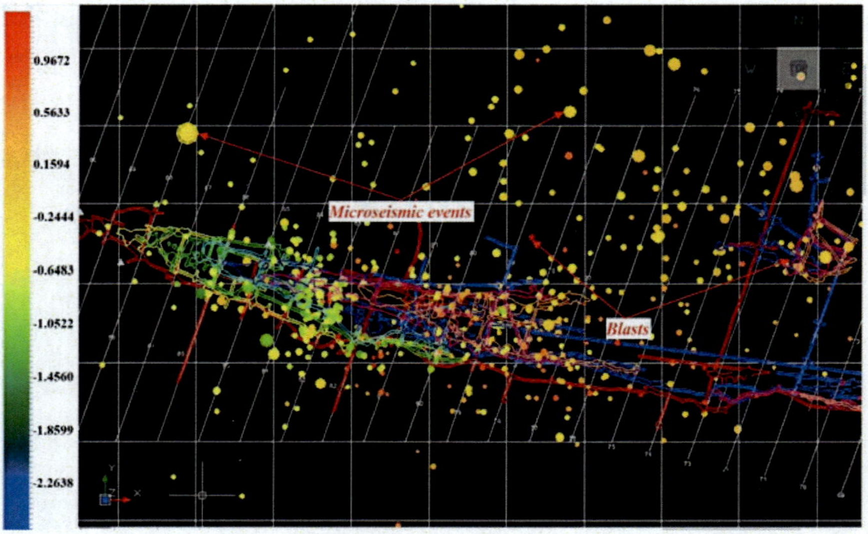

Fig. 12.5 The three-dimensional view of the roadway structure and the source location in the mining area (the third week)

In the first week, the interior acoustic event activities occur in many sections of the mining area. There are several gathering areas of geoacoustic events. Geoacoustic events are more concentrated in the region of the 85# line in 795 m, 82–83# line in 795 m, 85# line in 810 m, 85# line in 842 m, and 80# line in 860 m. In the second week, the gathering area of geoacoustic events has shifted with the mining activities, which becomes the region of the 83# line in 795 m, 83# line in 825 m, 84–85# line in 825 m, 85# and 81# line in 842 m, 83# line in 842 m, and 79# line in 860 m. In the third week, the geoacoustic events are concentrated in the region of the 88–89# line in 795 m, 80# line in 825 m, 81# line in 842 m, 83# line in 842 m, and 81# line in 860 m.

To verify the corresponding relationship between the geoacoustic event gathering area and the mining conditions, the events were compared with the field investigation (Figs. 12.6, 12.7, 12.8, and 12.9). The results showed that the daily mining activities of the mine were basically consistent with the distribution of geoacoustic events, and there were different degrees of rock failure in the event gathering area.

Fig. 12.6 The field condition of the area in 860 m 81# and 795 m 85# (In the first week)

Fig. 12.7 The field condition of the area in 810 m 85# (In the first week) and 825 m 83# (In the second week)

Fig. 12.8 The field condition of the area in 842 m 85# (In the second week) and 842 m 81# (In the second week)

Fig. 12.9 The field condition of the area in 795 m 88–89# (In the third week) and 860 m 81# (In the third week)

12.3 Tomography Analysis for Several Mine Layers

12.3.1 Data Processing

The signal quality is essential for the accuracy and resolution of tomography results. The collected signals are usually mixed with noise and abnormal signals, and need to be processed before the tomography calculation. The data processing is composed by signal discrimination, arrival picking, and event location. Considering the relationship between the quality and the signal-to-noise ratio, the blasting signals are selected for the tomography calculation. And the modified ratio [1] is selected as the arrival picking method. The arrival picking method is limited by the mine scale, the arrivals of S wave are mostly located at the end of P wave and there exists the obvious overlap phenomenon. Thus, the arrivals of P wave are relatively accurate and used for the tomography calculation. Since the location of blasting sources are unknown, it is necessary to obtain accurate event location using the probability-based velocity-free location method [2]. It avoids the impact of abnormal arrivals, and improves the robustness of location.

12.3.2 Tomography Analysis

The tomography application is conducted on the Shanxi Zhen'Ao mine, and 22 sensors installed on Level 795, 810, 825, 842, and 860 are utilized to acquire signals. The monitoring range of tomography technique is about 300 m × 520 m × 200 m based on the distribution of sensors and potential events. The data for tomography is from March 30 to April 22. The initial wave velocity for tomography model is set as 4500 m/s, corresponding to the velocity of the rock mass.

Considering the characteristics of high nonlinearity, and sensitivity to outliers for the tomography technique, the tomography data is divided into four parts according to date and quantity for iterative imaging. And each tomography iteration is based the prior tomography results. The tomography results and location results are shown in Figs. 12.10, 12.11, 12.12, and 12.13. The tomography results in Fig. 12.10 is obtained based on the initial model and signals from March 30 to April 3. There exist

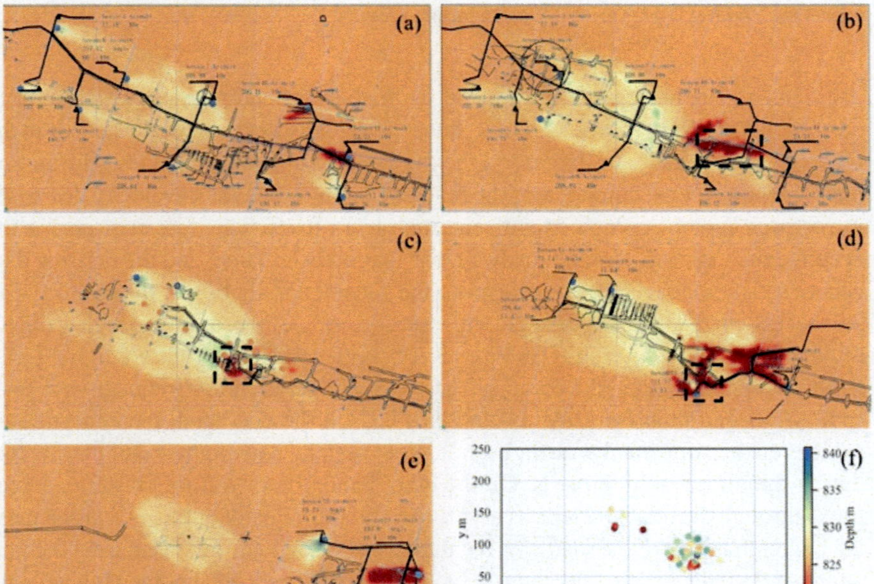

Fig. 12.10 Imaging results with blasting signals from March 30 to April 3. **a** Level A-A, **b** level B-B, **c** level C-C, **d** level D-D, **e** level E-E and **f** location results

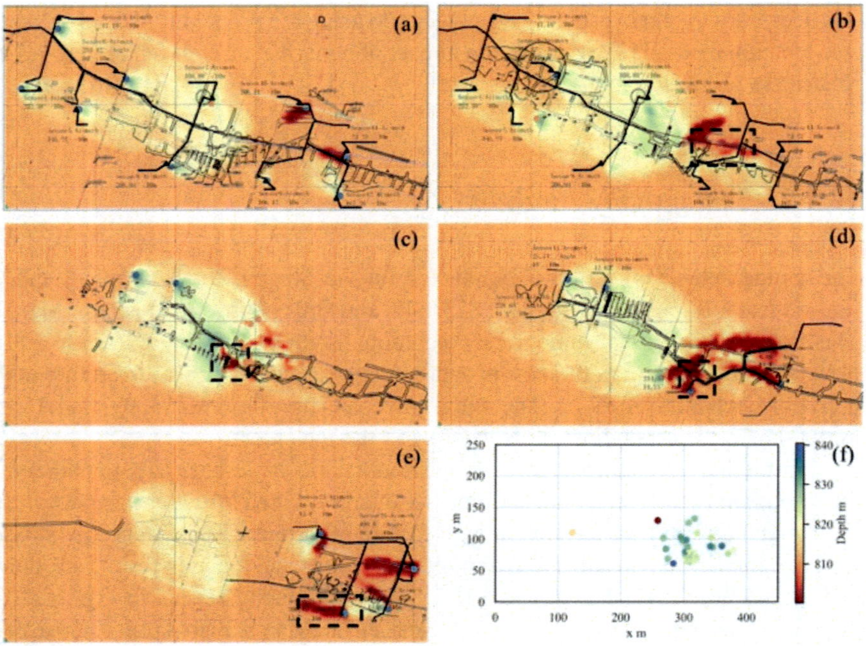

Fig. 12.11 Imaging results with blasting signals from April 4 to April 9. **a** Level A-A, **b** level B-B, **c** level C-C, **d** level D-D, **e** level E-E and **f** location results

distinguish velocity regions surrounded by the background regions, which refer to corresponding roadway in each level. Since the velocity difference between air in the roadway and rock mass is large, and the ray tracing path has the obvious diffraction phenomenon when it come across the roadway regions. It leads to velocity decreases in roadway regions compared with the rock mass itself. The data demand for the accurate reconstruction of the roadway distribution is far large, and the existing data cannot satisfy the requirement. However, the general distribution of the roadway can still be recognized from the tomography results. Besides, there are some high-velocity regions in tomography results. In Fig. 12.10b, high velocity regions are round Lines 81–83, where a mining site locates. The blasting operation are frequent with high source concentration from April 4 to April 29. On the one hand, the concentrated sources are accompanied with higher velocity gradient, and might accumulate errors during the tomography process. On the other hand, the rays from blasting sources travel through the intact rock mass to sensors, and the velocity of the intact rock mass are higher than that of the roadway. It enlarges the velocity difference during the iteration. And as the mining operation, it can be seen in Fig. 12.13 that the high velocity of level 825 decreases. The marked regions in Fig. 12.10c are corresponding to the hole and footage blasting regions around April 2. The high intensity blasting operation leads to high velocity regions. And then in the following tomography results, the velocity of these regions gradually decreases. Abnormal regions in Fig. 12.10d

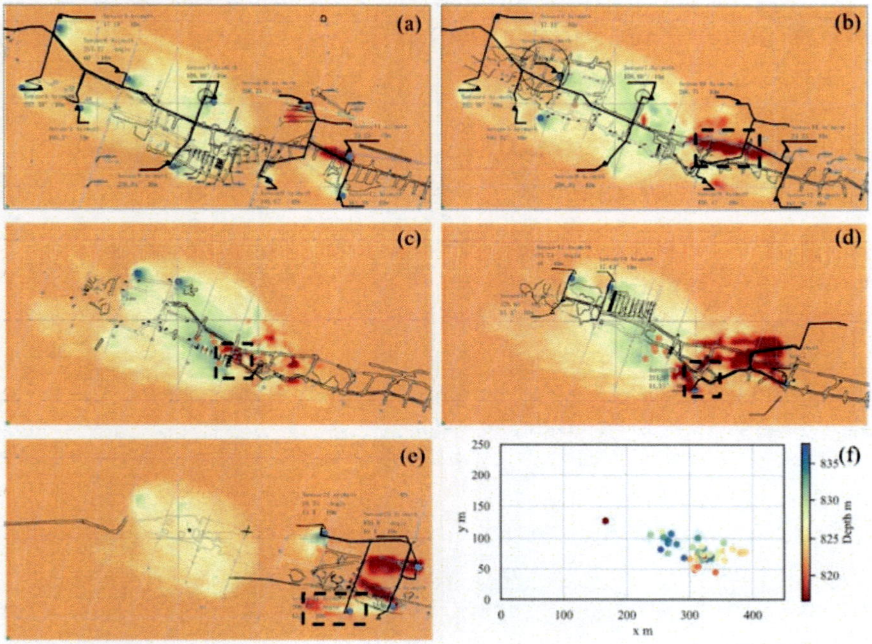

Fig. 12.12 Imaging results with blasting signals from April 11 to April 15. **a** Level A-A, **b** level B-B, **c** level C-C, **d** level D-D, **e** level E-E and **f** location results

are affected by the roof with low rock strength and sheet distribution, with collapses due to the large horizontal stress and frequent excavation blasting disturbance. The marked regions Fig. 12.10e belong to goaf, and are caused by the blasting operation on the bottom level. The velocity in marked regions gradually decreases in the subsequent imaging results.

12.4 Instability Risk Analysis of Mining Engineering

12.4.1 Tomographic Inversion

(1) Inversion parameters
The microseismic monitoring system went into service on 29 March 2022. In this section, the MS data were induced by mining activities from April to June 2022. Considering that the mining district in 795 level to 860 level, a total of $40 \times 20 \times 10$ voxel points with each voxel size of $20 \times 20 \times 10$ m in x, y, and z directions as input for tomographic calculation. All the events used in this section triggered more than nine sensors. According to the in-site surveys and the experiment results, average velocity of rock samples is 4,267.07 m/s (as

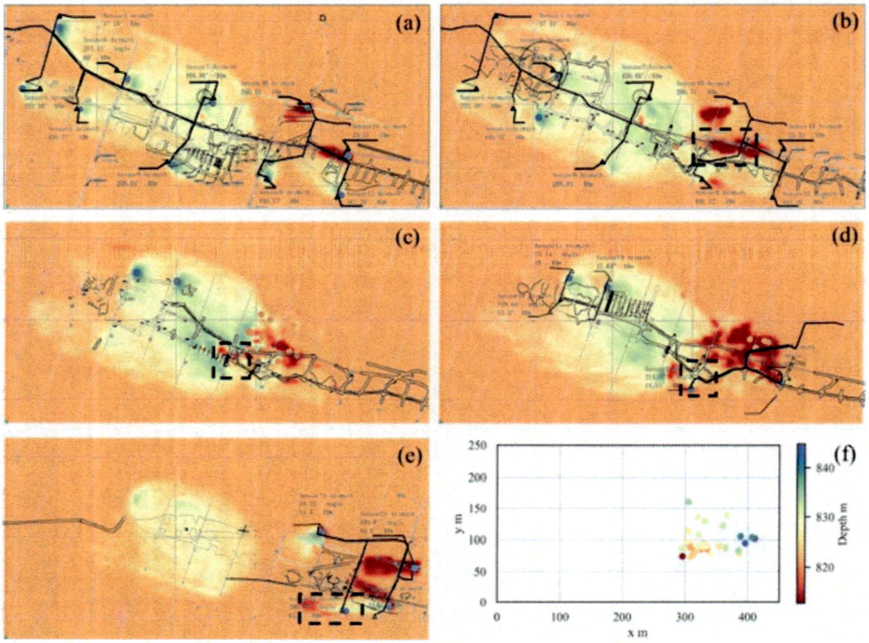

Fig. 12.13 Imaging results with blasting signals from April 16 to April 22. **a** Level A-A, **b** level B-B, **c** level C-C, **d** level D-D, **e** level E-E and **f** location results

shown in Fig. 12.14). Assume that velocity of P wave is 4,000 m/s in the first iteration. To reduce uncertainty, the speed is limited to a constant range between 2,000 and 6,000 m/s.

Plan views of velocity tomography results at the $z = 795, 810, 825, 842, 860$ m were generated to investigate anomaly seismic activity zones and analyzed seismic hazard risk for the next mining period about every 10 or 15 days (The

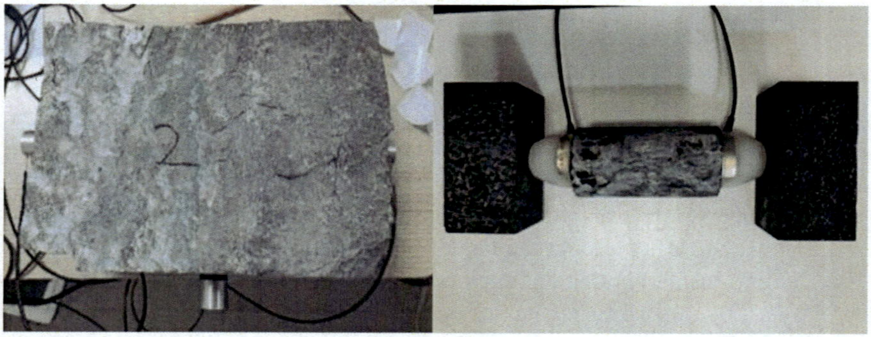

Fig. 12.14 Rock samples test of acoustic parameter

microseismic monitoring system was suspended due to the network malfunction from May 9 to May 15).

(2) Inversion result

According to Fig. 12.15, the exploratory line of No. 82 to No. 84 and No. 86 to No. 88 regions in the 795 level show high velocity level from April 1 to April 20, 2022. The main reason is that there were 12 blast works with explosive charges of 55 kg to 60 kg per work and a total of 804 kg of explosives being used. Fault structures may be affected by intense mining disturbances. These activities also affect the 810 level. Therefore, these regions are in danger in the next mining stage. From April 21 to April 30, to enhance production, the main works are ore hauling and drilling. The high-velocity levels of these regions have decreased by reducing the consumption of explosives during this period. Results show that high-velocity regions can be caused by high-magnitude mining disturbances and massive seismic rays.

From May 1 to May 15, blasting works in these main production regions are more frequently, especially at the 795 level and 810 level, with 30 blast works and a total of 1,770 kg explosives. It resulted in a wide range of high-velocity regions distribution. These activities also affect the adjacent layers. According to Fig. 12.15a–c, the tomographic images shows that 795 level, 810 level, and 825 level are generating a wide range of high-velocity regions from May 1 to May 8. These high-velocity regions are consistent with the blasting areas. At the same time, the seismic rays caused by blasting across the adjacent layers lead to some high-magnitude velocity disturbance in adjacent layers.

From May 16 to June 30, to enhance the production, the main work is to increase the ore hauling and reduce the blasting. In these 46 working days, 30 blast works were carried out at 795 level, with explosive charges of 20 kg to 30 kg per work and a total explosive weight of 684 kg. Fifteen blast works were carried out at 825 level, with explosive charges of 50 kg to 60 kg per work and a total explosive weight of 684 kg. Fifteen blast works were carried out at 842 level, with explosive charges of 40 kg to 55 kg per work and a total explosive weight of 732 kg. Twenty blast works were carried out at 860 level, with explosive charges of 50 kg to 60 kg per work, a total explosive weight of 1,104 kg. The reduced blasting work in spatial and temporal results in decreased high-velocity regions and low VG regions rate. Therefore, it can be seen that there is a good correlation between the high-velocity regions, the abnormal velocity regions, and the intense seismic activity regions.

According to the velocity matrix, The velocity-positive anomaly rockburst risk grade "middle" [3] areas include 810 level No. 82 line, 825 level No. 82 line, 842 level No. 86 to No. 88 line, 860 level No. 80 to No. 82 line, and 860 level No. 86 line. The velocity-positive anomaly rockburst risk grade "strong" areas include 795 level No. 82 to No. 84 line, No. 86 to No. 88 line, 810 level No. 86 line, and 825 level No. 84 line. The VG anomaly rockburst risk grade "middle" areas include 810 level No. 82 line, 825 level No. 82 to No. 84 line, 860 level No. 80 to No. 82 line, and 860 level No. 86 line. The VG anomaly rockburst risk grade "strong" areas include 795 level No. 82 to No.

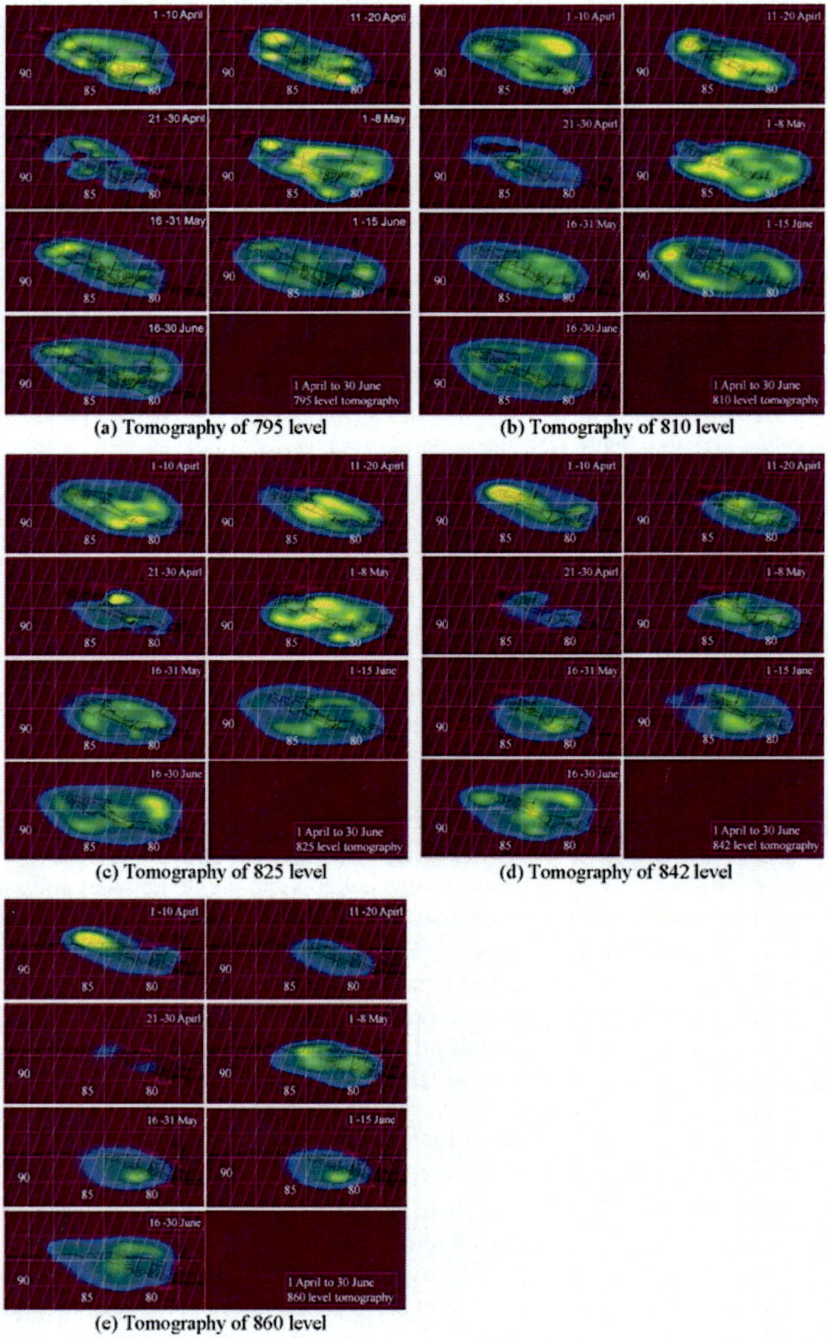

Fig. 12.15 Tomography images of **a** 795, **b** 810, **c** 825, **d** 842, and **e** 860 levels during the period from April 1, 2022 to June 30, 2022

84 line, 795 level No. 86 to No. 88 line, 810 level No. 86 line, and 842 level No. 86 to No. 88 line. The VG abnormal regions are close to the high-velocity areas. Velocity transition regions are distributed at the edge of the high-velocity areas. According to the experiments and rock strata, VG abnormally regions are also susceptible to rockburst risk. The stress concentration in the high-speed zone and the speed transition zone is due to the stress redistribution caused by mining disturbance. Mining activities may lead to instability, such as ore hauling, drilling, and blasting in adjacent regions. Therefore, we should pay more attention to the adjacent high-velocity area and velocity transition regions.

12.4.2 Variation of the Multi-parameter

(1) Sliding window
The S value and b value are comprehensive parameters of MS events, which can reflect the activeness and the magnitude ratio for a given time window. The results will be different in different time window sets. According to the three shifts mining management, we use one day as the sliding window to calculate the b value. Because the multi-parameters of mining working face may be influenced by disturbances of this layer and adjacent layer, the time window statistics are stratified by layer each day. Thus, we suggest that the rockburst risk grade analysis must combine with the tomography results and multi-parameters.

(2) Variation of the multi-parameter
Figures 12.16 and 12.17 show the 795 and 860 level variation characteristics of the multi-parameters from April 1 to June 30, 2022. From April 1 to April 20, the 795 level results show the S value fluctuated between 1.506 and 3.485, and the b value fluctuated between 0.313 and 1.248. The 860 level results show the S value fluctuated between 0.912 and 3.311, and the b value fluctuated between 0.192 and 0.978. Significantly, the multi-parameters in both levels sharply decreased to zero on April 18, the reason probably is the MS signal is too weak to trigger enough sensors. By definition of the S and b value, zero indicates low-frequency seismic activity and weak events. Thus, we set these parameters to zero on April 18. From April 1 to April 14, 256 events occurred in the 795 level release energy more than 1.00×10^3 J and one event release energy more than 1.00×10^4 J. This indicates that the microseismic was active. The energy of 16 events occurred in the 860 level exceeds 1.00×10^3 J (that of one event exceeds 1.00×10^4 J), which indicates that the microseismic was quiet relatively. Generally speaking, the multi-parameters are kept at a high level, which indicates that the micro-seismic activity was normal during this period relatively.

From April 21 to April 30, the main work is ore hauling and reducing the blasting. The results of the 795 level show the S value and b value have increased

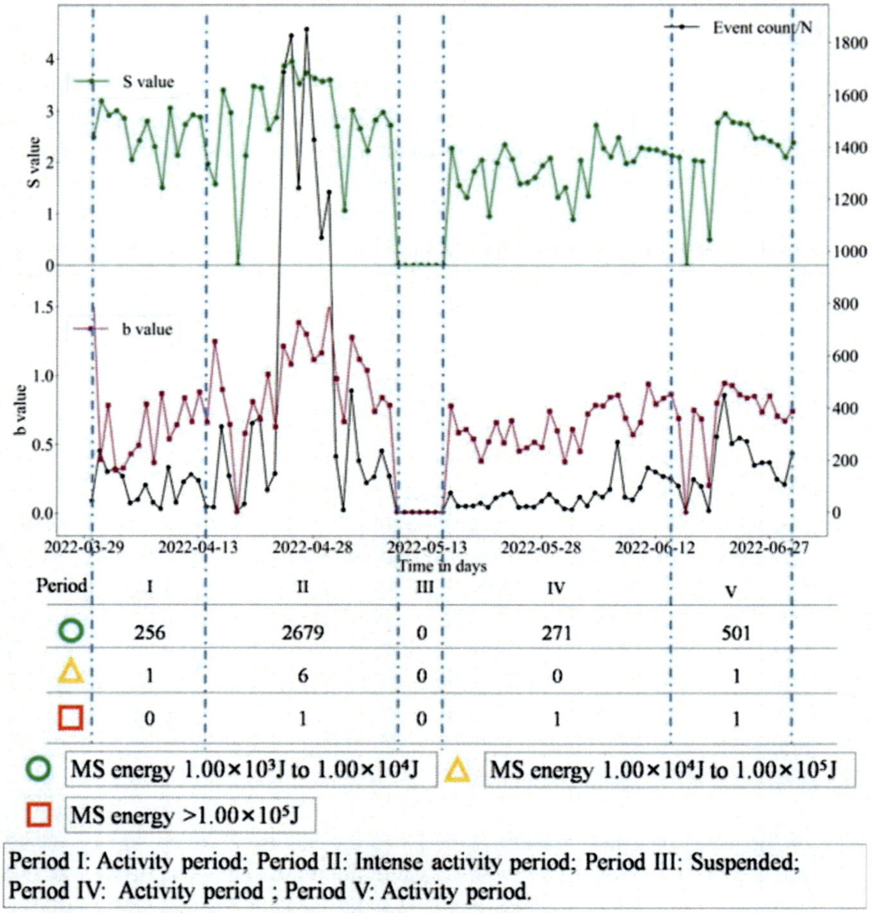

Fig. 12.16 The multi-parameter, MS event energy exceeding 1.00×10^3 J, and MS event count of 795 level: April 1 to June 30 2022

to a relatively high level of 3.95 and 1.598. There are more than 1,000 MS events per day, especially on April 27 with 1,851 event counts. The result of the 860 level show the S value and b value also increased to 3.77 and 1.262. There are more than 500 MS events per day, especially on April 25 with 1,248 event counts. Because the 795 level is the main production level, the multi-parameters of the 795 level are higher than the 860 level. Although the reduced blasting work decreased seismic activity regions, the rock mass stability is still affected by ore hauling and drilling. According to the database and in-site survey, numerous MS events were generated by these mining works. Furthermore, it makes the multi-parameters increase abnormally.

From May 1 to May 15, the average count of MS events has decreased from 1,000 to 180. The result of the 795 level show the S value fluctuated between

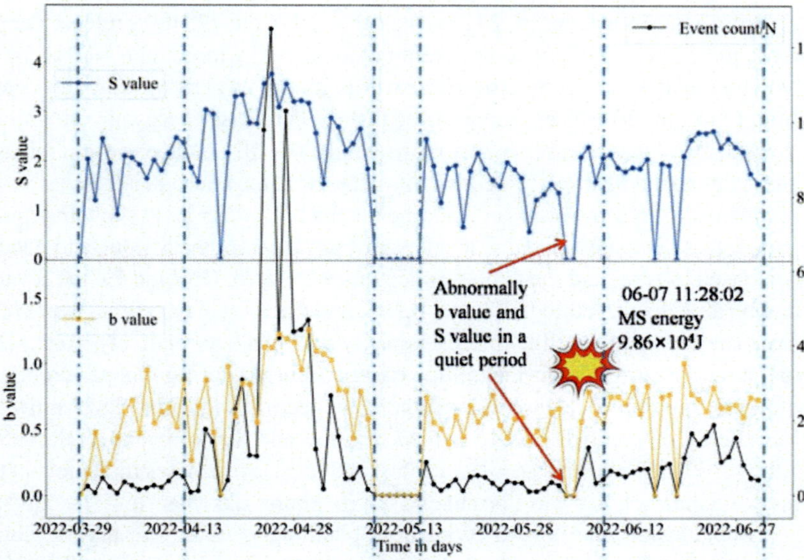

Period	I	II	III	IV	V
○	16	439	0	12	58
△	1	43	0	1+1(Rockburst)	24
□	0	0	0	0	4

○ MS energy 1.00×10^3J to 1.00×10^4J △ MS energy 1.00×10^4J to 1.00×10^5J

□ MS energy $>1.00\times10^5$J

Period I: Quiet period; Period II: Intense activity period; Period III: Suspended;
Period IV: Quiet period with abnormally b value and S value; Period V: Activity period.

Fig. 12.17 The multi-parameter, MS event energy exceeding 1.00×10^3 J, and MS event counts of 860 level: April 1 to June 30 2022

1.063 and 3.020, and the b value fluctuated between 0.668 and 1.278. The result of the 860 level show the S value fluctuated between 1.545 and 2.878, and the b value fluctuated between 0.438 and 1.099. From April 14 to May 8, 2679 events occurred in the 795 level release energy more than 1.00×10^3 J. Six events release energy more than 1.00×10^4 J, and one event releases energy more than 1.00×10^5 J. This indicates that the microseismic was strongly active. 439 events occurred in the 860 level release energy more than 1.00×10^3 J and 43 events release energy more than 1.00×10^4 J. This indicates that the microseismic was active. According to the results, from May 1 to May 8, 30 blast works were carried out at the 795 and 860 levels, with explosive charges of 55 kg to 65 kg per work and a total of 1,770 kg explosive. Therefore, more blasting will cause more high magnitude events and strong seismic activities, leading to high-level multi-parameters.

From May 15 to June 14, 271 events occurred in the 795 level release energy more than 1.00×10^3 J and one event releases energy more than 1.00×10^5 J, which indicates that the microseismic was in moderate activity relatively. From June 15 to June 30, 501 events occurred in the 795 release energy more than 1.00×10^3 J. One event releases energy more than 1.00×10^4 J, and one event releases energy more than 1.00×10^5 J, which indicates that the micro-seismic was active relatively. From May 16 to June 30, the results of the 795 level show the multi-parameters fluctuated at a high level evenly. On June 16, the S value and b value decreased to zero, and the total event counts were zero. On June 19, the S value and b value decreased to 0.496 and 0.204, and the total event counts were only five. They indicate that the microseismicity was inactive relatively. According to the records, some high-magnitude events occurred during this period due to the stress buildup or release. For example, the records of the 795 level show the MS energy of event ID 85,515 is 5.57×10^5 J and that of event ID 86,195 is 7.92×10^5 J. These stress concentrations and release processes indicate that the microseismic activity was normal and the rockburst risk was "low".

In conclusion, the mining production planning is mainly in the 795 level, where the S value fluctuated between 0.496 and 3.951, and the b value fluctuated between 0.204 and 1.59. Some high-magnitude events occurred during this period, and the MS energy fluctuated between 2.81×10^4 and 8.83×10^6. Due to the blasting work being carried out frequently in the 795 level, the stress has also been released. After the stress release process, the mining working face may reach a new stability condition. Due to the new equilibrium condition, although these works generated numerous MS events, the 795 level seismic was the inactivity and the rockburst risk was "low" during this period.

At 11:28:02 June 7, 2022, a rockburst occurred in the No. 80 line gob region of the 860 level. The records of the 860 level show that event ID is 84320 and MS energy is 9.86×10^4 J. The tomography results show this region is a velocity transition region with a "middle" rockburst risk. From May 15 to June 8, only 12 events occurred in the 860 level release energy more than 1.00×10^3 J and two events release energy more than 1.00×10^4 J, which indicates that the microseismic was quiet. According to the multi-parameters results, the S value increased from 1.36 on June 3 to 2.089 on June 6. After the rockburst, it decreased rapidly to 1.586 on June 8. The b value decreased from 0.661 on June 3 to 0.563 on June 6. After the rockburst, it increased to 0.788.

With the increased activity of MS events, the beginning of the quiet period of MS events was considered as the precursor of rockburst. In this case, June 4 and June 5 is the quiet period, in which the multi-parameters decreased rapidly to zero and event counts is zero. After the quiet period, the MS events activities increased relatively until a rockburst occurred. In this case, S value, b value, and event counts on June 6 are 2.089, 0.563, and 58, respectively. Then the rockburst occurred at 11:28:02 June 7. After the rockburst, S value, b value, and event counts on June 7 are 2.287, 0.788 and 128, respectively. From June 9 to June 30, 58 events occurred in the 860 level release energy more than 1.00×10^3 J. 24 events release energy more than 1.00×10^4 J, and four events release energy more than 1.00×10^5 J, which indicates that the microseismic was active and the stress has also been released. It should be noted that the relationship between rockburst and quiet period was an unnecessary but sufficient condition. This period might be a stage of original crack compaction. Thus, the multi-parameter precursors should be comprehensively used to determine the occurrence of a rockburst.

12.4.3 Prediction of Rock Burst Risk

(1) Rockburst Case study

According to the field records, the main reason for the rockburst at 11:28:02 on June 7 probably is that a blasting work charges of 60 kg was carried out in 860 level No. 82 line. The position of blasting work is 100 m apart from the rock-burst site. The blasting disturbance affects the compaction of the original crack. Furthermore, the compaction process will lead to the weakening or fracturing of the rock mass. As the tomography results show in Fig. 12.18a, b, No. 80 line gob region of the 860 level is a velocity transition region with velocity-positive anomaly rockburst risk grade "middle" and VG anomaly rockburst risk grade "middle".

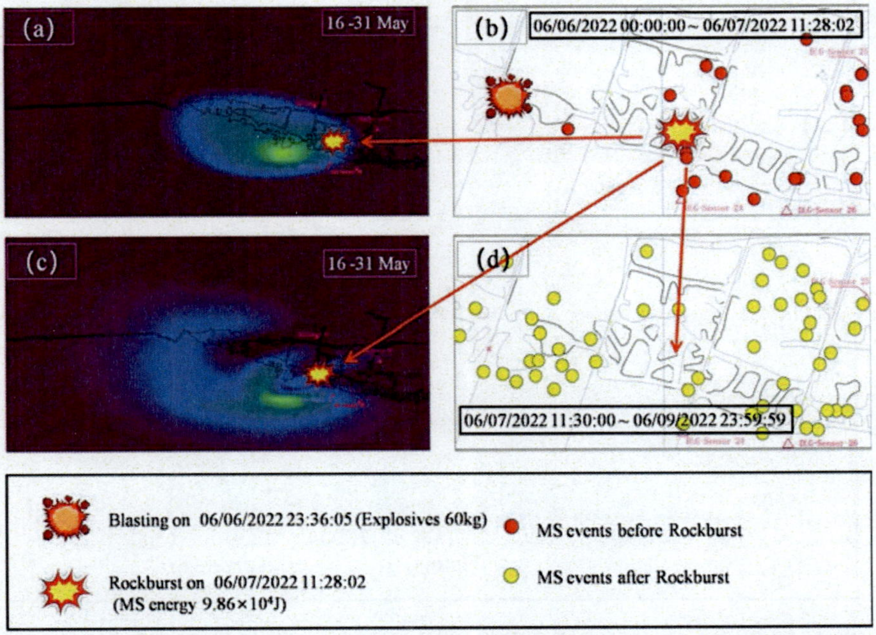

Fig. 12.18 Rockburst of 860 level 80 line gob area in June 7. **a** Tomography of 860 level velocity images, **b** MS events distribution before Rockburst, **c** Tomography of 860 level VG anomaly images and **d** MS events distribution after Rockburst

As shown in Fig. 12.18c–e, after the blasting work on June 6, the quiet period on June 4 and June 5 has been broken, and a lot of weak MS events began to appear in the regions. As shown in Fig. 12.19a, the MS energy fluctuated between 0.121 J and 349.22 J per hour before the rockburst on June 7. Most MS events are concentrated in the gob area but the number is relative. After the rockburst, MS event scattered around these regions, which indicates that the stress of the regions is released. According to the Omori's power-law decay characteristics [5, 6], the surrounding rock mass of the blasting area would return to a relatively stable state after approximately 20 days.

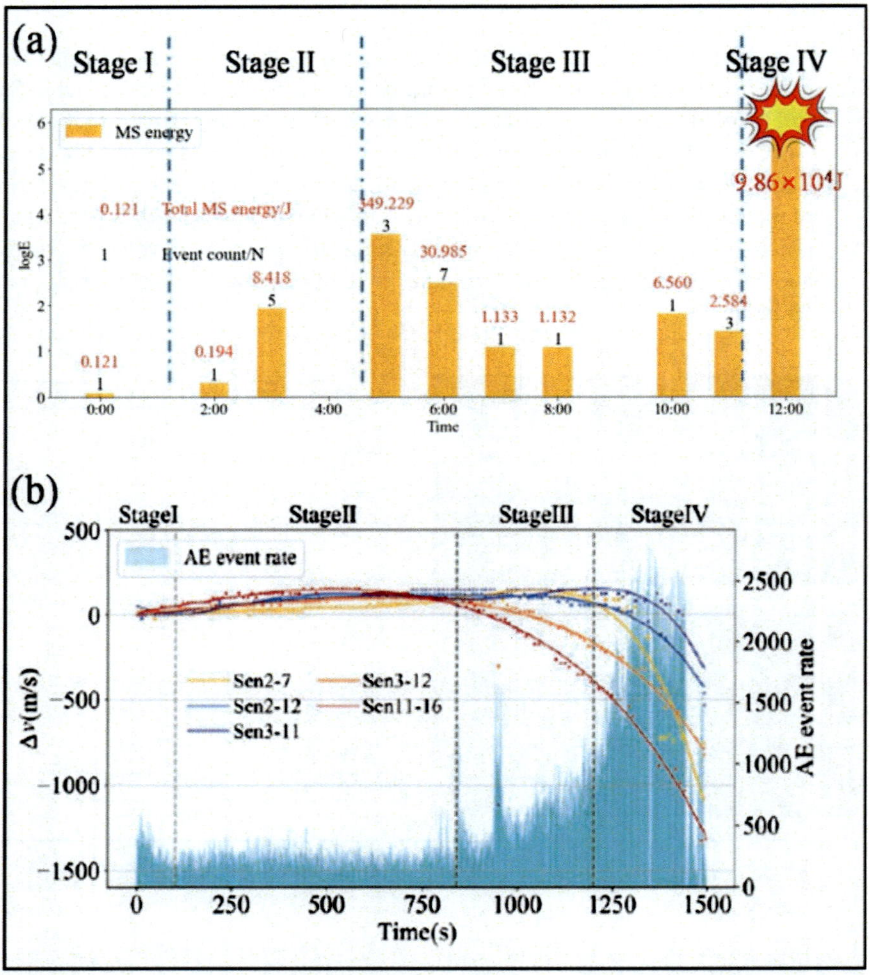

Fig. 12.19 Variation characteristics of MS energy and event in the study area after the blasting **a** MS energy and event count in the study area before the rockburst **b** AE event rate of granite sample 2 (from [4])

According to the experiment results [4] and the characteristics of MS events, the AE event rate gradually decreases and the rock sample tends to be compact in Stage I. The AE event rate is low and stable in Stage II. The rock is in the stage of elastic deformation. In this case, rock mass may be the stage of compact or elastic deformation from Stage I to II. Results show that MS events gradually appeared after the blasting, and these MS energies are no more than 10 J for the first four hours. In Stage III, the AE event rate increases slightly, and microcracks just initiate in the rock sample with the increase of stress. In Stage IV, the AE event rate increases sharply, and the rock sample breaks into pieces with different sizes and shapes at last. Frome Stage III to IV, more and more MS events occurred in the study area, especially at 04:43:54 when maybe a rib spalling occurred. The MS energy of event ID 84,229 is 323.87 J. That is similar to Stage II in the experiment. The rock is in the elastic deformation stage. After that, the MS event count increased more than the first four hours until the rock burst occurred.

Although the cut-off time point cannot be determined between each stage in this case, the trend of rock instability precursors will be the same. This characteristic is similar to the conclusion of Lu and Li [7, 8]. A process from rock deformation, failure to collapse is an irreversible process of energy dissipation, including both energy dissipation and energy release. The accumulation of MS events with abnormal multi-parameters can be regarded as a precursor of the rock burst risk. According to the experiment, rock mass under loading can divide into four stages: fracture closure, elastic deformation, fracture development, and rock burst damage. Microseismic signals mainly occur after the fracture closure. And high-magnitude events mainly occur in the fracture development and rockburst damage.

In summary, before the rock burst, MS events of each week gradually appeared and congregated around the tomography anomaly regions, and the multi-parameters changed suddenly. Thus, the tomography results, multi-parameters, mining plan, and the in-site survey can be used as a warning identification of a hazard. According to this information, the potential occurrence of different types of hazards can be determined.

(2) Determining Rockburst risk areas

According to the tomography results, multi-parameters, the mining plan, and the in-site survey, the following regions may contain a potential risk of rockburst:

(1) The room crosscut of No. 84 line in 795 level: the results of velocity tomography show the region rockburst risk grade is strong. The S value fluctuated between 0.496 and 3.951, and the b value fluctuated between 0.204 and 1.59. Some high-magnitude events occurred during this period, and MS energy fluctuated between 2.81×10^5 and 8.83×10^6. Recently, blast works have been carried out for the room crosscut frequently. The stress is also released, and the rockburst risk is "low". After the crosscut construction, accumulation of the redistribution stress may induce a serious

rockburst hazards. Some pressure release measures should be taken to prevent the occurrence of strong rockburst.

(2) In No. 82 and No. 86 line in 810 level, and No. 86 to No. 88 line in 842 level, the results of velocity tomography show that rockburst risk grade is "strong" in these transition regions or high-level velocity regions. The S value fluctuated between 1.316 and 3.627, and the b value fluctuated between 0.368 and 1.385. Recently, blasting works had been carried out for the mining works frequently. the stress is also released, and the rockburst risk was low. According to the Omori's power-law decay characteristics, the surrounding rock mass of the blasting area would return to a relatively stable state after approximately 20 days. After the open stope mining, accumulation of the redistribution stress may induce a serious rockburst hazard in the gob area. The backfilling should be taken in these regions as soon as possible.

(3) The goaf of No. 80 to No. 82 line in 860 level: the results of velocity tomography show that rockburst risk grade is "middle" in the transition region or high-level velocity region. After the rockburst on June 7, MS event scattered around these regions and the stress is released. However, the stability of the region will still be affected by blasting works carried out in the adjacent area. The backfilling should be taken in the goaf as soon as possible.

12.5 Conclusion

The microseismicity was analyzed in the Shanxi Zhen'Ao mine between April and June 2022. Considering the indication of the tomography results, multi-parameters, in-site surveys, mining parameters, and records of daily works as a warning identification of hazards, we provide a new method for early warning in underground mining. The main conclusions are as follows:

(1) Mining disturbance is the main reason for the instability of rock mass. The chapter reveals a good correlation between the Mining disturbance and MS information. Therefore, the sharp increase or decrease of MS event rate caused by mining disturbance in Spatio-temporal may be the precursor of surrounding rock instability.

(2) The high-velocity regions are consistent with the blasting areas. At the same time, when seismic rays pass through adjacent layers, strong velocity disturbance will be generated in adjacent layers. Therefore, it can be seen that there is a good correlation between the abnormal velocity regions and high seismic activity regions. Stress concentration is the redistribution of stress caused by mining disturbance in high-speed and speed transition areas. These disturbances may lead to the instability of rock mass.

(3) The changes of multi-parameters are consistent with the mining disturbances. The more sudden changes of the multi-parameters, the greater risk of the rock mass. The result shows that the S value fluctuates between 0.496 and 3.951,

and the b value fluctuates between 0.204 and 1.59. The high-magnitude events rate is also consistent with the mining disturbances. The more high-magnitude events occur, the faster stress releases.

(4) In the case of 7 June, the goaf is affected by the blasting disturbance of the adjacent area, which is a velocity transition region with velocity-positive anomaly rockburst risk grade "middle" and VG anomaly rockburst risk grade "middle". The S value increased from 1.36 to 2.089, and the b value decreased from 0.661 to 0.563. After the rockburst, the S value decreased from 2.089 to 1.586, and the b value increased from 0.563 to 0.788. The MS energy is less than 10 J for the first four hours.

(5) In-site survey and the results of MS event analysis for 795 level to 860 level include tomography results, MS event rate, b value, and S value. Before the rock burst, MS events of each week gradually appeared and congregated around the tomography anomaly regions, and the multi-parameters changed suddenly. Finally, the potential unsafe regions were discovered from the 795 to 860 levels.

The mining disturbance is closely related to the tomography results, MS multi-parameters, and the mining parameters. The combination of multi-parameters and seismic tomography has been proved to be a promising tool for evaluating rockburst risk during underground mining. The MS information can be used as a warning identification of a hazard and would have a wide application in the future.

References

1. Han L, Wong J, Bancroft J (2009) Time picking and random noise reduction on microseismic data. CREWES Res Rep 21:1–13
2. Dong L, Shu W, Li X, Han G, Zou W (2017) Three dimensional comprehensive analytical solutions for locating sources of sensor networks in unknown velocity mining system. IEEE Access 5:11337–11351. https://doi.org/10.1109/access.2017.2710142
3. Gong S-Y (2010) Research and application of using mine tremor velocity tomography to forecast rockburst danger in coal mine. Dissertation, China University of Mining and Technology
4. Dong L-J, Chen Y-C, Sun D-Y, Zhang Y-H (2021) Implications for rock instability precursors and principal stress direction from rock acoustic experiments. Int J Min Sci Technol 31(5):789–798. https://doi.org/10.1016/j.ijmst.2021.06.006
5. Shcherbakov R, Turcotte DL, Rundle JB (2004) A generalized Omori's law for earthquake aftershock decay. Geophys Res Lett 31(11). https://doi.org/10.1029/2004gl019808
6. Liu J-P, Xu S-D, Li Y-H, Lei G (2019) Analysis of rock mass stability based on mining-induced seismicity: a case study at the hongtoushan copper mine in China. Rock Mech Rock Eng 52(1):265–276. https://doi.org/10.1007/s00603-018-1541-y
7. Lu C-P, Liu G-J, Liu Y, Zhang N, Xue J-H, Zhang L (2015) Microseismic multi-parameter characteristics of rockburst hazard induced by hard roof fall and high stress concentration. Int J Rock Mech Min Sci 76:18–32. https://doi.org/10.1016/j.ijrmms.2015.02.005
8. Li Z-L, He X-Q, Dou L-M, Wang G-F (2018) Rockburst occurrences and microseismicity in a longwall panel experiencing frequent rockbursts. Geosci J 22(4):623–639. https://doi.org/10.1007/s12303-017-0076-7